T0179449

MICROARRAY INNOVATIONS

TECHNOLOGY AND EXPERIMENTATION

Drug Discovery Series

Series Editor

Andrew A. Carmen

illumina, Inc.
San Diego, California, U.S.A.

MICROARRAY INNOVATIONS

TECHNOLOGY AND EXPERIMENTATION

Edited by
GARY HARDIMAN

CRC Press
Taylor & Francis Group
Boca Raton London New York

CRC Press is an imprint of the
Taylor & Francis Group, an **informa** business

CRC Press
Taylor & Francis Group
6000 Broken Sound Parkway NW, Suite 300
Boca Raton, FL 33487-2742

First issued in paperback 2019

© 2009 by Taylor & Francis Group, LLC
CRC Press is an imprint of Taylor & Francis Group, an Informa business

No claim to original U.S. Government works

ISBN-13: 978-1-4200-9448-0 (hbk)
ISBN-13: 978-0-367-38581-1 (pbk)

This book contains information obtained from authentic and highly regarded sources. Reasonable efforts have been made to publish reliable data and information, but the author and publisher cannot assume responsibility for the validity of all materials or the consequences of their use. The authors and publishers have attempted to trace the copyright holders of all material reproduced in this publication and apologize to copyright holders if permission to publish in this form has not been obtained. If any copyright material has not been acknowledged please write and let us know so we may rectify in any future reprint.

Except as permitted under U.S. Copyright Law, no part of this book may be reprinted, reproduced, transmitted, or utilized in any form by any electronic, mechanical, or other means, now known or hereafter invented, including photocopying, microfilming, and recording, or in any information storage or retrieval system, without written permission from the publishers.

For permission to photocopy or use material electronically from this work, please access www.copyright.com (http://www.copyright.com/) or contact the Copyright Clearance Center, Inc. (CCC), 222 Rosewood Drive, Danvers, MA 01923, 978-750-8400. CCC is a not-for-profit organization that provides licenses and registration for a variety of users. For organizations that have been granted a photocopy license by the CCC, a separate system of payment has been arranged.

Trademark Notice: Product or corporate names may be trademarks or registered trademarks, and are used only for identification and explanation without intent to infringe.

Library of Congress Cataloging-in-Publication Data

Microarray innovations : technology and experimentation / editor, Gary Hardiman.
   p. ; cm. -- (Drug discovery series ; 11)
   Includes bibliographical references and index.
   ISBN 978-1-4200-9448-0 (alk. paper)
   1. DNA microarrays. 2. Gene expression. I. Hardiman, Gary, 1966- II. Title. III. Series.
   [DNLM: 1. Oligonucleotide Array Sequence Analysis--methods. 2. Gene Expression Profiling--methods. 3. Hybridization, Genetic. QZ 52 M625 2008]

QP624.5.D726M5125 2008
572.8'636--dc22   2008040407

Visit the Taylor & Francis Web site at
http://www.taylorandfrancis.com

and the CRC Press Web site at
http://www.crcpress.com

Contents

Preface

It has been almost six years since I published my first book *Microarray Methods and Applications—Nuts and Bolts*. The speed with which time has passed since then is hard to believe. More incredible though is the progress in the field of genomic technologies. In the past 12 months we have witnessed a new utility for DNA microarrays, where high-density oligonucleotide microarrays are repurposed as hybrid-selection matrices to capture defined genomic fragments as substrates for sequencing. This represents a paradigm shift from conventional approaches where DNA hybridization to a cognate array probe generates a coordinate signal and the resultant intensity is translated into biological information.

This book is very different from that inaugural volume, reflecting the rapid pace of this field. When *Nuts and Bolts* was published array technology was still a nascent field and its adoption by the scientific community heralded a new era, where assay throughput was increased by an order of magnitude, and accompanied by an exponential reduction in cost compared to pre-existing approaches.

Microarrays have evolved steadily over time from archetypal in-house complementary DNA (cDNA) arrays to robust commercial oligonucleotide platforms, a progression marked by migration to higher density biochips with increasing content and better analytical methodologies. This compendium is a cross section of the past five years in regard to microarray methods and applications and their usage in drug discovery and biomedical research. Improvements in automation (array fabrication and hybridization), new substrates for printing arrays, platform comparisons and contrasts, experimental design, data normalization and mining schemes, epigenomic array studies, electronic microarrays, comparative genomic hybridization, microRNA arrays, mutational analyses, clinical diagnostic arrays, and protein arrays and neuroscience applications are topics that are covered by experts in these areas.

Gary Hardiman
San Diego

Acknowledgments

As the editor of this compilation, it is a privilege to have been associated with this publication. I am grateful to those who have generously contributed material to this edition. Many people have helped (along the way) in making my involvement in this project possible.

I would particularly like to thank Dr. Andrew Carmen and Barbara Glunn.

I am grateful to all my colleagues at the University of California San Diego (UCSD), and particularly Dr. Chris Glass. I am indebted to the members of the The Biomedical Genomics Facility (BIOGEM) laboratory past and present, and particularly Colleen Eckhardt, James Sprague, Ivan Wick, Narimene Lekmine, and Dr. Roman Šášik. I would like to thank Deborah Seidle, Aurena Bacallan, and Cyndy Illeman at UCSD Core Biosciences.

I thank my family, my parents Maureen and Joe. I thank my wife Patricia and daughter Elena for their love, affection, and understanding.

Gary Hardiman
San Diego

Editor

Gary Hardiman, PhD has been, the director of The Biomedical Genomics Facility, since 2000, which specializes in high throughput genomic approaches, including DNA microarray technology and bioinformatics, at University of California San Diego, USA. He is a faculty member in the School of Medicine at UCSD. His research interests include pharmacogenomics and environmental toxicogenomics. Professor Hardiman serves on the editorial board of the journal *Pharmacogenomics* and has served on the executive committee of BioLink USA-Ireland. He is the editor of *Microarray Methods and Applications*, published by DNA Press, Inc. (2003), and *Biochips as Pathways to Drug Discovery*, published by CRC Press (2007).

Contributors

Christopher Adams
Invitrogen Corporation
Carlsbad, California

Mark Andersen
Invitrogen Corporation
Carlsbad, California

Shawn C. Baker
Illumina, Inc.
San Diego, California

Catalin Barbacioru
Applied Biosystems
Foster City, California

Shari L. Benson
Harvard Catalyst Laboratory for Innovative
    Translational Technologies
Harvard Medical School
Harvard University
Boston, Massachusetts

C. Ramana Bhasker
Department of Pediatrics, School
    of Medicine
University of California, San Diego
La Jolla, California

James D. Brenton
University of Cambridge
Cambridge, United Kingdom

Roger Canales
Applied Biosystems
Foster City, California

Connie L. Cepko
Department of Genetics
Harvard Medical School
Harvard University
Boston, Massachusetts

Frances Chan
Applied Biosystems
Foster City, California

Eugene Chudin
Illumina, Inc.
San Diego, California

Roberto Ciccocioppo
Department of Pharmacological Science
    and Experimental Medicine
University of Camerino
Camerino, Italy

Jacques Corbeil
University of Laval
Québec, Canada

C. Curtis
Cancer Research UK Cambridge
    Research Institute
Cambridge, United Kingdom

Gert Desmet
Vrije Universiteit Brussel
Brussels, Belgium

Ruediger Dietrich
SCHOTT Nexterion Microarray Solutions
SCHOTT Jenaer Glas GmbH
Jena, Germany

Douglas N. Gurevitch
Department of Bioengineering
University of California, San Diego
La Jolla, California

Daniel Haines
Regional Research and Development
A Division of SCHOTT North
    America Inc.
Duryea, Pennsylvania

Gary Hardiman
Department of Medicine
University of California, San Diego
La Jolla, California

Dalibor Hodko
Nanogen, Inc.
San Diego, California

Eivind Hovig
Department of Tumor Biology
Rikshospitalet-Radiumhospitalet
    Medical Center
Oslo, Norway

Kathryn Hunkapiller
Applied Biosystems
Foster City, California

Ashraf E.K. Ibrahim
Addenbrooke's Hospital
Hutchison/MRC Research Centre
Cambridge, United Kingdom

Sonia Jain
Department of Family and Preventive
    Medicine
University of California, San Diego
La Jolla, California

Tor-Kristian Jensen
PubGene AS
Oslo, Norway

Michael D. Kane
Department of Computer and
    Information Technology
Bindley Bioscience Center
Purdue University
West Lafayette, Indiana

Hio Chung Kang
UCSF Helen Diller Family
    Comprehensive Cancer Center
Cancer Research Institute
San Francisco, California

David Keys
Applied Biosystems
Foster City, California

Il-Jin Kim
UCSF Helen Diller Family
    Comprehensive Cancer Center
Cancer Research Institute
San Francisco, California

Kenneth Kuhn
Illumina, Inc.
San Diego, California

Winston Patrick Kuo
Harvard Catalyst Laboratory for
    Innovative Translational
    Technologies
Harvard Medical School

and

Department of Developmental Biology
Harvard School of Dental Medicine
Harvard University
Boston, Massachusetts

Fang Liu
Department of Tumor Biology
Rikshospitalet-Radiumhospitalet
    Medical Center
Oslo, Norway

Anbarasu Lourdusamy
Department of Pharmacological Science
    and Experimental Medicine
University of Camerino
Camerino, Italy

John Carlo Marioni
University of Cambridge
Cambridge, United Kingdom

Charles E. Massie
University of Cambridge
Cambridge, United Kingdom

Timothy K. McDaniel
Illumina, Inc.
San Diego, California

Adele Murrell
Cancer Research UK Cambridge
    Research Institute
Cambridge, United Kingdom

Steffen G. Oeser
Illumina, Inc.
San Diego, California

Lucila Ohno-Machado
Decision Systems Group
Brigham and Women's Hospital
Boston, Massachusetts

Kris Pappaert
Vrije Universiteit Brussel
Brussels, Belgium

Jae-Gahb Park
Seoul National University
Seoul, South Korea

Karen Poulter
Applied Biosystems
Foster City, California

Vardhman K. Rakyan
University of Cambridge
Cambridge, United Kingdom

Frédéric Raymond
University of Laval
Québec, Ontario, Canada

Rajendra Redkar
Regional Research and Development
A Division of SCHOTT North
   America, Inc.
Duryea, Pennsylvania

Alistair Rees
SCHOTT Nexterion Microarray
   Solutions
SCHOTT Jenaer Glas GmbH
Jena, Germany

Howard Reese
Nanogen, Inc.
San Diego, California

Richard J.D. Rouse
HTS Resources, LLC
San Diego, California

Barbara Ruggeri
Department of Pharmacological
   Science and Experimental
   Medicine
University of Camerino
Camerino, Italy

Raymond R. Samaha
Applied Biosystems
Foster City, California

Yaxian Shi
Full Moon BioSystems, Inc.
Sunnyvale, California

Dan Smolko
Nanogen, Inc.
San Diego, California

Wolfgang Sommer
National Institutes of Health
Bethesda, Maryland

Laura Soverchia
Department of Pharmacological
   Science and Experimental
   Medicine
University of Camerino
Camerino, Italy

Phillip Stafford
Biomining Worldwide
Tempe, Arizona

Yongming Sun
Applied Biosystems
Foster City, California

Paul Swanson
Nanogen, Inc.
San Diego, California

Simon Tavaré
University of Southern California
Los Angeles, California
and
University of Cambridge
Cambridge, United Kingdom

Natalie P. Thorne
The Walter and Eliza Hall Institute of
    Medical Research
Melbourne, Australia

Massimo Ubaldi
Department of Pharmacological Science
    and Experimental Medicine
University of Camerino
Camerino, Italy

Paul Van Hummelen
Flanders Interuniversity Institute for
    Biotechnology
Leuven, Belgium

Johan Vanderhoeven
Vrije Universiteit Brussel
Brussels, Belgium

Youxiang Wang
Full Moon BioSystems, Inc.
Sunnyvale, California

Yulei Wang
Applied Biosystems
Foster City, California

Steve Warrick
Invitrogen Corporation
Carlsbad, California

Whei-Kuo Wu
Autogenomics Inc.
Carlsbad, California

Shanshan Zhang
Full Moon BioSystems, Inc.
Sunnyvale, California

Yaping Zong
Full Moon BioSystems, Inc.
Sunnyvale, California

Summary of Chapters

CHAPTER 1: INTRODUCTION TO GENE EXPRESSION PROFILING WITH DNA MICROARRAY TECHNOLOGY

This chapter explores fundamental aspects critical to successfully utilizing microarray technology in the laboratory including (a) the role of DNA hybridizations in microarray technology, (b) the processes and logistics of creating a DNA microarray, (c) the common methods used to prepare samples for DNA microarray hybridization, (d) the critical components of DNA microarray hybridization and processing, and (e) an introduction to DNA microarray scanning.

CHAPTER 2: INTRODUCTION TO LARGE-SCALE GENE EXPRESSION DATA ANALYSIS

Common DNA microarray study objectives are discussed in this chapter, including (1) the identification of significant genes that are differentially expressed between two or more biological conditions, (2) the generation of possible hypotheses about the mechanisms underlying the observed phenotypes, and (3) the identification of gene expression patterns or "gene signatures" for classification purposes. An overview of different microarray platforms and data analysis methods including detailed quality assessment strategies, preprocessing methods, statistical methods for differential gene expression analysis, clustering analysis methods, and various functional annotation and interpretation methods and tools is presented.

CHAPTER 3: THE STATISTICAL DESIGN AND ANALYSIS OF GENE EXPRESSION DATA

The salient issues when designing and analyzing gene expression experiments are highlighted in this chapter. The key to a successful design is constructing an efficient and simple experiment that answers the biological question of interest within the framework of a statistical hypothesis. The design should ultimately balance experimental constraints such as the cost of chips, availability of RNA samples, and microarray platform precision to statistical considerations such as power and sample size.

CHAPTER 4: LABORATORY AUTOMATION FOR MICROARRAY EXPERIMENTATION

Microarray experimentation requires automation for creating and scanning the arrays. Both operations have two characteristics that demand automation: a large number of small features and very tight tolerances. These characteristics and the need for orderly arrays preclude manual array creation and data collection. The author discusses laboratory automation advances that have greatly aided microarray experimentation, namely the automation of array fabrication, sample preparation, and hybridization.

CHAPTER 5: HOW TO EVALUATE A MICROARRAY SCANNER

This chapter examines how the researcher can obtain the most reliable performance from a microarray scanner. After conducting extensive studies on a variety of commercial scanners, calibration tools that use a scanner calibration and validation slide were developed by the authors. These tools can be used to assist in the selection and evaluation of a microarray scanner.

CHAPTER 6: THE ROLE OF SUBSTRATES IN MICROARRAY EXPERIMENTATION AND HOW TO CHOOSE THE CORRECT COATING FOR MICROARRAYING

Open microarray systems are defined as those manufactured using robotic or manual deposition of probe molecules on coated substrates, followed by immobilization of probes. Open systems allow the design of custom arrays. The use of an appropriate substrate is an important factor for a successful microarray experiment. Microarray substrates have evolved over time and several formats are available for specific applications. Increased detection sensitivity provided by these substrates enable a more detailed analysis elucidating signals from poorly expressed mRNAs. This chapter guides the reader in the best choice of substrate for different microarray applications.

CHAPTER 7: TOWARD HIGHLY EFFICIENT OF AUTOMATED HYBRIDIZATIONS

A major problem with conventional or passive microarrays is that without active agitation, the number of target molecules available for hybridization is limited by molecular diffusion. Incomplete hybridization reactions are frequently found to be a systematic bias, which may in part explain the poor correlations sometimes observed among microarray platforms or laboratories. In this chapter, the authors discuss options for automated hybridization systems and how they improve array experimentation. Additionally they describe an approach they developed to circumvent diffusion, resulting in considerable reduction in hybridization time and sample volume.

CHAPTER 8: CHARACTERIZATION OF HYBRIDIZATION STOICHIOMETRY

In this chapter, the authors introduce a microarray system designed to measure the absolute abundance of target sequences in the hybridization milieu as opposed to the commonly reported relative abundance. This system uses a synthetic sequence control that hybridizes to every array feature and separate control sequences that are added to the hybridization mix in known concentrations to generate a set of values that represent the absolute amount of target present in the hybridization mix, across the dynamic range of expected target concentrations. The array design also facilitates quality control of the arrays since it allows the determination of the actual amount of probe DNA for each printed feature. The uniform signal also improves the accuracy of spot-finding and permits the easy identification of unusable features.

CHAPTER 9: DATA NORMALIZATION SELECTION

Classification of samples using microarray data refers to taking the expression array from the laboratory into the clinical setting. This approach has utility in grouping patients into responders and nonresponders to a therapeutic, guiding which patients would benefit from adjuvant chemotherapy, or delineating a sub-population that should report to their physician for more frequent follow-up based on their tumor expression profile. This chapter focuses on classification error, and in order to study this phenomenon, the author examined how normalization and platform affect classification error. Data are tested across three healthy human tissues, two commercial platforms, and nine different normalization methods using a simple sequential forward search feature selection. The results are presented in order to propose a normalization method that minimizes error and maximizes features with the highest predictive power.

CHAPTER 10: METHODS FOR ASSESSING MICROARRAY PERFORMANCE

This chapter describes a set of experiments routinely performed to assess the performance of Illumina bead arrays. These tests are utilized for validation whenever a new array design is implemented, changes occur to the manufacturing processes, or experimental protocols are modified. These experiments are of general interest, as the same approaches can be used to qualify the performance of any particular laboratory, laboratory personnel, reagent lots, and so on. In this chapter, the authors explore more broadly their rationale and describe community-wide efforts to standardize very similar tests, which should allow any research group or core lab to perform similar analyses in their own environment.

CHAPTER 11: COMPARISON OF DIFFERENT NORMALIZATION METHODS FOR APPLIED BIOSYSTEMS EXPRESSION ARRAY SYSTEM

While array technology has been widely used in discovery-based medical and basic biological research, its direct application in clinical practice and regulatory decision-making has been questioned. A few key issues, including the reproducibility, reliability, compatibility, and standardization of microarray analysis and results, must be critically addressed before any routine usage of microarrays in clinical laboratory and regulated areas can occur. Considerable effort has been dedicated to investigating these important issues, most of it focused on compatibility across different laboratories and analytical methods, as well as the correlation between different platforms. In this chapter, the authors investigate these issues and compare normalization approaches for a commercial array platform, the Human Genome Survey Microarray (Applied Biosystems).

CHAPTER 12: A SYSTEMATIC COMPARISON OF GENE EXPRESSION MEASUREMENTS ACROSS DIFFERENT HYBRIDIZATION-BASED TECHNOLOGIES

Many variants of high-throughput RNA-oriented technologies exist, and new implementations continue to be developed in the hope of providing more precise relative and absolute transcript-abundance measurements. In general, they can be grouped into two categories: hybridization- and sequencing-based approaches. Hybridization approaches include all forms of oligonucleotide-based microarrays and cDNA microarrays, whereas sequencing approaches include serial analysis of gene expression (SAGE), massively parallel signature sequencing (MPSS), and next-generation sequencing technologies. In recent years, data generated by the above technologies have been overwhelming. In this chapter, the authors discuss the problems associated with integrating cross-platform data. Additionally data sets from 10 different microarray platforms were compared and the results validated with quantitative real-time PCR. This study is unique in that the arrays were evaluated utilizing "actual" probe sequences.

CHAPTER 13: DNA METHYLATION ARRAYS: METHODS AND ANALYSIS

Mammalian DNA methylation describes a chemical modification that predominantly affects the cytosine base of CG dinucleotides. The regulation of DNA methylation is closely associated with other covalent modifications of the histone proteins on which DNA is assembled to form chromatin. These protein modifications include acetylation (transcriptionally activating) and methylation (transcriptionally repressive). Historically, epigenetics has played a role in cancer research, especially in the search for abnormally hypomethylated oncogenes or hypermethylated tumor-suppressor genes (i.e., genes promoting cancer that have become activated through hypomethylation, and genes suppressing cancer that have been deactivated through hypermethylation). While there are many publications exploring the biology of DNA methylation and the epigenome, and a large number of articles describing the development of approaches for studying DNA methylation, there are few articles that address the analytic issues involved in these new experiments. This chapter is aimed at the biologist who wants to understand the limitations in analyzing data obtained from different DNA methylation arrays, and the computational biologist wanting an entry point into this new and exciting area.

CHAPTER 14: ELECTRONIC MICROARRAYS: PROGRESS TOWARD DNA DIAGNOSTICS

Microarrays offer the advantage that many assays can be performed on a single device, an economy of scale that is functional at the microarray level. Multiplexing can be performed not only by analyzing a large number of genes or mutations in a sample but also by testing multiple patient samples. For example, Nanogen's 400-site microarray is capable of analyzing 23 genetic mutations for each of 64 samples on a single disposable microarray. Thus, costs are significantly reduced because many samples/assays are run on a single disposable microarray. In this chapter, the authors discuss electronic DNA microarrays and present examples of applications they have developed. Electronic microarray technology offers an unprecedented advantage of extremely rapid and controlled DNA hybridization that occurs within seconds compared to hours in passive microarrays.

CHAPTER 15: NEW APPLICATIONS FOR MICROARRAYS

Microarray-based comparative genomic hybridization (array CGH) and microRNA (miRNA) profiling are two of several emerging applications that enable effective, global interrogations of the genomic alterations, gene expression, and epigenetic mechanisms that underlie complex diseases. Mapping copy number gains and losses via CGH has led to potentially valuable diagnostic and prognostic markers for cancer and prenatal defects. MicroRNAs are an endogenous class of noncoding RNAs that play a significant role in gene regulation by acting as post-transcriptional inhibitors of gene expression. In this chapter, the authors describe these methods in detail.

CHAPTER 16: DEVELOPMENT OF AN OLIGONUCLEOTIDE MICROARRAY FOR MUTATIONAL ANALYSIS USING SINGLE OR MULTIPLE SAMPLE HYBRIDIZATION

Development of genotyping oligonucleotide microarrays that can analyze multiple samples in a single microarray will dramatically improve the sample throughput, time required, and cost of mutation analysis. The authors describe competitive DNA hybridization (CDH) as a method for the analysis of multiple samples in oligonucleotide microarrays. CDH facilitates the analysis of several samples on a single oligonucleotide microarray without the need for further sequence confirmation or other microarray experiments.

CHAPTER 17: DEVELOPMENT OF AN INTEGRATED MOLECULAR DIAGNOSTIC TEST TO IDENTIFY RESPIRATORY VIRUSES

From the avian flu to severe acute respiratory syndrome (SARS), respiratory viruses are currently the cause of great concern worldwide. Fear of an influenza pandemic is present through all strata of the population and the media devote a substantial amount of airtime to every new case of a suspicious respiratory disease. However, of equal importance are noninfluenza-related viral respiratory tract infections (VRTI), some of which result in death primarily among young infants and the elderly. This chapter describes the development of an integrated diagnostic assay to identify respiratory viruses. From the analysis of the problem to the design of the assay and its validation, this chapter walks the reader through all the design steps of a molecular diagnostic test involving primer design, PCR, primer extension, and microarray hybridization.

CHAPTER 18: MICROARRAYS IN NEUROSCIENCE

Microarray studies in neuroscience have covered a range of topics including the study of neurological and psychiatric disorders, substance abuse, and difficulties with memory and learning. Among the neurological disorders studied using microarray technology are Alzheimer's disease (AD), Parkinson's disease (PD), and multiple sclerosis (MS). Neurobiology poses several challenges to the application of microarray technology. The compartmentalization of the brain and the multitude of highly specialized cells result in highly heterogeneous samples and, consequently, expression data from brain samples represent averages of a variety of different transcriptional profiles. Furthermore, the majority of genes are expressed at low copy numbers, posing considerable demands on assay sensitivity and reproducibility. In this chapter, the authors review the issues associated with brain gene expression profiling and examine recent advances in neuroscience brought about through the use of microarrays.

CHAPTER 19: OPTIMIZATION OF PROTEIN ARRAY FABRICATION FOR ESTABLISHING HIGH-THROUGHPUT ULTRA-SENSITIVE MICROARRAY ASSAYS FOR CANCER RESEARCH

The microarray format is well suited for measuring biomarkers as it is a sensitive multiplex assay that requires small amounts of material. Antibody-based arrays require extensive optimization as proteins exhibit variable affinities and specificities. In this chapter, the authors outline the complexities in implementing reverse phase arrays. They describe the components of a high-capacity, sensitive protein microarray assay, consisting of the GeSiM Nano-Plotter piezoelectric spotter and the Zeptosens planar waveguide (PWG) imager.

1 Introduction to Gene Expression Profiling with DNA Microarray Technology

Michael D. Kane

CONTENTS

1.1 INTRODUCTION

Welcome to the emerging world of genomics and deoxyribonucleic acid (DNA) microarray technology. This remarkable and innovative technology has revolutionized how researchers determine changes in the expression levels of genes (actually messenger ribonucleic acids (mRNAs)) between two or more samples. This chapter is intended to introduce students and technical scientists to the fundamental concepts and logistics inherent to DNA microarray work, particularly those aspects relevant to conducting DNA microarray research in the laboratory. It also explores fundamental aspects critical to successfully utilizing microarray technology in the laboratory including (a) the role of DNA hybridizations in microarray technology, (b) the processes and logistics of creating a DNA microarray, (c) the common methods used to prepare samples for DNA microarray hybridization, (d) the critical components of DNA microarray hybridization and processing, and (e) an introduction to DNA microarray scanning. To begin our trek into this realm, we must briefly review the fundamental concepts of cellular and molecular biology.

It is important to remember that each living cell (and tissue) can express any gene in the genome at any time, at least theoretically. This notion is the rationale for utilizing high-density DNA microarray technology to study thousands of genes simultaneously, and reflects the fact that every living cell (nucleated eukaryotic cell) contains all the chromosomes (and therefore all the genes) within its nucleus. There are many genes that are utilized by all living cells (i.e., ubiquitous genes), and there are genes that are known only to be expressed in distinct cell types and/or at specific periods of time. For example, the cells in our retina contain all the genes needed to develop and function as a retinal cell, but they also contain the genes needed to function as a muscle cell. Yet retinal cells do not express muscle cell genes, and a scientist researching retinal cell biology would not expect to see muscle cell genes expressed in the retina (unless this was discovered as a component of a disease of the eye).

From a genomic perspective, the biology of the cell is the direct result of regulated expression (transcription) of specific genes, which encode specific proteins (translation) that confer the many different subcellular and biochemical processes within each cell. This cascade of "genomic information to biochemical function" begins with gene transcription in the cell's nucleus to form an mRNA (i.e., DNA to mRNA), which is then utilized at the cell's ribosome to synthesize proteins during translation (mRNA to protein). This is commonly described as an analogy to the construction industry, where the genes (DNA) in the nucleus are the "blueprints" for *all* known structures, and each mRNA is the specific "blueprint" taken to the job site (ribosome) to build our own protein. This dogma of cell biology provides the rationale for "gene expression profiling," where the detection (and amount) of each mRNA describes the biochemical processes and functions that are taking place in the cell. Extending our earlier analogy, gene expression profiling can determine which structures are under construction (i.e., what cellular proteins are being synthesized) by reading the blueprints being used at the job site (i.e., the mRNAs). As an example, if a gene expression profiling experiment shows that a drug-treated hepatic (liver) cell sample has increased the gene expression of glucose synthesis genes (i.e., mRNA levels) compared to an untreated sample, it can be inferred that the drug-treated hepatic cells are synthesizing more glucose. Utilizing DNA microarray technology, we can study the gene activity (mRNA levels) of essentially ALL genes within a sample, particularly when the genome of the model organism being studied has been well characterized.

This "genomic" approach to studying living systems (gene expression profiling) is much more efficient than determining the amount and activity of each protein or enzyme in the cell, simply because the methods for detecting DNA (and other nucleic acid polymers) are universal, and the detection of mRNA for a glucose synthesis gene is no different than the detection of mRNA for any other gene. Therefore, the strategy for DNA detection involves a universal set of assay conditions, simply by designing (and synthesizing) the complementary strand of DNA that will bind (hybridize) to our intended mRNA. This same strategy is utilized in polymerase chain reaction (PCR) primer design. Utilizing DNA microarray technology typically involves the conversion of mRNA to complementary DNA (cDNA) for stability and detection reasons, which will be discussed later.

Gene expression profiling is essentially the determination of changes in the abundance (amount) of each mRNA in a given biological sample. More specifically, the DNA microarray is measuring relative changes in the abundance of each mRNA in a given biological samples, compared to a control sample. This is an important distinction from other biochemical assays since we are not measuring the specific concentration of each mRNA in the sample. For example, clinical testing of cholesterol levels involves measuring the actual amount of the cholesterol within a volumetric unit of plasma (e.g., 220 ng/mL). Yet DNA microarray assays measure "relative" changes in mRNA for a number of reasons, but primarily since samples may vary tremendously in size and cell type(s). Extending our cholesterol testing example to an assay for "relative" changes in cholesterol, if we label the cholesterol with a fluorescent molecule (green fluorescence) in 1 mL of plasma taken before a human subject has eaten breakfast, and then fluorescence-label (red fluorescence) the cholesterol in 1 mL of plasma taken after our subject has eaten breakfast, we can determine how eating alters plasma cholesterol levels by measuring the amount of fluorescence. In this example, we can simply mix the green fluoro-labeled cholesterol (before breakfast sample) with the red fluoro-labeled cholesterol (after breakfast sample), and then simultaneously measure the amount of fluorescence (both red and green) in the mixed samples. If the green fluorescence is 12,000 and the red fluorescence is 24,000 (arbitrary fluorescence units), then the relative change in cholesterol after eating is twofold. In other words, the amount of cholesterol in the second sample has doubled compared to the first sample. This is exactly how a DNA microarray determines changes in the expression of each mRNA, where the mRNA in a test sample (e.g., drug-treated liver cells) and a control sample (e.g., untreated liver cells) are given green and red fluorescence labels, respectively. Changes in the amount of each mRNA in these samples involve "scanning" the DNA microarray after hybridization (i.e., taking a picture of the microarray to see both fluorescence colors), and measuring the amounts of red and green colors within each spot on the microarray. The details of this process will be discussed later.

Before discussing how DNA microarrays are made and analyzed, we must first review how scientists exploit "Watson–Crick" base pairing in the development of molecular biology methods and assays, such as DNA microarray technology.

1.2 DNA HYBRIDIZATIONS IN MICROARRAY TECHNOLOGY: SEQUENCE-SPECIFIC BINDING

Watson–Crick base pairing involves the interactions between each nucleotide within two different single strands of DNA (or any other nucleic acid polymer) that lead to the formation of a double-stranded DNA complex (i.e., the double helix). The four nucleotides or "bases" in DNA are guanine, cytosine, adenine, and thymine. DNA "base pairing," in its simplest terms, involves guanine (G) in one DNA strand binding to cytosine (C) in the other and, similarly, adenine (A) in one DNA strand binding to thymine (T) in the other. Figure 1.1 demonstrates that if a single strand of DNA contains the specific sequence of nucleotides that allows it to bind to another DNA strand, then these sequences are said to be "complementary." In this case, the sequences are such that each set of nucleotides forms an inter-strand base pair, and can therefore hybridize to form double-stranded DNA. It should be obvious that if you know the sequence of a gene you are interested in detecting, you can simply "design" the complementary sequence that will bind to that gene sequence. To extrapolate this sequence-specific interaction into the DNA microarray realm, the "probes" or "features" that are positioned on the surface of the microarray (single strands of DNA) have been designed (and synthesized) to be complementary to the genes they will detect, yet have little sequence similarity with any other gene in that genome. The design of DNA microarray probes is a complex process [1–3] and will not be discussed in this chapter, but it is important to understand that the DNA microarray utilizes the sequence-specific chemistry inherent to double-stranded DNA as the foundation of the assay. It is also important to note that DNA sequences inherently contain a direction (e.g., 5′ to 3′), which is important to designing complementary sequences, but in this introduction this aspect will not be discussed.

The DNA microarray harbors thousands of small "spots" or "features," and each of these spots is actually millions of strands of the same DNA sequence that have been covalently attached to the microarray surface. For example, if a spot on the microarray detects the gene "beta-actin," the spot is represented by millions of strands of the same DNA sequence that are complementary to beta-actin (these spots are essentially invisible to the naked eye). These strands of DNA within a single

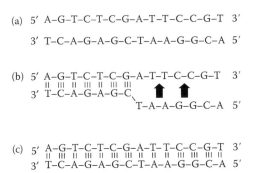

FIGURE 1.1 Sequence-specific DNA interactions. Two different DNA strands that are complementary (a) can form a double-stranded complex through a process called hybridization, where the two complementary strands of DNA bind through base pairing (b). This high-affinity, sequence-specific interaction (c) is a fundamental component of DNA molecular biology and DNA microarray technology.

spot on the microarray are called "probes." The reason why there are millions of strands of the same sequence within the spot is twofold. First, the amount of DNA present in the spot correlates with the overall binding capacity of the spot. The larger the binding capacity, the greater the amount of fluorescence signal that can be detected, thereby making the detection of the hybridized genes a more feasible process. Second, the binding capacity of the spot represents the detection range of the microarray assay. Since the DNA microarray is attempting to detect changes in gene expression levels, each spot must contain sufficient binding sites to adequately represent differences in expression levels (this is much more critical in single-color detection assays). The creation of a DNA microarray involves the attachment of these DNA sequence to the two-dimensional surface in such a manner that the sequences will not be removed during processing (i.e., washed off), yet allow the DNA sequence (probe) adequate molecular flexibility to facilitate binding (hybridization) with the complementary sequences derived from the biological sample.

1.3   CREATING A DNA MICROARRAY

The DNA microarray is created by placing DNA "probes" on the surface of a two-dimensional surface, typically a $1'' \times 3''$ glass slide with a functionalized surface. These probes come in two distinct forms: oligonucleotide and PCR probes ("cDNA probes"). Specifically, an oligonucleotide probe ("oligo-probe") is a single-stranded DNA that can range in size typically from 20 to 80 nucleotides in length and is synthesized using standard phosphoramidite chemistry. The synthesis of oligo-probes often involves the addition of a chemically active modifier to facilitate binding to the DNA microarray surface. The cDNA probe is essentially a PCR product (of almost any length) that is attached to the microarray surface using a specific attachment chemistry or simply ultraviolet cross-linking [3–7]. Since the term "cDNA" will be used later to describe the methods utilized for sample preparation, we will use the term "PCR probe" to avoid confusion. However, it is important to note that "cDNA microarrays" are prepared using PCR products rather than synthesized oligonucleotides.

The decision to utilize oligo-probes or PCR probes depends upon the amount of genomic information known about the organism or cell system under investigation. Since oligo-probes are designed using genomic information from the organism or cell system under investigation, it is nearly impossible to design oligo-probes for organisms where no genomic data are available (i.e., the genome and transcriptome have not been sequenced or characterized). Therefore, gene expression studies in "emerging" organisms (i.e., those with little genomic data available) often involve PCR products derived from a cDNA library.

It is important to note that there are several advantages to oligo-probes (compared to PCR probes), such as designing multiple oligo-probes to a single gene, targeting oligo-probe designs to specific exons or exon boundaries [8], and the fact that these can be designed to essentially avoid potential cross-hybridization with non-target genes [3]. In any case, we will explore the logistics of the creation of an oligo-probe microarray utilizing a functionalized microarray surface that reacts with primary amines, as well as a PCR probe microarray on a poly-lysine-coated microarray surface.

As mentioned earlier, the process of synthesizing oligo-probes (using standard phosphoramidite chemistry) often involves the addition of a chemically active modifier to facilitate covalent binding of the oligo-probe to the microarray surface (i.e., "spotting" the oligo-probe). In this example we will assume that the microarray surface harbors a functionalized surface to facilitate binding, which reacts with primary amines to create a Schiff's base through dehydration. Therefore, our oligo-probes will be designed and synthesized with an amino-modifier group at the 5′ end of the sequence. In other words, the DNA synthesis process will add a small molecule at the end of our oligo-probe sequence that contains a primary amine group. The amino-modifier can be added to either end of the oligo-probe sequence (5′ or 3′ end). Yet, since oligonucleotide phosphoramidite synthesis proceeds from the 3′ to the 5′ end, adding the modifier in the "last" synthesis step (i.e., 5′ end) will ensure that full-length oligo-probes will be attached to the microarray, and any synthesis failures in the reaction will be far less likely to attach. It is important to note that there are methods for oligo-probe spotting and attachment that do not require modifiers, which are not discussed in this chapter [9–11]. It should also be noted that PCR probes can be attached to amine-reactive surfaces using the amine groups native to DNA, or by adding a 5′ amino-modifier to the PCR primer(s) utilized in PCR probe amplification [3].

Spotting amino-modified oligo-probes simply involves delivering the probes to the microarray surface (i.e., DNA microarray spotting instrument) in buffers that lack amino groups, and incubating the spotted microarrays under conditions that facilitate Schiff's base formation [3]. This typically involves a specific amount of time (e.g., hours) at room temperature under controlled humidity conditions. Once the microarrays have been spotted, a subsequent step will be followed that quenches the reactivity of the remaining amino-reactive surface on the microarray (i.e., the spaces between the spots). Failure to effectively quench these amino-reactive groups will result in high background on the hybridized microarray or "black holes" where the spots are located [12,13], simply because the fluorescence-labeled sample will bind to this reactive surface.

Spotting PCR probes on a poly-lysine microarray surface is carried out in a similar manner, but steps are taken to link the double-stranded PCR product to the surface. Typically, this attachment method involves high heat (baking) and/or ultraviolet cross-linking. If the PCR probes are being spotted on a functionalized surface (e.g., amino-reactive surface), then the spotting protocol will be similar to the oligo-probe spotting process [3].

When spotting oligo- or cDNA probes, there are subsequent quality-control steps that can be utilized to verify that the DNA (probe) has been covalently attached and is hybridizing correctly. To verify that DNA attachment has occurred, protocols typically utilize a fluorescent dye that binds to DNA nonspecifically (e.g., sybergreen), and the fluorescent image derived from the "stained" microarray will show fluorescence where DNA (probes) have been successfully attached (i.e., bright spots). To verify that the spotted probes are capable of hybridization, protocols typically involve hybridizing the microarray with a short, fluorescence-labeled random DNA sequence (e.g., 5′-cyanine-3-labeled random 9-mer), and successful hybridization is demonstrated when the spots show fluorescence. The verification that the spotted probes are capable of hybridization is important, particularly when ultraviolet cross-linking is utilized (i.e., PCR probes), simply because excessive cross-linking will attach the probe to the microarray surface but prevent adequate hybridization performance [12,13]. In other words, the DNA has been excessively cross-linked to itself and is unavailable for hybridization.

Once the DNA microarrays have been prepared, they are stored at room temperature in a desiccated environment (often for months at a time) until needed for sample hybridizations.

1.4 SAMPLE PREPARATION FOR DNA MICROARRAY HYBRIDIZATION

Certainly, the first step in sample preparation for gene expression profiling is RNA isolation from the biological sample. There are a number of methods available for RNA isolation, and it is likely that your laboratory has established protocols for this step in sample processing. Once RNA is obtained from these samples, the mRNA is converted to cDNA using reverse transcription. This is much more stable than mRNA, and the biomolecular conversion of mRNA to cDNA provides the opportunity to add fluorescent groups to the cDNA. In this process mRNA serves as the sequence template for reverse transcription, which synthesizes the complementary strand of DNA from mRNA (cDNA synthesized from mRNA). Figure 1.2 demonstrates how reverse transcriptase (1) binds to a "primer" that is hybridized to the mRNA; (2) proceeds along the mRNA strand incorporating free nucleotides into the growing cDNA strand; and (3) the inclusion of free nucleotides, which are conjugated to fluorescent molecules, becomes incorporated into the growing cDNA strand, thereby creating a cDNA that harbors fluorescence for detection. Therefore, we have created one cDNA for each mRNA in the sample, and these fluorescence-labeled cDNAs represent the mRNAs in the original sample and will be hybridized to the microarray.

The key component of fluorescence labeling is the fact that one (or more) of the free nucleotides (ATP, CTP, GTP, and TTP) has a fluorescent dye conjugated to the nucleotide. Thus, when the

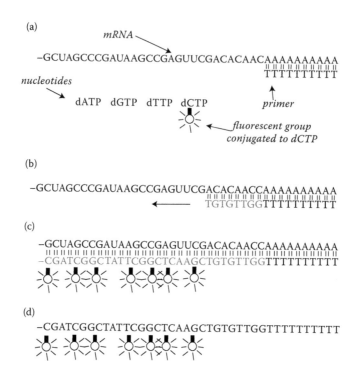

FIGURE 1.2 cDNA generation and fluorescence labeling using reverse transcription. mRNA is the single-stranded representation of a gene, which is the molecule being studied during gene expression profiling. mRNA is converted to cDNA through reverse transcription (a), which incorporates free nucleotides (including fluoro-labeled nucleotides) into the growing cDNA strand (b). The product of the reverse transcription reaction is a double-stranded cDNA : mRNA complex (c), and the mRNA strand is removed (e.g., RNAse H) to expose the fluoro-labeled cDNA that is ready for microarray hybridization (d).

dye-conjugated nucleotide is incorporated into the cDNA, the fluorescent dye is simultaneously added to the cDNA, thereby providing a fluorescent signal available for hybridization. As mentioned earlier, the DNA microarray is measuring the relative changes in the mRNA levels, and each microarray experiment involves two reverse transcription reactions (e.g., control and drug-treated). Therefore, the "control" (e.g., untreated) mRNA sample will be added to a reverse transcription reaction that includes a dye-conjugated nucleotide (green), whereas the "test" (e.g., drug-treated) sample will be added to a reverse transcription reaction that includes a different dye-conjugated nucleotide (red). The cDNAs derived from the two reactions will be mixed prior to microarray hybridization, creating a "two-color" sample. Therefore, the beta-actin gene in the samples (actually the fluoro-cDNA derived from the beta-actin mRNA) will be green in the control sample and red in the test sample, and both will hybridize to the beta-actin spot on the microarray (i.e., the probe that has been designed to detect beta-actin). In this example, if the test sample (i.e., drug-treated sample) causes beta-actin to increase the mRNA expression levels, then the beta-actin spot will appear more red than green (after color channel normalization). If the green fluorescence (control sample) from the beta-actin spot is measured at 10,000 relative fluorescence units (RFUs) and the red fluorescence (test sample) at 40,000 RFUs, then the test sample contains a fourfold increase in beta-actin expression (i.e., a 400% increase over the control). Figure 1.3 demonstrates how fluoro-labeled cDNA hybridizes to oligo-probes attached to the microarray surface.

Regarding the reverse transcription reactions used in sample labeling, it is important to understand that the initiation of a reverse transcription reaction requires a short double-stranded region (mRNA : DNA hybrid) to allow the reverse transcriptase to bind and synthesize to the

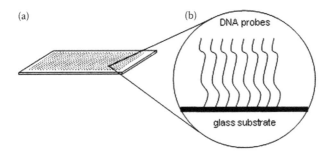

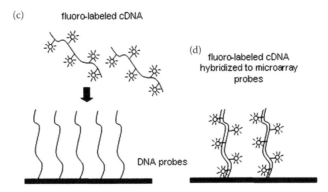

FIGURE 1.3 The DNA microarray hybridization process. The DNA microarray harbors thousands of "spots" or "features" (a), and each of these spots is actually millions of strands of a single DNA sequence (b) capable of binding to a specific gene (actually fluoro-labeled cDNA) in a sequence-specific interaction. During hybridization, the fluoro-cDNA derived from the reverse transcription reaction is placed on the microarray (c), and each fluoro-cDNA binds to the appropriate probe sequences (within the spots) during the hybridization incubation (d). The amount or intensity of fluorescence from each spot is detected during "scanning" or "imaging," which is the measure of each gene's expression level.

complementary strand (cDNA). This double-stranded region is created using a synthetic strand of DNA called a "primer." Typically, the reverse transcription primer utilized in these reactions is oligo-dT in eukaryotic samples (which hybridizes to the poly-A tail of eukaryotic mRNAs) or random primers in prokaryotic samples. It should be somewhat obvious that the cDNA is the molecular "representation" of the mRNA, and the cDNAs will hybridize to the microarray probes. Therefore, once the cDNA (fluoro-cDNA) has been derived from the mRNA, the mRNA is no longer needed and many protocols remove the mRNA from the mRNA:cDNA complex using ribonuclease H (RNAse H).

In eukaryotic samples, utilizing an oligo-dT primer will derive a 3' bias cDNA, whereas utilizing random primers alleviates this bias. In other words, the use of an oligo-dT primer, which will generate cDNAs from the 3' ends of the mRNAs (i.e., 3' bias), requires that the probes on the microarray are also designed to hybridize with 3'-biased cDNAs. If the oligo-probe design process allows the probe to be positioned in any location within the mRNA sequence, then random primers are required to generate cDNAs from the entire mRNA sequence.

There are extended methods for sample labeling that build on this simple example and subsequently generate cRNA (using in vitro transcription) from the cDNA to increase the overall fluorescence signal of the system, or incorporate a high-affinity binding site (instead of the direct incorporation of the fluorescence group) within the cDNA where the fluorescence group is added in a later step (i.e., "indirect" labeling methods). Yet the same principles described above apply to these methods, and the overall objective is to convert the mRNA in the sample to form cDNA that harbors fluorescence.

The two fluorescent dyes typically utilized in fluorescence labeling are cyanine-3 (Cy-3) and cyanine-5 (Cy-5), which are green- and red-colored dyes. Note that these colors describe the dyes' fluorescence (upon excitation), and not the color that appears to the naked eye. Cy-3 appears red to the naked eye, whereas Cy-5 appears blue to the naked eye.

Now that the mRNA has been converted to fluoro-cDNA, the sample is ready for hybridization with the microarray.

1.5   DNA MICROARRAY HYBRIDIZATION AND PROCESSING

The hybridization method(s) are aimed at placing the fluoro-cDNA on the two-dimensional surface utilizing a stringency conditions to facilitate sequence-specific binding. "Stringency" is a term used to describe the molecular (thermodynamic) energy required for binding two complementary, single-stranded DNA molecules, which is dependent largely on temperature, salt concentrations, and pH. High stringency conditions involve high temperatures and/or low salt concentrations, and DNA hybridizations proceed slowly but in a sequence-specific manner, whereas low stringency conditions involves cooler temperatures and/or high salt concentrations, and DNA can form double-stranded complexes even if their sequences are not complementary (i.e., nonspecific binding). As an example, distilled water's stringency is too high because it lacks salts and will quickly denature double-stranded DNA (i.e., do not put hybridized microarrays in distilled water, as this will quickly reverse the hybridization process!). Similarly, if you hybridize your DNA microarray at lower temperatures, you will get nonspecific binding of your fluoro-cDNA to all the probes, and every spot will appear to show nearly similar florescence content. There are many different hybridization buffers, which are associated with various hybridization temperatures, available for controlling the stringency of the hybridization.

Typically, hybridization involves placing the fluoro-cDNA in a specific buffer, and sandwiching a sample volume (50–500 μL) between the DNA microarray and a cover slip or blank glass slide. This assembly is then placed in a chamber where temperature, and sometimes humidity, is controlled. Typically, the hybridization needs more than 16–19 h (i.e., overnight) to allow sufficient time for the probes to bind to the fluoro-cDNAs in a sequence-specific manner. Once the incubation is complete, care should be taken while removing the excess sample through a series of buffer washes where stringency is controlled, and finally the microarrays (slides) are dried using centrifugation or airflow. The microarrays are now ready for scanning (i.e., fluorescence detection).

1.6   DNA MICROARRAY SCANNING

At this point the laboratory benchwork is essentially complete, and the microarrays are placed in a dedicated microarray scanning instrument. Recall that the cDNA sample placed on the microarray harbors fluorescent groups and, upon hybridization, the green and red fluorescence colors (representing the control and drug-treated samples, for example) are now bound to the probes (spots). Therefore, the spots will appear in varying colors from red to green to yellow (yellow is a rendering of mixed red and green fluorescence). If, for example, the control sample was labeled green (i.e., the reverse transcription reaction using the control sample included a green-conjugated nucleotide) and the drug-treated sample was labeled red, then a spot appearing red would indicate that that gene (mRNA) increased the expression level during drug treatment. Note that spots lacking any color (fluorescence) indicate that the gene (mRNA) was not expressed in the sample.

There are many different microarray scanning instruments; however, the overall objective of their use is similar. Essentially, the goal is to derive the best image of the microarray where the sensitivity of the detection platform is optimal for detecting "low-copy" genes (i.e., mRNAs that are not abundant in our samples), yet "high-copy" genes are not saturating the detector (i.e., the bright spots on the array are within the dynamic range of the system). This represents another challenge to DNA microarray technology utilization. Ideally, the imaging system will detect very rare or low-copy mRNAs while simultaneously providing high-quality data for abundant mRNAs, yet this is rarely the case. Fundamentally, mRNA levels within a tissue sample range from zero to millions of copies, yet the dynamic range of a typical fluorescence detection system is approximately four orders of magnitude. Thus the fluorescence detection range is smaller than the expected range of potential mRNA levels. With this disparity in mind, most microarray scanning instruments allow the user to adjust the sensitivity of the system (e.g., adjust the PMT gain) and it is the user's responsibility to adjust these variables for optimal performance. Therefore, increasing the sensitivity of the system to best detect dim or low-level spots increases the number of spots that will saturate the detector, and care must be taken to identifying these saturated spots since the effect of detector saturation is to underestimate the mRNA levels. Alternatively, decreasing the sensitivity of the imaging system to minimize spot saturation will decrease the number of mRNAs detected by omitting low-level fluorescent spots. The user must find a balance between these extremes to consistently derive the best data from a gene expression profiling—microarray study.

Once the microarray image has been derived using the scanner (typically this is actually two images representing the red and green images, and the scanner software displays an "overlay" of these images), raw data analysis is needed to (1) associate each spot with the gene (mRNA) that it is detecting; and (2) normalize the red and green channels to correct for any differences in initial RNA concentrations, labeling reaction efficiencies, and differences in the capabilities of each channel (red and green) within the scanner itself. This process can be time-consuming, but correct raw data analysis is crucial to deriving high-quality results for subsequent bioinformatics analysis.

1.7 CONCLUSION

The successful utilization of DNA microarray technology in gene expression profiling involves the integration of methods and includes aspects not discussed in this chapter, including study design, sample isolation methods, and data analysis. Therefore, scientists exploiting the capabilities of microarray technology must consider many technical and analytical components. There are many published reviews regarding the general utility of DNA microarrays [14–19], as well as published reports within specific scientific domains such as cancer, neuroscience, and pharmacogenomics to name a few, which should be explored with regard to the area of research under pursuit.

REFERENCES

1. Rimour, S, Hill, D, Militon, C, and Peyret, P. GoArrays: Highly dynamic and efficient microarray probe design. *Bioinformatics* 2005; 21: 1094–1103.
2. Haas, SA, Hild, M, Wright, APH, Hain, T, Talibi, D, and Vingron, M. Genome-scale design of PCR primers and long oligomers for DNA microarrays. *Nucleic Acids Res* 2003; 31: 5576–5581.
3. Kane, MD, Jatkoe, TA, Stumpf, CR, Lu, J, Thomas, JD, and Madore, SJ. Assessment of the sensitivity and specificity of oligonucleotide microarrays. *Nucleic Acids Res* 2000; 28: 4552–4557.
4. Dufva, M. Fabrication of high quality microarrays. *Biomol Eng* 2005; 22: 173–184.
5. Auburn, RP, Kreil, DP, Meadows, LA, Fischer, B, Matilla, SS, and Russell, S. Robotic spotting of cDNA and oligonucleotide microarrays. *Trends Biotechnol* 2005; 23: 374–379.
6. Park, CH, Jeong, HJ, Jung, JJ, Lee, GY, Kim, SC, Kim, TS, Yang, SH, Chung, HC, and Rha, SY. Fabrication of high quality cDNA microarray using a small amount of cDNA. *Int J Mol Med* 2004; 13: 675–679.
7. Dolan, PL, Wu, Y, Ista, LK, Metzenberg, RL, Nelson, MA, and Lopez, GP. Robust and efficient synthetic method for forming DNA microarrays. *Nucleic Acids Res* 2001; 29(21): e217.
8. Kane, MD, Dombkowski, AA, and Madore, SJ. The emerging utility of oligonucleotide microarrays. In: Q. Lu and M. Weiner (eds), *Gene Cloning and Expression Technologies*. Eaton Publishing, Westborough, MA, 2002, pp. 537–547.

9. Belosludtsev, Y, Iverson, B, Lemeshko, S, Eggers, R, Wiese, R, Lee, S, Powdrill, T, and Hogan, M. DNA microarrays based on noncovalent oligonucleotide attachment and hybridization in two dimensions. *Anal Biochem* 2001; 292: 250–256.

10. Call, DR, Chandler, DP, and Brockman, F. Fabrication of DNA microarrays using unmodified oligonucleotide probes. *Biotechniques* 2001; 30: 368–376.

11. Lipshutz, RJ, Fodor, SP, Gingeras, TR, and Lockhart, DJ. High density synthetic oligonucleotide arrays. *Nat Biotechnol* 1999; 14: 1675–1680.

12. Li, Q, Fraley, C, Bumgarner, RE, Yeung, KY, and Raftery, AE. Donuts, scratches and blanks: Robust model-based segmentation of microarray images. *Bioinformatics* 2005; 21: 2875–2882.

13. Hartmann, O. Quality control for microarray experiments. *Methods Inf Med* 2005; 44: 408–413.

14. Repsilber, D and Ziegler, A. Two-color microarray experiments: Technology and sources of variance. *Methods Inf Med* 2005; 44: 400–404.

15. Jares, P. DNA microarray applications in functional genomics. *Ultrastruct Pathol* 2006; 30: 209–219.

16. Hoheisel, JC. Microarray technology: beyond transcript profiling and genotype analysis. *Nat Rev Genet* 2006; 7: 200–210.

17. Fathallah-Shaykh, HM. Microarrays: applications and pitfalls. *Arch Neurol* 2005; 62: 1669–1672.

18. Trent, JM. Expression profiling using cDNA microarrays. *Nat Genet Supp* 1999; 21: 10–14.

19. Stoughton, RB. Applications of DNA microarrays in biology. *Annu Rev Biochem* 2005; 74: 53–82.

2 Introduction to Large-Scale Gene Expression Data Analysis

Anbarasu Lourdusamy, Massimo Ubaldi, Laura Soverchia,
Gary Hardiman, and Roberto Ciccocioppo

CONTENTS

2.1 INTRODUCTION

Recently, various deoxyribonucleic acid (DNA) microarray technologies to quantify the gene expression on a genome-wide scale have been developed to identify genes and pathways underlying complex disorders. DNA microarray approaches provide the ability to measure the expression levels of thousands of genes simultaneously. Common DNA microarray study objectives are (a) the identification of significant genes that are differentially expressed between two or more biological conditions, (b) the generation of possible hypotheses about the mechanisms underlying the observed phenotypes, and (c) the identification of gene expression patterns or "gene signatures" for classification purposes.

The use of DNA microarrays to study gene expression on a genome-wide scale results in large datasets, which are frequently difficult to analyze and could result in misleading conclusions. It is important to realize that several prerequisites need to be fulfilled to obtain meaningful results. For instance, the experimental design needs to fulfill some minimal criteria such as clearly defined hypotheses, and there should be an adequate number of samples representing different biological conditions and intrinsic variances. Furthermore, appropriate statistical data analysis methods should be determined, supporting the overall objectives of the study. For example, unsupervised methods such as clustering can help in finding similar gene expression patterns but do not provide statistically valid information on the magnitude of the transcriptional changes between different biological conditions.

In this chapter, we provide an overview of different microarray platforms and data analysis methods including detailed quality assessment strategies, preprocessing methods, statistical methods for differential gene expression analysis, clustering analysis methods, and various functional annotation and interpretation methods and tools.

2.2 MICROARRAY PLATFORMS

The high-density DNA microarray is one of the most powerful, versatile, and widely used tools for gene expression analysis. It is based on the principle of specificity and complementary hybridization of nucleotide sequences. Thousands of ordered individual nucleic acid species (*probes*) are hybridized with a complex mixture of labeled nucleic acids (the *targets*). The fluorescent-labeled targets bind to their appropriate probe partners arrayed on a solid surface (the *microarray*) and the intensity of emission can be assessed by an ion laser. The two main microarray technologies that are widely used are oligonucleotide and complementary DNA (cDNA) arrays.

2.2.1 Oligonucleotide Microarrays

High-density oligonucleotide microarrays were first manufactured on a large, commercial scale by Affymetrix, Inc. (Santa Clara, California). They are produced by chemical synthesis of oligonucleotides directly on a solid surface, using a photolithographic process [1,2]. In Affymetrix Genechips, the oligonucleotides are generally 25 bases long. As the sensitivity and specificity of a probe of 25-mers may not be high enough, each perfect match (PM) probe is paired with a mismatch (MM) probe, an identical probe except for a single base difference in a central position. PM and MM probes are called a *probe pair*. Each transcript is represented by 11–20 such probe pairs, called a *probe set*. The MM probes are used to estimate the degree of nonspecific hybridization and enable background subtraction. The strategy of multiple probes per gene greatly improves signal-to-noise ratios (SNRs) and reduces the rate of false positives.

In addition to Affymetrix, alternative platforms have emerged and are being widely used. Illumina (San Diego, California) has developed a bead-based technology for single nucleotide polymorphism (SNP) genotyping and gene expression profiling applications on two distinct substrates, the Sentrix LD BeadChip and the Sentrix Array Matrix (which multiplex up to 8 and 96 samples, respectively). The chemiluminescence-based Applied Biosystems Expression Array System (Foster City, California) employs standard phosphoramidite chemistry to synthesize 60-mer oligonucleotides, which are validated offline by mass spectrometry and are subsequently deposited onto a derivative nylon substrate. A platform widely used by the pharmaceutical industry is the CodeLink Bioarray previously from GE Healthcare (Piscataway, New Jersey) and currently available from Applied Microarrays, Inc. (Tempe, Arizona). The 30-mer oligonucleotides are synthesized *ex situ* using standard phosphoramidite chemistry in a similar manner to the ABI biochips. Agilent Technologies (Palo Alto, California) performs *in situ* synthesis of 60-mer probes via ink-jet printing using phosphoramidite chemistry. Combimatrix (Mukilteo, Washington) have developed a solid-phase oligonucleotide synthesis system using a method that electrochemically places monomers at defined locations on substrates. Nimblegen has synthesized microarrays containing 380,000 features using a digital light processor that creates digital masks to synthesize specific polymers.

The salient features of these different oligonucleotide platforms have been reviewed elsewhere [3–5]. A systematic comparison of gene expression measurements across different hybridization-based technologies is presented in Chapter 12.

2.2.2 cDNA Microarrays

The cDNA microarray platform was pioneered and made popular by Patrick Brown's laboratory at Stanford [6–8]. Genes to be spotted are generated with groups of sequenced clone libraries amplified by the polymerase chain reaction (PCR) ranging from 0.6 to 1.5 kb. Ribonucleic acid (RNA) is extracted and isolated from two different populations of cells or tissues from patients, and then reverse transcribed and amplified with the PCR. RNA is labeled with fluorescent pairs Cy5 (red) and Cy3 (green) and arrayed by contact printing with a robotic arraying machine onto glass slides. The hybridized microarray is scanned to acquire the fluorescent images.

2.3 MICROARRAY DATA ANALYSIS

Microarray data analysis is a complex process involving multiple steps and not only data generation, but also data analysis and mining using appropriate statistical methods. Systematic variations that largely influence gene expression measurements can occur at any step of the process: messenger RNA (mRNA) sample extraction and amplification, labeling, hybridization, and scanning. These variations can partly be reduced by background correction, normalization, and a summary of multiple probes per transcript or quality control (QC) measures.

Microarray experiments are frequently used for the purpose of identifying genes that are differentially expressed between multiple conditions. This amounts to comparing two or more groups and generating a ranked gene list with an appropriate test statistic of differential expression. In a further step, statistical significance is calculated for each gene and differentially expressed genes can be selected at specified cut-off values. Even for a simple case of comparing two groups, proper preprocessing and choice of gene selection method are crucial. This section will focus on various quality assessment methods, background correction and normalization, differential expression, clustering, and functional analysis methods.

2.3.1 QUALITY ASSESSMENT

Quality assessment is the first step and is critical for any subsequent microarray data analysis. Unwanted data variability introduced in each step of microarray data generation might affect data quality and mask the effect of interesting biological factors on the gene expression profile. Excluding low-quality chips in the analysis might increase the number of differentially expressed genes, while simultaneously reducing the expected false positives in the significant gene list.

For high-density oligonucleotide microarrays, the Affymetrix Microarray Suite expression analysis algorithms calculate several parameters relating to the quality of the expression chip, background intensity, noise, scaling factor, percent present calls, and 3′ to 5′ ratio. The background intensity, derived from the intensity values of the lowest 2% of cells on the chip, establishes an overall baseline intensity to be subtracted from all cells before gene expression levels are calculated. Noise is derived from the standard deviation of the background intensity measurement. Affymetrix suggests that, for best analytical results, background intensities and noise values should not exceed 100 and 4, respectively. The scaling factor is inversely related to the overall brightness of the chip and the values should be <3. The percent present call represents the percentage of probe sets called "Present" on an array, whereas the 3′ to 5′ ratio refers to the signal intensity ratios for the housekeeping genes such as *beta-actin* and *gapdh*. Although these quality metrics are useful for quality assessment, none of the recommended fixed threshold values have a connection to the expression values. In addition, these metrics ignore the within-chip variation and hence fail to detect spatial artifacts that occur within the chip. Reimers and Weinstein proposed a regional bias detection algorithm and also presented methods for visualizing and quantifying the levels of regional bias [9]. The quality measures can be obtained at the probe level for Affymetrix Genechips using probe-level models (PLM). Image plots of robust weights or residuals obtained from robust PLMs can identify aberrant chips [10]. In addition, exploratory data analysis methods and visualization can also be effectively used in quality assessment of Affymetrix Genechips. A simple image visualization of log intensity values, box plots of intensity distributions, and histograms of probe-level data can easily reveal potential artifacts and outliers among samples. Unsupervised methods such as hierarchical clustering and multidimensional scaling (MDS) can also be used to determine the overall relationship among samples [11].

For two-channel microarray data, quality assessment can be performed using spot quality, print-run quality, hybridization, and mRNA quality. Several diagnostics plots can be used to assess the spot and hybridization quality of individual arrays. One such diagnostic plot is shown in Figure 2.1, produced using the Bioconductor package "arrayQuality" for qualitative array assessment [12]. The diagnostic plot displays eight different panels including M/A plots, spatial plots, histograms, and dot plots. M/A plots display the log intensity ratio $M = \log_2(R/G)$ versus the mean log intensity $A = \log_2 \sqrt{RG}$. This plot type is widely used to visualize array data because it directly displays the red-to-green ratios, which are often the quantities of interest in most experiments. Furthermore, M/A plots make it easy to identify intensity-dependent biases in the data. The spatial plots are the graphical representations of microarray data according to the locations on the slide. Three different plots are generated showing raw M values, normalized M values, and raw A values. These plots can be used to visually detect uneven hybridization and missing spots, and also to identify spatial effects on the hybridized arrays. The diagnostic plot also generates histograms of the signal-to-noise log-ratio for comparison of the distribution of intensities between the two channels.

2.3.2 PREPROCESSING

Preprocessing oligo arrays generally involves three steps: background correction, normalization, and summarization. Recently, more methods have become available for preprocessing the Affymetrix Genechips. The Affymetrix system uses Microarray Suite 5.0 (MAS5) in which the background is

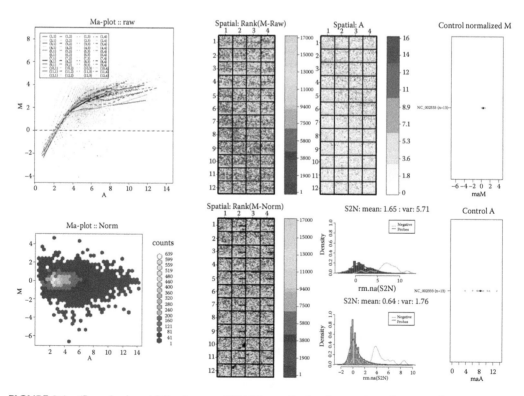

FIGURE 2.1 (See color insert following page 138.) Diagnostic plots for two-channel array quality assessments.

calculated on the basis of the second percentile signal in each of the 16 subsections of the chip [13]. This method corrects both PM and MM probes. The Robust Multichip Average (RMA) algorithm, which is implemented in the Bioconductor package "Affy," ignores the MM probes altogether for background correction [14]. It subtracts a background value that is based on modeling the PM signal intensities as a convolution of an exponential distribution of signals and a normal distribution of nonspecific signals. Li and Wong proposed model-based expression indexes (MBEI) that comprise an error model for PM and MM probes [15]. This method is implemented in the freeware program dCHIP and the option is also available for PM, but only for correction. Recently, Wu et al. introduced a sequence-based statistical model in the modified version of RMA, and it is implemented in the Bioconductor package "GCRMA" [16].

Normalization in microarray experiments is carried out based on the assumption that only a small proportion of genes will be differentially expressed among the thousands of genes present in the array and/or that there is symmetry in the up- and down-regulation of genes. The Affymetrix MAS5 uses a trimmed mean-based scaling method in which all arrays are scaled to a chosen baseline array. Both RMA and GCRMA use quantile normalization, which imposes the same empirical distribution of intensities to each array. The quantile normalization approach first ranks data on each array and substitutes data of the same rank across all arrays by the mean of the data [17]. The dCHIP program uses the invariant set normalization method in which a subset of genes that are stable across all conditions are selected as references rather than using standard housekeeping genes. Other normalization methods for oligo arrays include cyclic Loess, variance stabilization [18], contrast normalization [19], and splines smoothing.

The final summarization step generates an expression value for each gene from the set of probe values. There are several summarization methods available for Affymetrix Genechip arrays. MAS5 uses the one-way Tukey's biweight method to get a gene expression summary. dCHIP uses a model to estimate the signal based on the original scale. Both RMA and GCRMA use median polish to obtain a single gene expression value for each probe set on the log-transformed intensities. It has been shown that RMA and dCHIP (Figure 2.2) generate comparable expression values with less dispersion than other methods.

Likewise, for two-channel arrays, there exist various methods for preprocessing and normalization. Most standard image processing algorithms extract the signal intensities for each spot and from the surrounding background. The measurement of background intensities can be averaged over entire arrays or taken from the area adjacent to a spot. The simplest form of background correction is to subtract the background intensities from the foreground intensities. The other sophisticated background correction methods such as "normexp" and "morph" are implemented in the Bioconductor package "limma" [20]. These methods adjust the foreground intensities adaptively for the background intensities and result in positive adjusted intensities only. To obtain reliable data for the normalization procedure, the poor-quality spots should be removed using different filtering criteria such as the spot's size, SNR, and the difference between the spot's mean and median intensities.

There are several methods for normalizing of two-color arrays. The most common is the global normalization, which ensures that every array has the same median intensity. The global normalization method works better when there is no intensity effect and no block effect. However, it was soon realized that dye biases were dependent on spot intensity and the global normalization method was not adequate to remove these biases. Yang et al. proposed locally weighted linear regression (Loess) normalization, an intensity-dependent method to account for dye biases [21]. This iterative normalization method works on the assumption that the average of up-regulated expression profiles is approximately the same as the down-regulated expression profiles at each intensity level. This assumption is not necessarily valid for customized arrays. A similar approach, using a different local regression method rather than Loess, was proposed by Kepler et al. [22]. Other normalization techniques include methods based on analysis of variance (ANOVA) proposed by Kerr et al. [23] and Wolfinger et al. [24]. All the intensity-based normalization methods described earlier use all or most of the genes on the array. An alternate approach is to use a smaller subset of genes such as

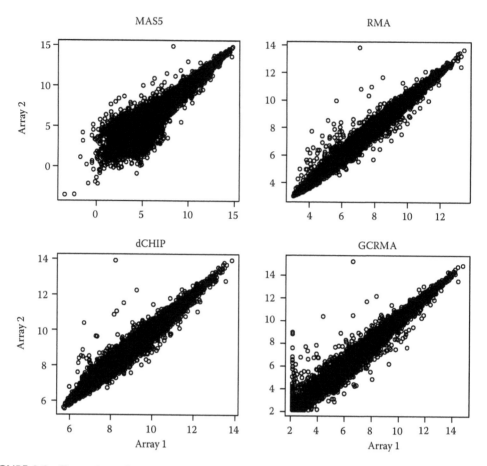

FIGURE 2.2 Expression values generated from four different methods. Expression levels detected on the Array 1 (*x*-axis) are plotted against levels detected on the Array 2 (*y*-axis); dCHIP and MAS 5.0 are shown on the log scale for compatibility with RMA and GCRMA.

housekeeping and invariant genes. Although often assumed to be constantly expressed, the expression of housekeeping genes can vary substantially. As a result, using housekeeping genes normalization might introduce another potential source of error. Another similar approach is to find genes that are constantly expressed across different experimental conditions. One such approach is the rank-invariant set of genes proposed by Tseng et al. [25]. Other useful methods include the two-way semilinear model and robust normalization.

2.3.3 DIFFERENTIAL EXPRESSION

Identifying genes that show reproducible differences in mRNA abundance between sample classes is the most basic and important step in microarray data analysis. A huge variety of methods have been used for this task and the choice of method is highly dependent on the experimental setting. In the early days of microarray experiments, when no or only a very limited number of replicates were included in the experimental design, a fixed fold-change cut-off (usually twofold) was used to define differentially expressed genes. This approach does not take into account the biological and experimental variability in the data, and thus many genes with high fold-changes but poor-quality data were mistakenly identified as being differentially expressed, whereas genes with reproducible data but low fold-changes were missed. In addition, the choice of fold-change threshold is very subjective and is very much dependent on the experiment and technological variables [26]. To avoid these problems, many sophisticated statistical approaches have been developed based on different models. These statistical tests usually involve three steps: calculating a test statistic, assigning the significance, and choosing a cut-off value for the statistical significance.

For a simple two condition experiment, one way to identify the differentially expressed genes is to use an ordinary t-test, $t = M/(s/\sqrt{n})$, where M is the average log-ratio, s is the standard deviation of the M-values, and n is the number of replicates. In this approach, a gene with a very small s value due to its low expression level will have a large t-value regardless of the M-value, and thus this gene can be selected as the differentially expressed gene even though it is not truly differentially expressed. To overcome this problem of the traditional t-test, various methods have been proposed using a penalized t-test, where a constant is added in the denominator of t, thus making it less sensitive to small variances. Among these methods, there are significant analysis of microarrays (SAMs) by Tusher et al. [27], nonparametric empirical Bayes by Efron et al. [28], and B-statistics proposed by Lönnstedt et al. [29], and further developed by Smyth et al. [30]. The empirical Bayes methods work by effectively borrowing information across genes to obtain the variance estimate, thus avoiding the problem of many genes and few replicates. The resulting test statistic is a moderated t-statistic, where the weighted average of variance estimates and the constant factor are used instead of single-gene estimated variances. Related to Bayesian approaches, Baldi and Long [31] suggested a Bayesian probabilistic framework that uses a parametric Bayesian method (Bayes t-test). It has been shown that the Bayes t-test works effectively in analyzing microarray data with fewer samples. The t-test approach could be further generalized for analyzing more than two samples by using ANOVA-based approaches. Kerr et al. [23] proposed ANOVA models that take care of specific effects such as dye, slide, treatment, gene, and their respective interaction effects. Smyth et al. [30] generalized the moderated t-statistic to a multiconditional case and this approach is implemented in the Bioconductor package "limma."

After a test statistic has been selected, the next step is to compute the significance (P-value) of the test statistic and to choose a cut-off value, above which the genes will be considered as differentially expressed. Typically, P-values are calculated by looking up normal or t-tables that rely predominantly on mathematical assumptions. Resampling methods, such as permutation or bootstrapping, are also used often to calculate the P-values. With estimated P-values, a statistical task is then to find significantly differentially expressed genes. An important aspect here is the need to control for simultaneous testings of hypotheses. Since thousands or tens of thousands of genes are examined simultaneously, there is a risk of identifying false positives. The most conservative approach to multiple testing is to control the family-wise error rate (FWER)—the probability of at least one false positive among significant genes. The Bonferroni single-step correction is an extremely stringent approach of FWER, where the significance levels are divided by the number of tests that are performed. The Holm's step-down correction is less stringent and makes successively smaller adjustments for higher P-values. Recently, less conservative methods based on false discovery rates (FDRs) have been proposed as an appropriate alternative to FWER-based methods. The FDR is the expected proportion of false positives in the list of results. Benjamini and Hochberg proposed the FDR control procedure to control false positives in the significant gene list [32]. Storey and Tibshirani suggested using q-values, which are FDR equivalent [33].

2.3.4 CLUSTER ANALYSIS

The underlying principle of applying clustering to expression data is the assumption that genes with similar expression patterns are likely to be functionally related. A wide variety of methods are available to organize the genes into groups with similar expression patterns. These methods depend on defining a measure of similarity between expression profiles, and each measure can reveal different features in the data. The two most widely used measures are Euclidean distance and Pearson's correlation coefficient. The Euclidean distance between two genes G_1 and G_2 across n samples can be defined as

$$D(G_1, G_2) = \sqrt{\sum_{i=1}^{n}(g_{1i} - g_{2i})^2}$$

Euclidean distance is best used when the magnitude of expression level is important. The Pearson's correlation coefficient distance measure can be calculated as

$$D(G_1, G_2) = 1 - r(G_1, G_2) = 1 - \frac{\sum_{i=1}^{n}(g_{1i} - \overline{g_1})(g_{2i} - \overline{g_2})}{\sqrt{\sum_{i=1}^{n}(g_{1i} - \overline{g_1})^2 \sum_{i=1}^{n}(g_{2i} - \overline{g_2})^2}}$$

where g_1 and g_2 are mean expression values for genes G_1 and G_2 across n samples. Pearson's correlation coefficient measure is useful when the pattern of expression in the genes or samples is more important. It has been suggested that Euclidean distances are most suitable for absolute expression values as in Affymetrix Genechips, whereas Pearson's correlation coefficient measures are more appropriate for ratios. The clustering methods described here are based on distance measures and include hierarchical clustering, k-means clustering, and self-organizing maps (SOMs).

Hierarchical clustering is a most widely used method that does not require any prior information and works by successively grouping together very similar pairs of genes [34,35]. The algorithm starts by comparing each gene with every other gene. It selects the two genes with the most similar expression patterns, groups them together into a "node," and then repeats the procedure with the remaining genes. The algorithm stops once all the genes have been placed in a node. The grouping of genes into a node can be done in different ways. Single-linkage clustering uses the minimum distance between a member of one cluster and a member of other cluster, for all members. Complete-linkage clustering uses the maximum distance, whereas average-linkage clustering uses the average distance. The results of hierarchical clustering can be easily visualized as a heat map with ordered clusters so that patterns in gene expression become visually accessible (Figure 2.3).

The hierarchical clustering method has been successfully used in clustering microarray data and is especially useful in ordering genes to visualize global patterns. The method, however, suffers from the assumption that the internal structure of the data is basically hierarchical—that the genes can be accurately ordered in terms of their similarity to one another. If the genes are not correlated across all conditions, hierarchical clustering will not perform well. In addition, at each iterative stage, the merging of two nodes is based on the pair-wise distances of all nodes at that stage instead of any global criterion. When applied to noisy microarray data, more sophisticated clustering methods such as k-means clustering and SOMs have been shown to outperform hierarchical clustering.

The k-means clustering algorithm works by first randomly placing every gene into one of the preset clusters [36]. The user is required to specify the number k of clusters a priori. The "centroid" of each cluster is calculated by taking the mean expression level of the genes in that particular cluster. The iterative algorithm assigns each gene to the cluster with the smallest distance to the next centroid. It then calculates the new centroid from the genes that belong to the cluster and replaces the old centroid. The computation process is terminated when there is no further change in the assignment of genes to centroids. The k-means algorithm usually produces good clustering results if k is correctly chosen. The method is, however, unstable and highly affected by the presence of scattered genes in the expression data. In addition, the calculation of centroids requires the data be in the Euclidean space with Euclidean distance as the similarity measure.

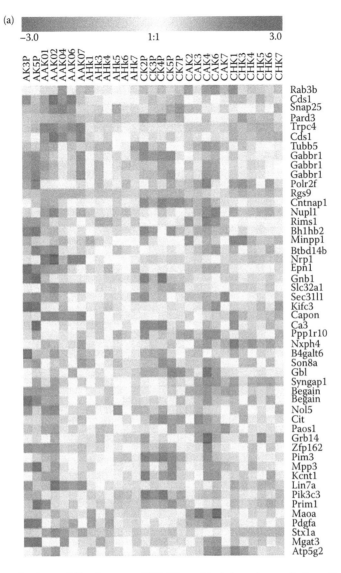

FIGURE 2.3 **(See color insert following page 138.)** Hierarchical clustering analysis applied to gene expression data. Hierarchical clustering method (a) has been applied to an unordered dataset (b) the hierarchical clustering clearly reveals underlying patterns that can help identify coexpressed genes.

SOMs are clustering algorithms that are similar to *k*-means clustering but have been shown to work better in both robustness and accuracy when applied to microarray data [37,38]. The algorithm first maps *k* nodes in a low-dimensional (usually two-dimensional) grid space from the *d*-dimensional space, in which *d* is the number of features, and then the nodes are adjusted iteratively. A gene is chosen randomly at each time, compared to samples, and moved in *d*-dimensional space, toward the sample to which it is closest. The movement of the nodes in the *d*-dimensional space depends on their distance to the chosen gene and the two-dimensional geometry of the nodes. The magnitude of the movement decreases as the number of iterations increases. After a number of iterations, the nodes in the grid become organized in a manner that represents the topological structure of the input samples. Thus clusters generated from nodes close to each other in the two-dimensional grid geometry will have similar expression patterns.

(b)

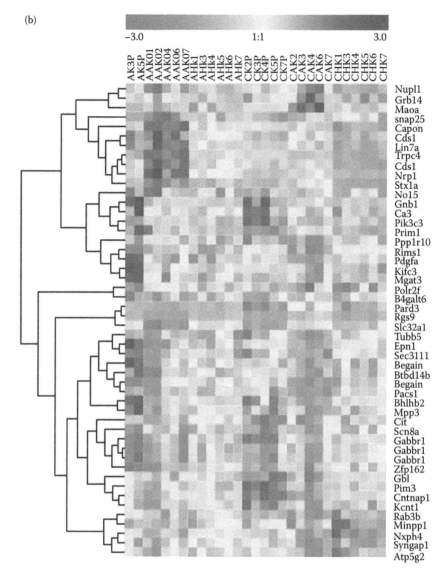

FIGURE 2.3 *(Continued)*

2.3.5 FUNCTIONAL ANALYSIS AND INTERPRETATION

Differential expression analysis of microarray data typically generates a list of significant probe sets or probes that are differentially expressed across experimental conditions. With or without cluster analysis, the next important task is to add biological knowledge to the selected identifier lists either from existing literature or from such databases as Entrez Gene, Unigene, UniProt, Gene Ontology (GO), and Kyoto Encyclopedia of Genes and Genomes (KEGG) pathways. Most of these databases classify genes into biological categories or classes that represent their function. A further step is to estimate the statistical significance of association between the classes and probes of the obtained list. Several automated web accessible databases and tools are being implemented to enrich the lists of candidate differentially expressed probes with the biological information available for the correspondent genes, and most of them use information from either the GO or the KEGG pathways database.

TABLE 2.1
List of Tools for Functional Analysis and Interpretation

Tool	Statistical Test	Functional Categories	Source	References
BiNGO	Binomial, hypergeometric	GO	http://www.psb.ugent.be/cbd/papers/BiNGO/	[44]
CLENCH	Binomial, hypergeometric, χ^2	GO (only for *Arabidopsis thaliana*)	http://www.stanford.edu/~nigam/Clench/	[45]
DAVID/EASE online	Fisher's exact test	GO, KEGG, and BioCarta pathways, protein interactions, gene–disease associations, protein functional domains, and motifs	http://david.abcc.ncifcrf.gov/	[39,40]
eGOn	Binomial	GO	http://www.genetools.microarray.ntnu.no/egon/index.php	[46]
FatiGO+	Fisher's exact test	GO, KEGG pathways, Interpro motifs, Transfac motifs, CisRed motifs	http://fatigo.bioinfo.cipf.es/	[43]
FuncAssociate	Fisher's exact test	GO	http://llama.med.harvard.edu/cgi/func/funcassociate	
GeneMerge	Hypergeometric	GO, KEGG, chromosomal location	http://genemerge.bioteam.net/	[47]
GoMiner	Fisher's exact test	GO	http://discover.nci.nih.gov/gominer/	[48]
Gostat	Fisher's exact test, χ^2	GO	http://gostat.wehi.edu.au/	[42]
GoSurfer	χ^2	GO	http://bioinformatics.bioen.uiuc.edu/gosurfer/	[49]
GO::TermFinder	Hypergeometric	GO	http://search.cpan.org/dist/GO-TermFinder/	[50]
GOTM	Hypergeometric	GO	http://bioinfo.vanderbilt.edu/gotm/	[51]
GOToolBox	Binomial, hypergeometric, Fisher's exact test	GO	http://burgundy.cmmt.ubc.ca/GOToolBox/	[52]
Ontology Traverser	Hypergeometric	GO	http://vortex.cs.wayne.edu/ontoexpress/	[53]
Onto-Tools	Binomial, hypergeometric, χ^2	GO, KEGG, chromosomal location		[41]

The GO project comprises over 19,000 terms in molecular function, biological process, and cellular component. GO has been used to annotate whole genomes and find enriched functional categories in differentially regulated gene lists or clustered genes in microarray experiments. This two-step approach simply compares the number of differentially expressed genes observed in each GO category with the number of expected genes in the same category by chance. If the observed number is substantially different from the expected one (by chance), then the category is reported as significant using some statistical model. Several statistical models such as hypergeometric, χ^2, binomial, and Fisher's exact test have been used to calculate *P*-values and are implemented in different available tools. Among these tools, the most popular ones are DAVID [39,40], Onto-express [41], GOstat [42], and FatiGO [43]. Some of the tools use GO, KEGG, and other biologically relevant categories or terms for enrichment analysis. Table 2.1 shows a list (not an exhaustive one) of the most widely used tools.

2.4   CONCLUSION

Microarray technology combined with bioinformatics will have a great impact on biological discovery. Although gene expression profiling using DNA microarray technologies has brought exciting changes in many biological disciplines, it remains a challenge to transform the large amount of raw data into reliable and meaningful biological sense. In this chapter, we have discussed various methods that are being used in each step of microarray data analysis, starting from quality assessment through to inferring and assigning biological themes. With emerging different statistical methods, web-based software tools, and ongoing efforts in developing open source software for data analyses, there are many reasons to be optimistic that microarray results will be robust and analyses will be relatively straightforward.

REFERENCES

1. Lipshutz RJ, Fodor SP, Gingeras TR, and Lockhart DJ. 1999. High density synthetic oligonucleotide arrays. *Nat Genet* 21(Suppl. 1): 20–24.
2. Lockhart DJ, Dong H, Byrne MC, Follettie MT, Gallo MV, Chee MS, Mittmann M, Wang C, Kobayashi M, Horton H, and Brown EL. 1996. Expression monitoring by hybridization to high-density oligonucleotide arrays. *Nat Biotechnol* 14(13): 1675–1680.
3. Hardiman G. 2004. Microarray platforms—comparisons and contrasts. *Pharmacogenomics* 5(5): 487–502.
4. Wick I and Hardiman G. 2005. Biochip platforms as functional genomics tools for drug discovery. *Curr Opin Drug Discov Devel* 8(3): 347–354.
5. Hardiman G and Carmen A. 2006. DNA biochips—past, present and future; an overview. In: *Biochips as Pathways to Discovery* (A Carmen and G Hardiman, eds). Taylor & Francis, New York, pp. 1–13.
6. Schena M, Shalon D, Davis RW, and Brown PO. 1995. Quantitative monitoring of gene expression patterns with a complementary DNA microarray. *Science* 270(5235): 467–470.
7. Shalon D, Smith SJ, and Brown PO. 1996. A DNA microarray system for analyzing complex DNA samples using two-color fluorescent probe hybridization. *Genome Res* 6(7): 639–645.
8. DeRisi J, Penland L, Brown PO, Bittner ML, Meltzer PS, Ray M, Chen Y, Su YA, and Trent JM. 1996. Use of a cDNA microarray to analyse gene expression patterns in human cancer. *Nat Genet* 14: 457–460.
9. Reimers M and Weinstein JN. 2005. Quality assessment of microarrays: visualization of spatial artifacts and quantitation of regional biases. *BMC Bioinformatics* 6: 166.
10. Bolstad BM, Collin F, Brettschneider J, Simpson K, Cope L, Irizarry RA, and Speed TP. 2005. Quality assessment of Affymetrix GeneChip data. In: *Bioinformatics and Computational Biology Solutions Using R and Bioconductor* (R Gentleman, VJ Carey, W Huber, RA Irizarry, and S Dudoit, eds). Springer, New York, pp. 33–47.
11. Reimers M, Heilig M, and Sommer WH. 2005. Gene discovery in neuropharmacological and behavioral studies using Affymetrix microarray data. *Methods* 37(3): 219–228.
12. Paquet A and Hwa Yang JY. Available at http://arrays.ucsf.edu/archive/software/2004.

13. Affymetrix. 2001. *Statistical algorithm reference guide*. Technical report, Affymetrix.
14. Gautier L, Cope L, Bolstad BM, and Irizarry RA. 2004. Affy—analysis of Affymetrix GeneChip data at the probe level. *Bioinformatics* 20(3): 307–315.
15. Li C and Wong WH. 2001. Model-based analysis of oligonucleotide arrays: expression index computation and outlier detection. *Proc Natl Acad Sci USA* 98: 31–36.
16. Wu Z. and Irizarry RA. 2004. Preprocessing of oligonucleotide array data. *Nat Biotechnol* 22: 656–658.
17. Bolstad BM, Irizarry RA, Astrand M, and Speed TP. 2003. A comparison of normalization methods for high density oligonucleotide array data based on variance and bias. *Bioinformatics* 19(2): 185–193.
18. Huber W, von Heydebreck A, Sueltmann H, Poustka A, and Vingron M. 2003. Parameter estimation for the calibration and variance stabilization of microarray data. *Stat Appl Genet Mol Biol* 2(1), article 3. Available at: http://www.bepress.com/sagmb/vol2/iss1/art3.
19. Chen YJ, Kodell R, Sistare F, Thompson KL, Morris S, and Chen JJ. 2003. Normalization methods for analysis of microarray gene-expression data. *J Biopharm Stat* 13(1): 57–74.
20. Smyth GK. 2004. Linear models and empirical Bayes methods for assessing differential expression in microarray experiments. *Stat Appl Genet Mol Biol* 3(1): Article3.
21. Yang IV, Chen E, Hasseman JP, Liang W, Frank BC, Wang S, Sharov V. et al. 2002. Within the fold: assessing differential expression measures and reproducibility in microarray assays. *Genome Biol* 3(11): RESEARCH0062.
22. Kepler TB, Crosby L, and Morgan KT. 2002. Normalization and analysis of DNA microarray data by self-consistency and local regression. *Genome Biol* 3(7): RESEARCH0037.
23. Kerr MK, Martin M, and Churchill GA. 2000. Analysis of variance for gene expression microarray data. *J Comput Biol* 7(6): 819–837.
24. Wolfinger RD, Gibson G, Wolfinger ED, Bennett L, Hamadeh H, Bushel P, Afshari C, and Paules RS. 2001. Assessing gene significance from cDNA microarray expression data via mixed models. *J Comput Biol* 8(6): 625–637.
25. Tseng GC, Oh MK, Rohlin L, Liao JC, and Wong WH. 2001. Issues in cDNA microarray analysis: quality filtering, channel normalization, models of variations and assessment of gene effects. *Nucleic Acids Res* 29(12): 2549–2557.
26. Yang YH, Dudoit S, Luu P, Lin DM, Peng V, Ngai J, and Speed TP. 2002. Normalization for cDNA microarray data: a robust composite method addressing single and multiple slide systematic variation. *Nucleic Acids Res* 30(4): e15.
27. Tusher VG, Tibshirani R, and Chu G. 2001. Significance analysis of microarrays applied to the ionizing radiation response. *Proc Natl Acad Sci USA* 98(9): 5116–5121.
28. Efron B, Tibshirani R, Storey JD, and Tusher V. 2001. Empirical Bayes analysis of a micoarray experiment. *J Am Stat Assoc* 96: 1151–1160.
29. Lönnstedt I, and Speed T. 2002. Replicated microarray data. *Stat Sin* 12: 31–46.
30. Smyth GK, Michaud J, and Scott H. 2005. The use of within-array duplicate spots for assessing differential expression in microarray experiments. *Bioinformatics* 21(9): 2067–2075.
31. Baldi P and Long AD. 2001. A Bayesian framework for the analysis of microarray expression data: regularized *t*-test and statistical inferences of gene changes. *Bioinformatics* 17(6): 509–519.
32. Benjamini Y and Hochberg Y. 1995. Controlling the false discovery rate: a practical and powerful approach to multiple testing. *J R Stat Soc Ser* B 57: 289–300.
33. Storey JD and Tibshirani R. 2003. Statistical significance for genomewide studies. *Proc Natl Acad Sci USA* 100(16): 9440–9445.
34. Eisen MB, Spellman PT, Brown PO, and Botstein D. 1998. Cluster analysis and display of genome-wide expression patterns. *Proc Natl Acad Sci USA* 95(25): 14863–14868.
35. Spellman PT, Sherlock G, Zhang MQ, Iyer VR, Anders K, Eisen MB, Brown PO, Botstein D, and Futcher B. 1998. Comprehensive identification of cell cycle-regulated genes of the yeast *Saccharomyces cerevisiae* by microarray hybridization. *Mol Biol Cell* 9(12): 3273–3297.
36. Tavazoie S, Hughes JD, Campbell MJ, Cho RJ, and Church GM. 1999. Systematic determination of genetic network architecture. *Nat Genet* 22(3): 281–285.
37. Tamayo P, Slonim D, Mesirov J, Zhu Q, Kitareewan S, Dmitrovsky E, Lander ES, and Golub TR. 1999. Interpreting patterns of gene expression with self-organizing maps: methods and application to hematopoietic differentiation. *Proc Natl Acad Sci USA* 96(6): 2907–2912.
38. Toronen P, Kolehmainen M, Wong G, and Castren E. 1999. Analysis of gene expression data using self-organizing maps. *FEBS Lett* 451(2): 142–146.
39. Dennis G Jr, Sherman BT, Hosack DA, Yang J, Gao W, Lane HC, and Lempicki RA. 2003. DAVID: database for annotation, visualization, and integrated discovery. *Genome Biol* 4(5): P3.

40. Hosack DA, Dennis G Jr, Sherman BT, Lane HC, and Lempicki RA. 2003. Identifying biological themes within lists of genes with EASE. *Genome Biol* 4(10): R70.

41. Draghici S, Khatri P, Bhavsar P, Shah A, Krawetz SA, and Tainsky MA. 2003. Onto-Tools, the toolkit of the modern biologist: Onto-Express, Onto-Compare, Onto-Design and Onto-Translate. *Nucleic Acids Res* 31(13): 3775–3781.

42. Beissbarth T and Speed TP. 2004. GOstat: Find statistically overrepresented Gene Ontologies within a group of genes. *Bioinformatics* 20(9): 1464–1465.

43. Al-Shahrour F, Diaz-Uriarte R, and Dopazo J. 2004. FatiGO: a web tool for finding significant associations of Gene Ontology terms with groups of genes. *Bioinformatics* 20(4): 578–580.

44. Maere S, Heymans K, and Kuiper M. 2005. BiNGO: a Cytoscape plugin to assess overrepresentation of Gene Ontology categories in biological networks. *Bioinformatics* 21(16): 3448–3449.

45. Shah NH and Fedoroff NV. 2004. CLENCH: a program for calculating Cluster ENriCHment using the Gene Ontology. *Bioinformatics* 20(7): 1196–1197.

46. Berriz GF, King OD, Bryant B, Sander C, and Roth FP. 2003. Characterizing gene sets with FuncAssociate. *Bioinformatics* 19(18): 2502–2504.

47. Castillo-Davis CI and Hartl DL. 2003. GeneMerge—post-genomic analysis, data mining, and hypothesis testing. *Bioinformatics* 19(7): 891–892.

48. Zeeberg BR, Feng W, Wang G, Wang MD, Fojo AT, Sunshine M, Narasimhan S, et al. 2003. GoMiner: A resource for biological interpretation of genomic and proteomic data. *Genome Biol* 4(4): R28.

49. Zhong S, Storch F, Lipan O, Kao MJ, Weitz C, and Wong WH. 2004. GoSurfer: A graphical interactive tool for comparative analysis of large gene sets in Gene Ontology space. *Appl Bioinformatics* 3(4): 1–5.

50. Boyle EI, Weng S, Gollub J, Jin H, Botstein D, Cherry JM, and Sherlock G. 2004. GO: TermFinder—open source software for accessing Gene Ontology information and finding significantly enriched Gene Ontology terms associated with a list of genes. *Bioinformatics* 20(18): 3710–3715.

51. Zhang B, Schmoyer D, Kirov S, and Snoddy J. 2004. GOTree Machine (GOTM): A web-based platform for interpreting sets of interesting genes using Gene Ontology hierarchies. *BMC Bioinformatics* 5: 16.

52. Martin D, Brun C, Remy E, Mouren P, Thieffry D, and Jacq B. 2004. GOToolBox: Functional investigation of gene datasets based on Gene Ontology. *Genome Biol* 5(12): R101.

53. Young A, Whitehouse N, Cho J, and Shaw C. 2005. OntologyTraverser: An R package for GO analysis. *Bioinformatics* 21(2): 275–276.

3 The Statistical Design and Analysis of Gene Expression Data

Sonia Jain

CONTENTS

3.1 INTRODUCTION

Over the past 10 years, the development of microarray biotechnology has become a popular and powerful tool in understanding biological processes at a molecular level [1]. Microarray technology allows for the simultaneous study of thousands of genes in an automated setting. These high-throughput assays are critical in a number of biomedical applications such as cancer classification [2], disease diagnosis [3], and drug discovery [4]. At the same time, there has been a call for innovation in the statistical science discipline to develop a new methodology to analyze these complex and high-dimensional genomic assays. Several statistical areas where we have witnessed increased development include multiple comparisons and testing [5,6], high-dimensional data analyses [7], machine-learning techniques [8], Bayesian methods [9,10], and the design of experiments [11].

Microarray experiments have typically been considered exploratory discovery tools or an intermediate step to determining a subset of biologically interesting genes. However, one of the first steps in creating a successful gene expression microarray study is the development of a statistically rigorous experimental design to address a scientific hypothesis. Careful statistical design can provide both cost-saving benefits and efficient use of a specific microarray platform. The scientific community now expects that a gene expression experiment be reproducible [12] so that spurious results are diminished.

The intent of this chapter is to provide an overview of designing microarray experiments with a focus on the statistical issues that may arise. In particular, we will describe the general steps in a typical microarray experiment, principles of designing an experiment, highlight sources of variability, and consider sample size and statistical power.

3.2 STEPS IN A MICROARRAY EXPERIMENT

In this section, we provide a list and description of steps that should be considered when conducting a microarray experiment. We focus on the role of the statistician. Refer to Draghici [13] and Simon et al. [14] for variations on this list.

3.2.1 DEVELOPING A RESEARCH HYPOTHESIS

The first step prior to beginning an experiment is to define the scientific question of interest. Often, this can be framed in terms of a statistical hypothesis that can be tested. For example, in an experiment comparing gene expression profiles of two groups of patients (disease versus controls), we may ask, "Are there genes that are differentially expressed (up- or down-regulated) between the disease group versus the control group?" At this initial stage of the study, it is critical to enlist a statistical collaborator to help define or narrow the research hypothesis into a "testable" statistical hypothesis.

3.2.2 DESIGN OF MICROARRAY EXPERIMENT

The design of a microarray study is an integral component of the experimental process. There are two fundamental types of design in a gene expression experiment: the actual design of the array and the allocation of ribonucleic acid (RNA) samples to arrays. The first type of design determines which deoxyribonucleic acid (DNA) probes will be printed on the array, the spatial configuration of the probes on the array, and the selection of controls. This will not be discussed in this chapter. Instead, we focus on the second type of experimental design and will delve into the principles of statistical design in Section 3.3.

The key to a successful microarray study is to obtain input on the design prior to any data collection or array processing. Careful planning on the number of arrays to be included can be cost-effective, since a single array on a commercial platform can be expensive. Issues that should be determined at the design stage include selecting the array platform best suited for the experimental problem (e.g., large, encompassing arrays versus specific gene pathway arrays), deciding which hypotheses to test (e.g., selecting factors to study, time-points, and so on), setting the statistical operating characteristics (i.e., level of statistical power, type I error rate, false discovery rate tolerance, and so on), choosing the number of biological samples, number of biological and technical replicates, and methods to control for sources of variation.

3.2.3 CONDUCT THE EXPERIMENT

Once the statistical design has been formalized, the samples can be collected, and the array experiments can be processed. It is important to take note of any irregularities in the conduct of the experiment, as this may introduce bias into the experimental system. Since the experimental procedure involves many steps from sample collection to the final stages of array processing, it is often difficult to ascertain the sources of experimental variation.

3.2.4 IMAGE ANALYSIS

After microarray hybridization, an image file of pixels is generated that represents the gene expression intensities. The purpose of image analysis is to convert and process pixel data into numeric measures of gene expression intensity. This numeric matrix of data can then be used for analyzing the experiment. There are several image analysis methods that are frequently employed in microarray analysis depending on the platform. The discussion of image analysis is beyond the scope of this chapter, but details can be found in Simon et al. [14].

3.2.5 DATA PREPROCESSING AND NORMALIZATION

Prior to data analysis to address the research question of the experiment, microarray data requires preprocessing and normalization. The type of preprocessing that occurs is often dependent on the array platform. For example, methods have been developed specifically for two-color versus one-color systems [15–17]. One common preprocessing procedure is to take the natural logarithm of intensity values [15].

Normalization, another mode of preprocessing, is a technique to scale the genes across arrays in order to identify and reduce systematic variation stemming from differences in genetics, experimental conduct, or calibration issues. Normalization is also used within an array to minimize noise and poor-quality probes. Methods to normalize microarray data are numerous, and much of the statistical advancement in microarray methodology involves the development of normalization techniques (see e.g., Refs. [18–20]). There is no general consensus on the "best" method to utilize yet, and differences in normalization methods can lead to differences in findings [21]. Some popular normalization methods include RMA [18], dChip [19], and SNOMAD [22]. For further discussion, see the following Refs. [23,24].

3.2.6 STATISTICAL AND BIOLOGICAL INFERENCES

The data analysis of gene expression experiments is a complex task performed after preprocessing and normalization. The type of data analysis required will depend on the scientific research question. In general, for microarray analysis, research questions can fall into two major categories: (i) gene expression profile comparisons and (ii) class discovery. Further, we briefly outline analytic methods commonly used for these tasks.

3.2.6.1 Gene Expression Profile Comparisons

One purpose of comparing gene expression profiles is to identify genes that are biologically different among prespecified groups. This is often referred to as determining differential gene expression. Here, gene expression is treated as the response variable in a statistical model, while an indicator variable that distinguishes classes is the independent variable. Methods to determine differential gene expression range from simple (and sometimes naïve) fold changes to sophisticated supervised machine-learning techniques.

In a simple two-group comparison, for example comparing cancer versus control groups, to detect differential gene expression, statistical methods often fall under the umbrella of performing a statistical test for each gene and generating a p-value to ascertain statistical significance. One fundamental test to do this is the two-sample Student's t-test. However, when many genes are being tested simultaneously, the probability that a type I error (i.e., a false-positive error) will occur increases dramatically as the number of gene tests increases. This is referred to as a multiplicity problem, which is not unique to microarray analysis. However, much attention and development of statistical methodology for multiple testing to combat this problem in the context of gene expression has intensified over the years. Traditional multiple testing correction method, such as the Bonferroni method, which controls the family-wise error rate (i.e., the probability that at least one false-positive error will be committed), is deemed too conservative when dealing with thousands of genes [25]. An alternative framework to handle multiplicity is the "false-discovery rate" (FDR) approach [5,26], in which one controls the proportion of false positives among the genes that are identified as differentially expressed. For a review of multiplicity, refer to Ref. [25]. Other methods that address the differential expression problem include Cyber-T [27] and SAM [6].

3.2.6.2 Class Discovery

Class discovery addresses exploratory molecular questions that are not framed as formal statistical hypotheses. This mode of analysis is considered a first step in microarray analysis that may yield insight into designing future studies. Class discovery is used to identify patterns in gene expression profiles, in which samples can be clustered together based on similar profiles to define classes. These clusters of samples may be hierarchical in that sub-classes may exist within classes.

In a microarray experiment, clustering can be conducted in two ways. The first concerns the clustering of tissue samples (i.e., human, mouse, and cell line) based on gene intensities. Tissue clusters have been instrumental in the discovery and characterization of new sub-classes of diseases [2]. The second type is clustering the genes based on the tissue samples. Gene clusters have been used to identify genetic pathways or families of genes that are co-regulated [28].

Clustering techniques fall under a larger class of statistical unsupervised learning methods. These methods allow the data to find natural clusters without using information such as tissue phenotypic traits or gene function. Typical clustering methods include k-means clustering [29], hierarchical clustering [30], self-organizing maps [31], model-based clustering methods such as mixture models [32], and Bayesian approaches to clustering [33,34]. Although clustering gene expression data are very popular, it should be noted that many of these distance-based clustering methods were not developed to handle the high dimensionality that is inherent in microarray data. Since microarray data typically involve gene-to-tissue ratios that are 100-fold and genes that are not independently distributed, the cluster analysis problem in this context is not standard. Methods to reduce the dimensionality of the data have been considered in conjunction to clustering, such as principal components analysis (PCA) [35], but this is nonideal. One key difficulty with PCA is identifying a reasonable number of principal components to use. Another difficulty is attempting to biologically interpret the principal components, since a single component could be composed of hundreds of genes. Multidimensional scaling [36] is another dimension-reduction technique utilized in microarray analysis.

Caution should be taken when applying clustering methods, since these techniques are often not robust to outliers and variations in samples. Normalization and data transformations have a pronounced effect on clustering [37]. It is recommended that sensitivity analysis be performed to confirm findings.

3.3   PRINCIPLES OF EXPERIMENTAL DESIGN

In any study, the purpose of the experimental design is to separate noise from signal, so that the factors that influence the signal can be observed and identified. Often in microarray studies, this aspect of the statistical process is unfortunately ignored. The development of good experimental design is grounded in statistical theory; some classical references include Cochran and Cox [38], Montgomery [39], and Neter et al. [40].

A good experimental design will address sources of variability in the microarray study and will reduce the experimental variability of the expression intensities. As Draghici [13] points out, there are three fundamental concepts in experimental design: replication, randomization, and blocking. We will briefly describe these concepts and how they relate to designing microarray studies.

3.3.1 REPLICATION

Replication is a confusing term in microarray analysis, since there are a number of levels of replication. The purpose of replication is to average out or reduce the amount of experimental variability by increasing the sample size. Replication is required to estimate experimental variability. Microarray replicates belong to two categories: technical or biological replicates. Technical replicates are made from the same sample of RNA and shared across replicate arrays. In contrast, biological replicates are made from the RNA of biologically independent samples. The types of biological replicates depend on whether we are considering controlling variation within an individual, within-group biological replicates, or between-group biological replicates. For example, variation exists (i) between individuals within the same class (e.g., patients in a control group will have person-to-person variability); (ii) between samples from the same individual (e.g., repeated blood draws taken from a patient over a month); and (iii) between individuals in different diagnostic classes (e.g., patients in the control group versus patients in the disease group).

The selection of types of replicates most appropriate for a study will depend on the type of microarray platform, type of tissue being studied, the scientific question of interest, and finally, financial considerations. In general, when resources are limited, it is best to optimize the number of biological replicates between groups.

3.3.2 RANDOMIZATION

Randomization is a fundamental statistical technique to minimize bias. It is a method to equalize factors that have not already been accounted for in the experimental design. These factors could potentially influence the outcome of the experiment and create confounding. One example of randomization is the location of replicate probes on the array. If replicate probes are located side-by-side, then a processing malfunction that damaged one side of the array could yield intensities that appear to be the result of differential gene expression instead of experimental error. Another example of randomization is the order in which arrays are processed. Consider an experiment comparing controls to diseased patients. One way to reduce bias would be to process the arrays at random across the two groups, instead of processing one group first and then the other.

3.3.3 BLOCKING

Blocking is a procedure by which experimental units are grouped into homogeneous clusters in an attempt to improve the comparison of treatments by randomly allocating the treatments within each cluster or "block". Here, factors of interest are allowed to vary, whereas noninteresting factors are held constant. Blocking diminishes the variability due to the differences between blocks [13]. In microarray experiments, the physical array is a block, in that all probes on the array are subject to the same factors during the experimental process. Within a slide, the intensities are expected to be more homogeneous than across slides. Blocking is possible only when the factors are under the experimenter's control. Otherwise, randomization is the only option to control for these nuisance factors.

3.3.4 DESIGNS

The type of experimental design utilized in a microarray study will depend on the array platform, factors under consideration, and cost. Well-designed experiments can lead to increased efficiency and cost-savings. We outline three basic experimental designs and refer readers to Montgomery [39] for details.

3.3.4.1 Completely Randomized Design

The completely randomized design is the most basic experimental design. Here, we assume that there is only one factor with m levels in which the data are collected for the various levels of the factor at random. The total number of measurements for each level of the factor is n. If the number of observations per factor level is identical, then this is known as a balanced design. The statistical model can be described as

$$X_{ij} = \mu + \lambda_i + \varepsilon_{ij}$$

where μ is the overall mean, λ_i the effect of factor i, and ε_{ij} the variability unexplained by the model. Here, $i = 1, \ldots, m$ and $j = 1, \ldots, n$. This design can be analyzed by a one-way analysis of variance (ANOVA) statistical model.

3.3.4.2   Randomized Block Design

The randomized block design is a modification of the completely randomized design. Here, individual groups of homogeneous measurements are blocked together so that factors are assigned randomly to the blocks. If every block is assigned to every factor, this is a randomized block design. The statistical model for the block design is

$$X_{ij} = \mu + \lambda_i + \beta_j + \varepsilon_{ij}$$

where μ is the overall mean, λ_i the effect of factor i, β_j the effect of block j, and ε_{ij} the random noise. Here, $i = 1, \ldots, m$ and $j = 1, \ldots, b$, where b is the number of blocks. This design can be analyzed by a two-way ANOVA statistical model for blocks.

3.3.4.3   Factorial Designs

The factorial design is an experimental design that allows for at least two factors, say A and B, and allows for all possible combinations of A and B levels to be considered. Further, the interaction between factors A and B can also be analyzed. The statistical model for a factorial design with factors A and B is

$$X_{ijk} = \mu + \lambda_i + \beta_j + (\lambda\beta)_{ij} + \varepsilon_{ijk}$$

where μ is the overall mean, λ_i and β_j represent factors A and B, $(\lambda\beta)_{ij}$ the interaction of factors A and B, and ε_{ijk} the random noise. This design can be analyzed by a two-way ANOVA statistical model.

Other basic statistical designs include incomplete block design, Latin square design, and split-plot design [39]. For two-color arrays, special designs have been developed to address the issues of imbalance due to dyes. These designs include reference design, loop design, and dye-swap design. Please refer to Refs. [13,14] for details on these designs.

3.4   SAMPLE SIZE AND POWER CONSIDERATIONS

A fundamental concern in the designing of microarray experiments is the allocation of samples to arrays that will produce results that are statistically valid or have reasonable statistical power. Poor statistical power has been cited as one of the reasons for lack of reproducibility in some array studies.

Statistical power measures the test's ability to reject the null hypothesis when it is actually false. In other words, it measures the power to make a correct decision. Power is also known as the probability of not committing a type II error (i.e., when the null hypothesis is not rejected when it is, in fact, false). The maximum power for a test is 1, while the minimum is 0. Ideally, we desire a test to have high power and be close to 1. One common way to increase power is to increase the sample size. However, due to financial constraints, increasing the sample size in a microarray study is not usually feasible. Other factors that influence power include variance of individual measurements, acceptable false-positive rate, and unknown true difference we wish to detect.

In array studies, it is often difficult to assess statistical power. To perform power calculations, assumptions are required which are not always realistic in a microarray context. For example, for basic two-sample calculations using a t-test structure, we assume that the samples have a Gaussian distribution. Another common assumption is that variances of relative expression levels across hybridizations are constant, which is unlikely to be true.

There is a trade-off of power against the false-positive rate. Since array results are typically intermediary and validated via other experimental procedures (e.g., reverse transcription-polymerase chain reaction (RT-PCR)), a higher false-positive rate may be tolerable. A few references for power and sample size considerations include Simon et al. [14], Zien [41], and Lee [12].

3.5 CONCLUSION

In this chapter, we have highlighted some of the salient issues when designing and analyzing gene expression experiments. The key to a successful design is to construct an efficient and simple experiment that answers the biological question of interest in the framework of a statistical hypothesis. The design should ultimately balance experimental constraints such as cost of chips, availability of RNA samples, and microarray platform precision to statistical considerations such as power and sample size. A microarray study is strengthened by having a statistical collaborator join the study in the planning stages to guide hypothesis generation, sample size allocation, and troubleshoot sources of bias in the conduct of the experiment.

REFERENCES

1. Schena M. *Microarray Biochip Technology*. Eaton Publishing, Westborough, MA: BioTechniques Press, 2000.
2. Golub TR, Slonim DK, Tamayo P, Huard C, Gaasenbeek M, Mesirov JP, Coller H, et al. Molecular classification of cancer: Class discovery and class prediction by gene expression monitoring. *Science* 1999; 286: 531–537.
3. Stremmel C, Wein A, Hohenberger W, and Reingruber B. DNA microarrays: A new diagnostic tool and its implications in colorectal cancer. *International Journal of Colorectal Disease* 2002; 17: 131–136.
4. Debouck C and Metcalf B. The impact of genomics on drug discovery. *Annual Review of Pharmacology and Toxicology* 2000; 40: 193–208.
5. Storey JD. A direct approach to false discovery rates. *Journal of the Royal Statistical Society, Series B* 2002; 64: 479–498.
6. Tusher VG, Tibshirani R, and Chu G. Significance analysis of microarrays applied to ionizing radiation response. *Proceedings of the National Academy of Sciences of the USA* 2001; 98: 5116–5121.
7. Gordon AD. *Classification*. New York: Chapman and Hall, 1999.
8. Mertens BJA. Microarrays, pattern recognition and exploratory data analysis. *Statistics in Medicine* 2003; 22: 1879–1899.
9. Medvedovic M and Sivaganesan S. Bayesian infinite mixture model based clustering of gene expression profiles. *Bioinformatics* 2002; 18: 1194–1206.
10. Do K-A, Mueller P, and Tang F. A nonparametric Bayesian mixture model for gene expression. *Journal of the Royal Statistical Society, Series C—Applied Statistics* 2005; 54: 1–18.
11. Kerr MK and Churchill GA. Statistical design and the analysis of gene expression microarray data. *Genetical Research* 2001; 77: 123–128.
12. Lee M-LT, Kuo FC, Whitmore GA, and Sklar J. The importance of replication in microarray gene expression studies: Statistical methods and evidence from repetitive cDNA hybridizations. *Proceedings of the National Academy of Sciences of the USA* 2000; 97: 9834–9839.
13. Draghici S. *Data Analysis Tools for DNA Microarrays*. New York: Chapman and Hall/CRC Press, 2003.
14. Simon RM, Korn EL, McShane LM, Radmacher MD, Wright GW, and Zhao Y. *Design and Analysis of DNA Microarray Investigations*. New York: Springer, 2003.
15. Yang YH, Buckeley MJ, Dudoit S, and Speed TP. Comparison of methods for image analysis of cDNA microarray data. *Journal of Computational and Graphical Statistics* 2000; 11: 108–136.

16. Kerr MK and Churchill GA. Experimental design for gene expression microarrays. *Biostatistics* 2001; 2: 183–201.
17. Dobbin K, Shih J, and Simon R. Statistical design of reverse dye microarrays. *Bioinformatics* 2003; 19: 803–810.
18. Irizarry RA, Hobbs B, Collin F, Beazer-Barclay YD, Antonellis KJ, Scherf U, and Speed TP. Exploration, normalization, and summaries of high density oligonucleotide array probe level data. *Biostatistics* 2003; 4: 249–264.
19. Li C and Wong WH. Model-based analysis of oligonucleotide arrays: Expression index computation and outlier detection. *Proceedings of the National Academy of Sciences of the USA* 2001; 98: 31–36.
20. Yang YH, Dudoit S, Luu P, Lin DM, Peng V, Ngai J, and Speed T. Normalization for cDNA microarray data: A robust composite method addressing single and multiple slide systematic variation. *Nucleic Acids Research* 2002; 30: e15.
21. Ding Y and Wilkins D. The effect of normalization on microarray data analysis. *DNA and Cell Biology* 2004; 23: 635–642.
22. Colantuoni C, Henry G, Zeger S, and Pevsner J. SNOMAD (Standardization and Normalization of MicroArray Data): Web-accessible gene expression data analysis. *Bioinformatics* 2002; 18: 1540–1541.
23. Quackenbush J. Microarray data normalization and transformation. *Nature Genetics* 2002; 32 (Suppl.): 496–501.
24. Park T, Yi SG, Kang SH, Lee S, Lee YS, and Simon R. Evaluation of normalization methods for micro-array data. *BMC Bioinformatics* 2003; 4: 33.
25. Dudoit S, Shaffer J, and Boldrick JC. Multiple hypothesis testing in microarray experiments. *Statistical Science* 2003; 18: 71–103.
26. Storey JD. The positive false discovery rate: A Bayesian interpretation and the q-value. *Annals of Statistics* 2003; 31: 2013–2035.
27. Baldi P and Long AD. A Bayesian framework for the analysis of microarray expression data: Regularized t-test and statistical inferences of gene changes. *Bioinformatics* 2001; 17: 509–519.
28. Jensen ST, Liu XS, Zhou Q, and Liu JS. Computational discovery of gene regulatory binding motifs: A Bayesian perspective. *Statistical Science* 2004; 19: 188–204.
29. Hartigan JA and Wong MA. A k-means clustering algorithm. *Applied Statistics* 1979; 28: 100–108.
30. Ward JH. Hierarchical groupings to optimize an objective function. *Journal of the American Statistical Association* 1963; 58: 236–244.
31. Kohonen T. *Self-Organizing Maps*. Berlin: Springer, 1995.
32. McLachlan GJ, Bean RW, and Peel D. A mixture model-based approach to the clustering of microarray expression data. *Bioinformatics* 2002; 18: 413–422.
33. Medvedovic M, Yeung K, and Bumgarner R. Bayesian mixture model based clustering of replicated microarray data. *Bioinformatics* 2004; 20: 1222–1232.
34. Jain S and Neal RM. A split-merge Markov chain Monte Carlo procedure for the Dirichlet process mix-ture model. *Journal of Computational and Graphical Statistics* 2004; 13: 158–182.
35. Yeung KY and Ruzzo WL. Principal component analysis for clustering gene expression data. *Bioinformatics* 2001; 17: 763–774.
36. Kruskal JB. Multidimensional scaling by optimizing goodness of fit to a nonmetric hypothesis. *Psychometrika* 1964; 29: 1–27.
37. Kim SY, Lee JW, and Bae JS. Effect of data normalization on fuzzy clustering of DNA microarray data. *BMC Bioinformatics* 2006; 7: 134.
38. Cochran WG and Cox GM. *Experimental Designs*. New York: Wiley, 1992.
39. Montgomery DC. *Design and Analysis of Experiments*, 6th edition. New York: Wiley, 2005.
40. Neter J, Kutner MH, Nachtscheim CJ, and Wasserman W. *Applied Linear Statistical Models*, 4th edition. Chicago: Irwin, 1996.
41. Zien A, Fluck J, Zimmer R, and Lengauer T. Microarrays: How many do you need? *Journal of Computational Biology* 2003; 10: 653–667.

4 Laboratory Automation for Microarray Experimentation

Douglas N. Gurevitch

CONTENTS

4.1 WHY LABORATORY AUTOMATION?

Owing to the very nature of microarray experimentation, once researchers decided to use this technology, they also decided to use laboratory automation. Microarray experimentation requires automation for printing and scanning the arrays. Both operations have two characteristics that demand automation: a large number of small features and very tight tolerances. These characteristics and the demand to make orderly arrays in order to track features and data preclude manual array creation and data collection.

Other uses of laboratory automation that can greatly aid microarray experimentation include automated hybridization, sample preparation, and probe synthesis or preparation. All of these can greatly aid the reproducibility of microarray experiments. Most of these systems, as well as the systems to create the microarrays, are based on robotics.

4.1.1 WHAT IS A ROBOT?

A robot is a flexible, automated tool. The official definition from the Robotics Institute of America is "... a reprogrammable multi-functional manipulator designed to move materials, parts, tools or specialized devices, through variable programmed motions for the performance of a variety of tasks." For laboratory automation, the majority of robots are designed to move small volumes of liquid very precisely (microliter scale).

4.1.2 USES OF AUTOMATION BY TASK

One experimenter task that requires laboratory automation is the robotic manufacturing of custom arrays. Other microarray technologies, especially the preprinted, *in situ*-synthesized or lab-on-a-chip technologies are created by automation at the site of their manufacturers. Custom arrays are typically spotted or printed in-house. Another task that can be automated is the assay/hybridization process itself. This can be done using automated hybridization systems, which consist of computer-controlled pumping and thermal cycling systems. These systems can help eliminate experimenter-to-experimenter assay process differences. Data collection also requires automation to scan the arrays under high magnification and with the appropriate light sources and filters to measure the fluorescence resulting at each spot due to the sample binding to the probe. Lastly, many liquid handling, assay, and sample preparation processes can be automated using more general liquid handling robots.

4.2 CREATING CUSTOM-SPOTTED MICROARRAYS

All in-house microarray spotting or "printing" systems are based on gantry or "Cartesian" style robots. These robots move a "print head" from source microtiter plates to print onto the substrates, usually specially coated microscope slides. The actual spotting or printing is accomplished by one of the two techniques: (1) contact or transfer printing and (2) noncontact or "ink-jet" style printing.

4.2.1 CONTACT PRINTING

Contact or transfer printing is the dominant method used for spotting or printing microarrays. This consists of finely manufactured pins that dip into the source liquids in a microtiter plate and then touch off this liquid onto the surface of the destination substrates, which can be glass slides, nylon membranes, or other materials. Contact printing usually prints volumes in the 0.5–2 nL range depending on the liquid and pin diameter. The vast majority of the pins being used are "quill" designs, emulating the quill pens of yesteryear (Figure 4.1). A finely manufactured slot in the pin wicks up the liquid to be printed due to capillary action. Some manufacturers can even tailor these surfaces to affect the surface energy and the pins' ability to wick the source fluid. This slot then acts

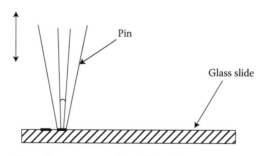

FIGURE 4.1   Quill pin touching off spots onto a glass slide substrate.

as a reservoir, allowing the pin to print many times either as replicates on the same slide, or on multiple slides. A few pins also have a reservoir at the top of the slot to increase the capacity for multiple prints. "Quill" or slotted pins are best for lower viscosity or aqueous liquids. Unfortunately, many slotted pins do not have circular cross sections at the contact surface, and poor spot shape can frequently result. Hitachi (Alameda, California) has developed a variation on this theme for their pin, which has a complex machined "dimple" at the tip's end. This "dimple" acts as the reservoir, holding the liquid by surface tension for spotting.

The solid pin is better suited for nonaqueous and higher viscosity fluids. This pin transfers only that liquid which is attached by surface tension on the external surface of the pin. Given the low volume attached to the surface, these pins effectively have a one-spot reservoir of liquid. This requires that solid pins be dipped for every print and for a slower overall printing process. Ideally, especially with higher viscosity fluids, these pins can print more circular spots more consistently. This is not the case with more aqueous and lower viscosity fluids, where surface energy levels usually cause ringed or lopsided, undersized spots.

Both types of pins can be made out of a variety of materials. Currently, the most common is stainless steel. Although stainless steel has good corrosion resistance and compatibility with biomaterials, at the dimensions currently desired for microarray printing, it is easily deformed and damaged and is worn by use quickly. It is also difficult to machine; therefore, the pins tend to be expensive. Another material currently in use is titanium. This is both stronger and lighter than stainless steel with equally good corrosion resistance and biocompatibility. Titanium pins also last longer due to the material's strength. Unfortunately, it is difficult to manufacture pins in this material, and therefore they are expensive. Tungsten, an additional material, machined by electro-chemical processes has found widespread use over the past few years. Tungsten is also highly corrosion-resistant and biocompatible. It is also the strongest of the three and the pins are much more difficult to damage and therefore last much longer. The main drawback is the increased weight of the pins due to tungsten's high density. This increased weight can increase wobble of the pins during the printing process.

Wobble of the pins is an issue that is sometimes glossed over by manufacturers of spotting and printing robots. The pins must have low enough friction to slide up and down easily to ensure the best printing. This requires that the holes for the pins in the print head be slightly oversized, which allows the pins to wobble slightly. A number of manufacturers claim ± 1–$10\,\mu$m positional accuracy for their robots. This is the accuracy of the actual motion components, or of the robot, in the system. This is not the final positional accuracy of the printed spots. Owing to the wobble and manufacturing tolerances of the pins, total positional accuracy is in the order of ± 15–$25\,\mu$m. This is usually not a problem, thanks to the spot-finding algorithms in the analysis software, but can limit how many spots can be printed per slide.

4.2.2 Noncontact Printing

All microarray noncontact printing systems are based on "ink-jet" printers used for printing text and images. Unfortunately, true ink-jet printers print inks with incredibly consistent viscosity from cartridge to cartridge, optimized to prevent clogging. Moreover, the ink is designed for minimal spread on an absorbent substrate (paper), yielding smaller spots that spread very little. Unfortunately, this is not true for noncontact printing microarrays, as the viscosity of each probe to print may be different, with quite a bit of spread on the substrate (slide), and a high rate of clogging. This differing viscosity can thus cause each probe to print different volumes and different size spots.

Ink-jet printing comes in two types: piezo-electric actuator and pressurized micro-valve type (Figure 4.2). Neither of these should be confused with bubble-jet technology. Bubble jets were the first of the "jet" printing systems. Pioneered by Hewlett–Packard, a bubble-jet print head actually boils a small plug of ink to achieve a single bursting bubble. The force of the bubble bursting propels an ink drop through a micro-nozzle and onto the paper. Obviously, this would be very detrimental to any biological materials.

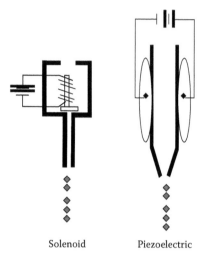

Solenoid   Piezoelectric

FIGURE 4.2   Examples of ink-jet print heads.

For microarray printing, the piezo-electric actuator takes the following two incarnations. Both use a piezo-electric crystal to generate a pressure wave that provides the motive force for a drop to shoot out of the exit nozzle as:

- A glass capillary wrapped by a blanket of piezo-electric ceramic that gives the capillary a sharp squeeze when driven by a voltage pulse
- A micro or silicon lithography machined diaphragm pump in which the diaphragm is actuated by a piezo-electric ceramic block

Both print volumes of 200–700 pL per drop. Both clog very easily due to bubbles, particles, or sticky, high-viscosity fluids, with the diaphragm system showing more reliability. Both can print extremely quickly, but there are severe cleaning requirements between aspirations of probe, usually including a step to blot off any remaining liquid from the cleaning. None of the readily available systems can aspirate from a 1536-well plate, as the actual nozzles are too wide to go deeply into the wells of such plates.

The other main ink-jet-based technology is the pressurized, micro-valve released design. This consists of a capillary nozzle/tip linked to a syringe pump with a micro-solenoid valve in between. The sample is aspirated (sucked up) into the nozzle, the syringe pump preloads the liquid column with pressure, and the micro-valve opens to expose the liquid in the nozzle to a sharp pressure wave to drive the release of a droplet. These systems can also bulk dispense reagents for high-throughput screening (HTS) applications, especially for 1536 well and higher density (3456) applications. In this application, ink-jet-based dispensing can really speed up processing.

4.2.3   Spot Morphology and Placement Issues

All printing techniques designed for in-house use suffer from either spot morphology (shape) or placement issues. For contact printing, the prevalent issue is spot shape. Quill pins usually do not have a circular pin-end shape, and are frequently rectangular in shape. This leads to oval-shaped printed spots. Both solid and quill pins can print "incomplete" circle spots in which the spots can lose crescent-shaped portions of their profiles. Furthermore, solid pins, as well as quill pins running out of liquid, can print doughnut or ringed spots. These either have less or no liquid at their centers, or have a center spot with a ring of liquid surrounding it. This is typically due to liquid spreading out during contact and retracting during separation of the pin from the substrate. These problems can also be caused by clogging due to either excessive particles in the liquid or the sample liquid being too viscous.

Noncontact printing also poses problems. The most serious issue is the placement of the spot on the substrate. The drops ejected from the ink-jet dispensing heads can be diverted from the optimal location by air currents, electrostatic fields, and by liquid retained on the head (leading to the need to blot off the heads after cleaning).

4.3 AUTOMATED HYBRIDIZATION SYSTEMS

Another good use of automation for microarray experimentation is for automating the hybridization assay. A number of reasons support the use of automation for running the assay:

- Automation removes the experimenter-to-experimenter variation normally seen when complex biochemical experiments are run manually.
- Automated systems usually can run 12 slides at the same time, leading to greatly increased throughput.
- Most use gasket/seal systems that are better at preventing leakage during the assay versus manual cover-slip methods.
- One of the systems (the Lucidea ASP from Amersham Biosciences, Piscataway, New Jersey) allows for reuse of the sample by allowing it to be pulled off of the slide after the assay is over.

These systems also have a major drawback in that they are very expensive. However, most of the users of these systems agree that the improved consistency and repeatability of the assay afforded by these systems is worth the cost. If the highest consistency or quality of data is the goal of your microarray group and you are processing a large number of arrays, automating the hybridization assay should be a serious consideration. A comparison of automated hybridization stations is discussed in Chapter 7.

4.4 READING THE DATA: SCANNERS

4.4.1 TYPES OF OPTICS AND SENSORS

Given the dimensions of features on most microarrays, the sheer number of spots, and the accuracy required to make the most of the data, an automated scanner is required. A number of manufacturers make scanners to detect fluorescence of microarray features at a number of wavelengths. Details on various scanners technologies can be found in the book *Microarray Image Analysis—Nuts & Bolts* by Shah and Kamberova [1]. A methodology on how to evaluate scanner performance is presented in Chapter 5.

Most scanners fall into two main categories. The first of these is based on confocal microscope optics with a photomultiplier tube (PMT) sensor to detect the intensity of the fluorescence. PMTs also act as photon amplifiers to increase the signal from dim sources. The confocal optics collects photons from a very thin focus plane. This combination rejects background noise while having high sensitivity. Unfortunately, due to the very thin focal plane of the confocal optics, these systems are very sensitive to the microarray slide not being flat or consistently the same thickness. This problem is most often caused by the stage that holds the slide in the scanner not being flat.

The second type of scanner is based on standard microscope optics with a charge-coupled device (CCD) camera as the sensor. These systems are usually less expensive to produce, but they suffer from the following problems. They usually reject less background noise, resulting in noisier data and because CCDs do not amplify the signal like PMTs, they are usually less sensitive. However, they are much better at ignoring stage and slide flatness problems.

4.4.2   Excitation (Illumination) Types

There are three types of excitation or illumination technologies for causing the tag molecules in the spots of a microarray to fluoresce. The method with the highest stability, illumination intensity (higher power for higher sensitivity), and cost is the gas laser. All the most expensive scanners use gas lasers for their consistency and power. Gas lasers are also fairly large, so the scanner must also be large. The next most common type uses diode or "solid-state" lasers. These systems have less stability and intensity/power than the gas laser-based systems, but they also cost less. Solid-state lasers are also smaller and allow for the scanner to be smaller than if it were based on gas lasers. A few systems are in use based on controlled halogen bulb light sources. This type of illumination is the least expensive of the three and has the lowest power requirements, but it also has the least stability and illumination intensity (leading to the lowest sensitivity). Unfortunately, all these excitation options are very difficult to manufacture, such that they output the same intensity from system to system.

Zhang and his group from the University of Texas M.D. Anderson Cancer Center carried out a detailed comparison of several available microarray scanners [2]. They scanned the same microarray slides with three commercial scanners. The images were analyzed with independent microarray image analysis software. The correlation coefficients for the various comparison experiments between scanners fell within the range 0.90–0.96. When images and data were obtained with the same scanner at different times, the correlation coefficient was about 0.93. This means that any variability observed by the authors was not a result of the instrument used. Therefore, data produced by different laser scanners can be reliably compared. It is important however to use a reliable image analysis software and compare data analyzed with the same feature extraction algorithm. The final decision on a scanner depends on the throughput in the lab and the specific requirements of the research project and microarray techniques applied by the team.

4.4.3   More about Scanners

Many of the high-end scanner manufacturers have options or systems that include automated slide loading. This allows the systems to be "walk-away," to run the scans on their own without an instructor. Unfortunately, nearly all scanners also have slide stage flatness issues that can add as much as 20% variability to the results (more normally about 5%). The LS series scanner from Tecan (San Jose, California) eliminates this problem by holding the slides in a four-at-a-time slide holder that has the outer dimensions of a microtiter plate. This holder is then loaded onto a stage that has a full five degrees of freedom (DOF) of motion and control to actively control for stage flatness. This system can additionally scan microtiter plates (albeit upside down). It has the advantage that standard plate handling robots will be able to load the scanner and feed a large number of microarrays for scanning. This system uses best-of-breed gas lasers and PMT technology. Unfortunately, it is more expensive than competing systems due to the cost of the five DOF stage. The advantage the control of the stage for flatness gives is significant. In side-by-side tests, the actively controlled stage reduced coefficient of variance (CV) of read spots by 4% compared to a standard gas laser/PMT system with the same sensitivity. And as stated earlier, truly out-of-flat stages can severely affect data reliability.

4.4.4   Automation of Microarray Image Analysis

Automation in microarray image analysis is a necessity in large-scale experiments or clinical trial projects. There are two major requirements that are met by automating this step:

- High throughput of image processing
- Unbiased analysis of the images, which guarantees high-quality data.

One example is the automated module in ImaGene from BioDiscovery (Marina del Rey, California). An important characteristic of this module is its autonomous operation, which offers batch-mode (overnight) processing of multiple images. An operator will load the images for analysis and experimental configuration and start the system (Figure 2.3). This walk-away approach allows one person to set many images for analysis in lesser time.

The system offers consistent quality because results do not vary with experience of the operator. The program correctly identifies signal pixels from background, and throws away the contamination pixels. After quantification, the files are saved for further analysis. ImaGene's batch module has been designed as a shared resource to be executed on a server or as part of a microarray Laboratory Information Management System (LIMS). Users are able to review saved data from their personal computers with ResultsReviewer™ at a convenient time.

4.5   GENERAL LABORATORY AUTOMATION

Why should the microarray experimenter consider other processes to automate? First, the biology of microarray experimentation is highly variable, requiring the need to decrease all other sources of potential variability to obtain optimal results. Second, experimenter-to-experimenter error can add between 1% and 3% to the overall CV. Third, samples and reagents are precious, and automation can reduce the volumes used and lost due to human error. General laboratory automation aids in reducing the amount of variability in the experimental processes. Automation of interest to the user of microarrays falls into two categories: material handling and chemical/DNA synthesis systems.

4.5.1   MATERIAL HANDLING LABORATORY AUTOMATION

Liquid material handling laboratory automation is based on "Cartesian" or gantry style robots. These robots are most often used as dispensers or liquid handlers (liquid transfer). Dispensers pump from a reservoir using a microprocessor-controlled peristaltic, diaphragm, or syringe pump. These are used for bulk reagent distribution or dilution tasks. Liquid handling/transfer systems aspirate liquid from one place and dispense it to another. For fine work they use either syringe pumps attached to the working tips via tubing or syringe-at-the-tip systems, most frequently seen in the 96- and 384-channel systems. Working volumes are in the microliter to milliliter range. These systems are useful for preparing and purifying samples, reagents, and other assay materials. Another type of material transfer robot is the clone picker/gridder. These systems combine active vision systems that look for bacterial colonies on growth plates with a picking or gridding tool. Once a colony is identified, the robot stabs it with a transfer pin. This pin is then touched off onto a new growth plate or into a microtiter plate (inoculation) for further growth of the colony. These systems can also be mounted with gridding pin tools for creating arrays either on membranes, growth plates, or other, more novel, substrates. These systems are more expensive due to the complexity of having vision systems and complex tool-cleaning processes, but they also can increase the throughput substantially. Material handling systems are important for high-volume, highly repetitive tasks, as they can reduce assay errors and potential repetitive motion injuries to laboratory personnel.

Labware material handling systems are of two types. First is the sample transport system. This can be a robot arm, track mounted or not, or a conveyor system. One new entry combines simple pick-and-place robot arms with a flexible, easy to reconfigure conveyor system. These systems move microtiter plates, tubes, and sometimes reservoirs around. Most are based on industrial automation and are not for the faint of heart to attempt to program on their own. A number of vendors offer preintegrated systems with easy to reconfigure software systems that prevent the user from having to learn how to program each robot independently. The second material handling system for labware is the automated storage system (some with environmental control). These are either plate stackers or automated carousels in an enclosure with a robot arm to get the plates out to a single access point. A plate stacker system, arm, or conveyor based, attached to a single plate reader, can greatly increase the throughput by providing a walk-away solution to time-consuming sample retrieval or empty microtiter plate feeding. Automated storage systems become much more necessary with large libraries of compound or samples, especially if they need to be stored in chilled or incubator conditions. These systems range from quite reasonable simple stacker/arm combinations to multi-million dollar environmentally controlled automated storage warehouses. The biggest issue with using labware handling automation is to judge the need correctly (including possible growth) in order to get the most cost-effective solution.

4.5.2 CHEMICAL/DNA SYNTHESIS AUTOMATION

Automated chemical/DNA synthesis systems come in two types. The first type is based on liquid handling robots. In this incarnation the synthesizers pipette the chemicals into tube- or plate-based solid-phase extraction columns. These systems provide a great deal of flexibility, including the ability to switch chemical techniques, but tend to be slower than the next type due to the limited number of pipette motions that can be done concurrently (based on the number of tips).

The second type of synthesis system uses valves and pumps to simultaneously pump the chemicals through flow-through columns. These systems are much less flexible, but much faster due to their multi-channel approach. The use of either type of these systems for the creation of probes can give the experimenter much greater knowledge as to the quality of the probes used on the microarray. However, these systems require a great deal of knowledge in the use of hazardous chemicals. Therefore, the experimenter should weigh carefully the advantage that the control of the process provides versus the ease of purchasing the materials. If the composition of the probes needs to be kept secret for intellectual property concerns, then using one of these systems to create the probes is required.

4.6 CONCLUSION

As stated at the beginning of this chapter, microarray experimentation requires automation in order to make and scan the arrays. Furthermore, due to issues of scale, sensitivity, and variability in the microarray process, the serious researcher needs to analyze the various components and consider whether or not automation can help reduce experimental variability.

REFERENCES

1. Shah S and Kamberova K (Eds). *Microarray Image Analysis—Nuts & Bolts*. Abington, PA: DNA Press, LLC, 2002.
2. Ramdas L, Wang J, Hu L, Cogdell D, Taylor E, and Zheng W. Comparative evaluation of laser-based microarray scanners. *BioTechniques* 2001; 31: 546–552.

5 How to Evaluate a Microarray Scanner

Yaping Zong, Shanshan Zhang, Youxiang Wang,
and Yaxian Shi

CONTENTS

5.1   INTRODUCTION

Microarray technology has become an important tool for the study of gene expression, drug discovery, single nucleotide polymorphism (SNP) detection, and other applications [1–3]. Even though the technology has become more prevalent, many variables that affect the performance of experiments and the reproducibility of results remain unresolved. To obtain the best microarray results, one needs to identify and understand the main factors influencing microarray data, then concentrate on minimizing variations related to these factors, and finally obtain reliable data with confidence. Microarray processes generally include

- Sample preparations, such as sample extractions from cell lines
- Amplification and purification, and probe labeling and purification
- Spotting or printing
- Hybridization
- Image and data acquisition
- Data analysis and interpretation.

Each step involves a number of variables that are discussed throughout this book. In this chapter, we will focus our attention on the imaging system for microarrays.

From a microarray user's point of view, one may ask—how can we select a good and reliable scanner that generates consistent images with superior quality? In other words, how do we know that a scanner is operating accurately and properly? After conducting extensive studies on a variety of commercial scanners in the market, scientists developed several effective tools, including the Scanner Calibration Slide, Scanner Validation Slide, and All-Purpose Evaluation Slide, for evaluating and calibrating microarray scanners (Full Moon BioSystems, Inc., Sunnyvale, California, www.fullmoonbiosystems.com). The straightforward and easy-to-use products, incorporating Full Moon BioSystems' (FMB') proprietary coating technologies that enhance the reliability and accuracy of the calibration tools, provide users with extra benefits for evaluating and comparing microarray scanners.

5.2   SCANNER EVALUATION

Currently, there is a wide selection of microarray scanners available in the market [4]. Manufacturers publish a long list of hardware and software specifications, many of which come from design engineers' prospective. As end users, researchers are primarily concerned about the scanner's actual performance in real experiments. Since the engineers' specifications may not always be realized in true microarray experiments, how do we interpret the numbers beyond the hardware point of view and translate them into real performance specifications? Do two scanner units produce the same results? How do we identify the differences and correct them? In short, how do we evaluate the performance of a scanner and make sure it fits our needs?

Microarray scanner evaluation involves two aspects, intra-system evaluation and inter-system calibration. In intra-system evaluation, one examines the scanner's uniformity, alignment and focus, noise, dynamic range, repeatability, stability, channel cross-talk, and other important attributes of a microarray scanner. This process ensures that the scanner is functioning properly, the data acquired are consistent, and the concentrations of targets and probes are adequately configured so that they respond within the scanner's linear dynamic range. An inter-system calibration is required when multiple scanner units are used. Owing to the complexity of microarray scanners, each scanner may not necessarily produce the same readouts even when the settings seem to be the same. If one wishes to compare two sets of microarray results, he/she must first ensure that the two scanners from which the results were acquired produce comparable results. Inter-system calibration allows the users to use the same tool to obtain the results from multiple scanners, compare the data, and adjust the scanners' settings accordingly.

5.2.1   INTRA-SYSTEM EVALUATION

In general, commercial microarray scanners can be classified into two categories based on the light source and the detection system used [4]. One type of scanner is charged-coupled device (CCD)

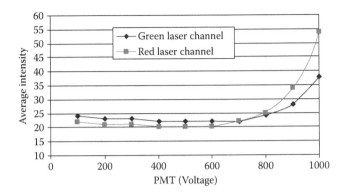

FIGURE 5.1   Scanner noise varies at different PMT settings.

based, where a white light source is used for illumination. The other type is photomultiplier tube (PMT) based, which utilizes lasers as excitation source. Regardless of which light source is used, we can evaluate the performance of a scanner from the following aspects.

5.2.1.1   Scanner Noise

Scanner noise refers to both the dark current noise and the shot noise. The dark current noise is produced by the PMT or by the photon detection devices, whereas the shot noise comes from the particle nature of light sources [1].

A scanner's noise can be easily estimated by performing several scans at different PMT voltages while leaving the slide holder empty. Figure 5.1 shows the noise variations at different PMT voltage settings. The results show that the noise virtually remains constant when the PMT setting is <700 V. On this particular scanner, the noise level dramatically increases as the PMT voltage reaches 800 V. Please keep in mind that these data are specific to the scanner being tested; other commercial scanners may exhibit different responses.

If the dark noise needs to be estimated separately, one can block the excitation laser beam, and then carry out a scan without a slide. The average signal obtained is the scanner's dark noise.

5.2.1.2   Scanner Uniformity

Scanner uniformity usually refers to the scanning consistency in X- and Y-directions. In most situations, during the course of scanning, the stage that carries the slide holder moves in the Y-direction (Y-axis) and the light source or scanning head with lens moves in the X-direction.

To determine X- or Y-direction's scanning uniformity, one can create a special fluorescent slide by carefully placing one piece of adhesive fluorescent tape (such as Kapton adhesive tape)* on the surface of a glass slide. Make sure that the tape rests completely flat and uniform on the slide surface. Load the slide on a scanner; perform a quick scan to optimize the scanner's PMT setting to ensure that the signals from both green and red channels are not saturated. To verify the fluorescent slide's uniformity and eliminate any interfering factors coming from the slide, scan the entire slide, and calculate the average intensities at different positions across the entire slide. Then rotate the slide 180° and scan again. Perform the same analysis. Plot a graph of normalized intensities against the positions on the slide surface. Figures 5.2 and 5.3 show two plots as illustrative examples.

---

* Kapton© is a trademark of DuPont. Kapton film tapes are used extensively in printed circuit board production. They are transparent, which allows observation of the circuit boards during processing. See http://www.kaptontape.com.

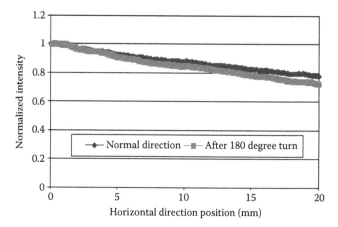

FIGURE 5.2 *X*-direction scanning uniformity. The two plotted lines provide a similar trend—normalized signal intensities in both directions (normal direction and 180° reverse direction) continuously decrease toward the right side of the graph while the intensity level remains relatively consistent; hence, the Kapton slide's uniformity did not affect the scanning uniformity. Therefore, the nonuniformity (decrease in intensity) exhibited here can be the result of the *X*-stage being out of alignment.

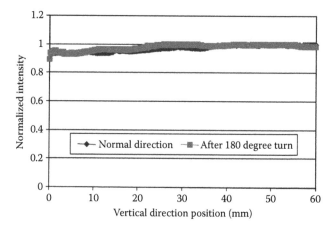

FIGURE 5.3 *Y*-direction scanning uniformity. Similar to Figure 5.2, the normalized intensity remains consistent on both scans, which confirms that the slide is relatively uniform. The left portion of the graph, where the intensities were lower, indicates that this particular area of the scanner requires optimization.

5.2.1.3 Scanner Alignment and Focus

When scanning an array with both Cy3 and Cy5 channels, weak signals, blurry spots, or un-superimposed images are sometimes observed. In this case, the scanner's alignment and focus may require further attention. Typically, alignment problems occur when scanner's mechanical stages fail to return to its origins correctly, or when lasers alignment is not optimized.

FMBs' Scanner Validation Slide can be used to diagnose a scanner's problems with alignment and focus. The Validation Slide contains arrays of spots that produce exceptional signals of Cy3 and Cy5 fluorescent dyes in dilution series. The arrays consist of two identical blocks of 29 unique samples printed in 29 columns, and each column contains 12 repeats of each sample. Among the arrays, there are five columns that contain Cy3 and Cy5 dye mixtures, which produce yellow spots when scanned. Figure 5.4 shows the layout of an FMB Scanner Validation Slide.

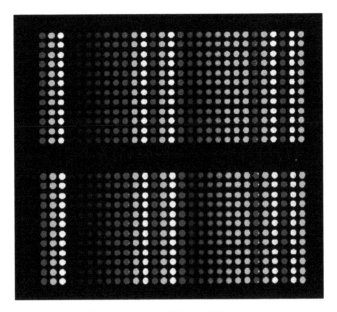

FIGURE 5.4   Array image from an FMB Scanner Validation Slide.

When it is necessary to check the alignment or focus, scan an FMB Scanner Validation Slide, after optimizing the other settings on the scanner. If the alignment of the scanner is at the optimal condition, the spots in the five columns with a mixture of Cy3 and Cy5 dyes will appear to be perfectly superposed, without any overflow of red or green colors at the edges of each spot. At the same time, the spots with single color dyes will appear in single color, red or green, looking sharp and well defined.

In addition, there is an alternative method to detect a problem with focus, where FMB's Calibration Slide serves as a useful tool. Our market research on major commercial microarray scanners has shown that most scanners' actual limit of detection (LOD) for Cy3 or Cy5 dye is ~2–6 molecules/μm^2 when a coated slide is used. Upon a FMB Calibration Slide at the optimized settings on a scanner, and when the scanner focus is properly adjusted, the actual LOD can be determined by locating the last column with visible spots and cross-referencing the number of molecules per micron square for that particular column. (Concentration data is provided with the Calibration Slide.)

The third method to determine whether a scanner is out of focus is through testing the scanner's capability for a slide's backside-signal rejection. Normally, when a scanner's focus is properly calibrated, the signals detected should only come from the fluorescent dyes on the slide surface. It should not detect anything from the backside of a slide. FMB's Calibration Slide or Validation Slide may be used to examine whether the scanner's backside signals can be rejected. To do so, simply turn the slide upside down and scan the back surface of the slide (the side without arrays). Examine the image to see if you can observe any signals from fluorescent arrays on the other side. If backside signals are observed, there may be a problem with the scanner's focus.

5.2.1.4 Dynamic Range

The dynamic range of a scanner is referred to as the system's linear response range [4]. While scanner manufacturers usually focus on the electronic dynamic range, as end users, it is essential for us to consider the dye molecule linear response range and the biological experiment dynamic range, which is closely related to the scanner system, biological samples used, and the amount of targets retained on the slide (slide retention). Additionally, the LOD is also regarded as a very important factor related to dynamic range. Commonly, the LOD is defined as the minimum detectable signal for which the signal-to-noise ratio (SNR) is 3. When the LOD is lower, the sensitivity of the scanner is better. To calculate SNRs, the following formula is commonly used:

$$SNR = \frac{\text{Average signal intensity} - \text{Average background intensity}}{\text{Standard deviation of background signals}}$$

When setting up a microarray experiment and optimizing sample concentrations, one should consider whether the experimental results will produce signals that fit within the linear response range of the scanner. For scientists who want to get maximum information out of a set of experiments where cellular material is limited, such as examining low expressions of messenger ribonucleic acids (mRNAs), obtaining results with high SNRs using a scanner with high sensitivity and low LOD is critical.

How is the linear dynamic range evaluated? FMB Scanner Calibration Slide is available as a tool to help users verify a scanner's linear dynamic range and LOD. The Calibration Slide contains arrays that produce exceptional signals of Cy3 and Cy5 fluorescent dyes. Each block consists of 28 sets of two-fold dilutions of Cy3 or Cy5, coupled with three sets of blanks and one set of position markers. Each column contains 12 repeats of each sample.

Follow this step-by-step instruction to determine the linear dynamic range of your microarray scanner.

1. Load an FMB Calibration Slide on the scanner's slide holder.
2. Optimize PMT settings and make sure that the spots in the last 4–5 columns on the far right side of the image with high dye concentrations produce saturated signals.
3. Scan the slide with both red and green lasers.
4. Analyze the image and calculate SNR.
5. Plot SNR against dye moluecules per square micrometer (these reference data are provided with the product).
6. Determine the linear dynamic response range and LOD based on SNR > 3.

The detection limit and linear dynamic range can be determined from Figure 5.5. The linear dynamic range in this experiment is approximately 3–3.5 orders of magnitude for fluorescent dyes on silanized glass surface. Typically, the scanner dynamic range claimed by most manufacturers is 5 orders based on hardware specifications. However, in normal biological experiments, due to the fluorescent background of the slide and the scanner's self-generated noise, the actual detection limit for dyes can only reach 3–4 orders of magnitude.

5.2.1.5   Repeatability

The stability and repeatability of a scanner is essential for obtaining reliable microarray data with confidence [5–7]. A scanner's repeatability is a function of the stability of the system's power supply, laser, PMT responses, and electronics. How does one determine scanner's repeatability? First, a good fluorescent sample must be selected. The sample must be able to produce strong signals but remain stable under continuous laser illumination. Fluorescent dyes usually would not be a good choice because they degrade and quench under continuous illumination. Therefore, a slide with fluorescent tapes, such as the Kapton-taped slide we discussed earlier, is a good choice for this experiment. Load the slide onto a scanner and optimize appropriate settings. Scan the slide 10 times and then analyze the images. Plot signal intensity against the number of times of repeat scans, as shown in Figure 5.6. The variation for the 10 scans may also be calculated based on the data obtained.

5.2.1.6   PMT Responses

The green and red channels usually possess different sensitivities for Cy3 and Cy5 dyes, which may result from differences in laser power and PMT sensitivity [7]. When you scan arrays carrying both

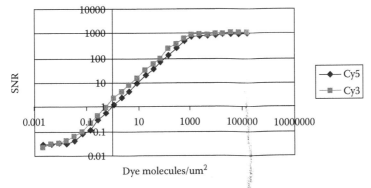

FIGURE 5.5 Correlations between SNR and dye density (dye molecules per square micron).

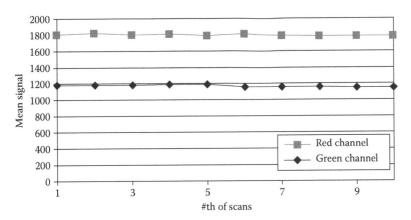

FIGURE 5.6 Scanner repeatability for 10 repeated scans.

Cy3 and Cy5 dyes, it may be useful to match Cy3 and Cy5 signal intensities. This may be accomplished by balancing PMT voltages so that the signal ratio of Cy3 and Cy5 is close to 1. To achieve the balance, one would first determine the differences in response from two channels (red and green), and then modify the PMT voltage settings of both channels to obtain desired results. FMB Calibration Slide can be used for balancing the signals from two channels. The slide contains Cy3 and Cy5 arrays in separate blocks, while the concentrations for each block are identical (Figure 5.7). By simply conducting a scan of a Calibration Slide at the same PMT voltage, the sensitivity difference from the two channels will be revealed. Then, you can adjust PMT voltages accordingly to balance the signals.

Another very important feature of the Calibration Slide is its ability to measure the linear response range of the PMT for Cy3 and Cy5 in different concentrations, which in turn will provide critical information for choosing the optimal PMT setting and for maximizing the SNR for your samples. To determine the linear response range, scan a FMB Calibration Slide at different PMT voltages, analyze the images, and plot the data to determine the appropriate PMT setting.

5.2.1.7 Channel Cross-Talk

Normally, when the red laser is illuminated alone, only Cy5 signals are detected; similarly, when the green laser is activated, only Cy3 signals are acquired. However, when scanning a slide hybridized with a mixture of Cy3- and Cy5-labeled probes, and the red laser is turned on and the green laser is

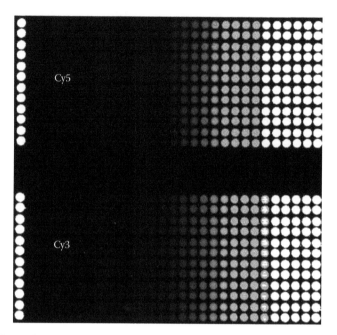

FIGURE 5.7 An image from FMB Calibration Slide showing two blocks of arrays. The upper block contains Cy5 arrays, and the lower block contains Cy3 arrays.

off, sometimes one may observe some Cy3 signals in addition to Cy5 signals, which are supposed to be the only signals detected. This phenomenon is called channel cross-talk, and it results from signals leaking from one channel into the other. Because cross-talk is concealed in dual-color images, this effect is less noticeable when both red and green lasers operate simultaneously to detect a mixture of Cy3- and Cy5-labeled samples. One function of the FMB Scanner Calibration Slide is to help users detect channel cross-talk and estimate variations. Scan both sets of arrays by illuminating only one laser, red or green, then immediately scan again with the other laser. Analyze the image, calculate the intensity of Cy3 and Cy5 spots for both scans, and determine the percentage of the channel cross-talk.

5.2.1.8 Tail Effect

Occasionally, an image may show "spots with tails," where the spots appear to have been smudged in the horizontal direction. This phenomenon is usually observed around spots with high signal intensities and has been identified as tail effect or pixel-to-pixel cross-talk. It results from the residual signals left in the electronic circuitry systems causing the PMT to generate false responses and produce fictitious signals. The FMB Calibration Slide containing 8-order dilution series of Cy3 and Cy5 can be used to verify this phenomenon. By examining images acquired from such slides, you can determine whether the tail effect exists.

5.2.2 INTER-SYSTEM CALIBRATION

The inter-system calibration is especially important when more than one scanners are used and when users wish to compare the results derived from different scanners. Regardless whether the different scanner units are manufactured by the same manufacturers, each scanner may produce different readouts even when all the settings are identical. Therefore, comparing data in terms of signal intensity (counts) will lead to false conclusion.

To accurately compare microarray results, one must first create a correlation between scanner's readouts and sample concentrations so that the results are comparable. For each scanner, a correlation curve must be established. Section 5.2.1.4 provides detailed instructions on how to establish such working curve. Once the curve is set up, the scanner's linear response range can be determined. Within this linear range, based on the SNR results, one can extrapolate samples concentrations. This data is suitable for comparisons.

5.3 SUMMARY

The information provided here on microarray scanners offers simple methods to evaluate the most important features of a microarray scanner. Our goal is to develop some simple, yet robust methods that everyone can manage to use to evaluate and calibrate microarray scanners. We hope the information provided here will help you select or maintain your scanner system(s) in the optimal working conditions and help you obtain consistent and reliable microarray results.

APPENDIX A: MICROARRAY SCANNER CALIBRATION SLIDE CATALOG NUMBER: FMB DS 01

DESCRIPTION

FMB Microarray Scanner Calibration Slide is developed for users to perform quantitative evaluations of their microarray scanners. It is designed for determining the dynamic range, LOD, and uniformity of microarray scanners. This is an excellent tool for detecting laser channel cross-talk and laser stability.

This tool contains two separate array blocks of dilution series of Cy3 and Cy5 fluorescent dyes. Each block consists of 28 sets of twofold dilutions of Cy3 or Cy5, coupled with three sets of blanks and one set of position markers. Each column contains 12 repeats of each sample.

APPLICATIONS

- Analyze and calculate scanner's dynamic range and detection limit
- Detect and analyze variations in performances among different microarray scanner units
- Detect channel cross-talk
- Verify laser alignment
- Determine scanning uniformity

SPECIFICATION

- Slide size: 75 ± 0.3 mm $\times 25 \pm 0.3$ mm $\times 1.0 \pm 0.02$ mm
- Array dimensions: two blocks; each block contains 32 columns $\times$ 12 rows
- Array layout: see Figure A.1
- Spot center-to-center distance: 320 μm

STORAGE

The slides are individually packaged in air and moisture-resistant foil bags, and protected from light. Once opened, the slides should be kept away from light. When not in use, please store the slides desiccated in the dark.

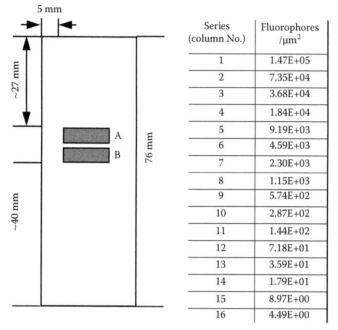

Series (column No.)	Fluorophores /μm^2	Series (column No.)	Fluorophores /μm^2
1	1.47E+05	17	2.24E+00
2	7.35E+04	18	1.12E+00
3	3.68E+04	19	5.61E−01
4	1.84E+04	20	2.80E−01
5	9.19E+03	21	1.40E−01
6	4.59E+03	22	7.01E−02
7	2.30E+03	23	3.50E−02
8	1.15E+03	24	1.75E−02
9	5.74E+02	25	8.76E−03
10	2.87E+02	26	4.38E−03
11	1.44E+02	27	2.19E−03
12	7.18E+01	28	1.10E−03
13	3.59E+01	29	0
14	1.79E+01	30	0
15	8.97E+00	31	0
16	4.49E+00	32	Position marker

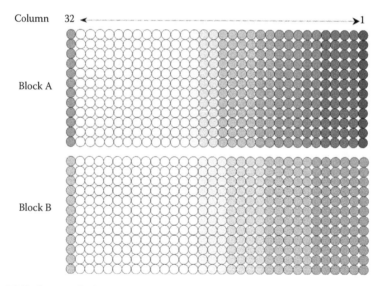

FIGURE A.1   FMB Scanner Calibration Slide layout.

APPENDIX B: MICROARRAY SCANNER VALIDATION SLIDE CATALOG NUMBER: FMB SCV 01

DESCRIPTION

FMB Microarray Scanner Validation Slide is designed for general validation and evaluation of microarray scanners. This Validation Slide also operates as a demonstration slide, a useful tool for demonstrating the overall performances and functions of microarray scanners.

The slide contains arrays that produce exceptional signals of Cy3 and Cy5 fluorescent dyes in dilutions series. The arrays consist of two identical blocks of 29 unique samples printed in 29 columns, and each column contains 12 repeats of each sample.

APPLICATIONS

- Verify whether the scanner operates properly by assessing the accuracy and correctness of responses from scanning lasers, PMT, and laser filters
- Detect channel cross-talk
- Detect and analyze variations in performances among different microarray scanner units
- Operates as a demo slide during microarray scanner demonstrations
- Verify laser focus and alignment

SPECIFICATION

- Slide size: 75 ± 0.3 mm $\times 25 \pm 0.3$ mm $\times 1.0 \pm 0.02$ mm
- Array dimensions: two identical blocks; each block contains 29 columns $\times$ 12 rows
- Array layout: see Figure B.1
- Spot center-to-center distance: 300 μm

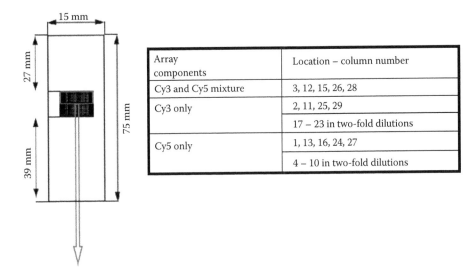

Array components	Location – column number
Cy3 and Cy5 mixture	3, 12, 15, 26, 28
Cy3 only	2, 11, 25, 29
	17 – 23 in two-fold dilutions
Cy5 only	1, 13, 16, 24, 27
	4 – 10 in two-fold dilutions

Column number   1 2 3 4 5 6 7 8 9 10 11 12 13 14 15 16 17 18 19 20 21 22 23 24 25 26 27 28 29

FIGURE 5B.1   (See color insert following page 138.) FMB Scanner Validation Slide layout.

STORAGE

The slides are individually packaged in air and moisture-resistant foil bags, and protected from light. Once opened, the slides should be kept away from light. When not in use, please store the slides desiccated in the dark.

REFERENCES

1. Gershon D. Microarray technology. An array of opportunities. *Nature*, 2002; 416(6883): 885–891.
2. Templin MF, Stoll D, Schrenk M, Traub PC, Vohringer CF, and Joos TO. Protein Microarray Technology. *Trends Biotechnol.*, 2002; 20(4): 160–166.
3. Afshari CA. Perspective: microarray technology, seeing more than spots. *Endocrinology*, 2002; 143(6): 1983–1989.
4. Becker KG. The sharing of cDNA microarray data. *Nat. Rev. Neurosci.*, 2001; 2: 438–440.
5. Seidow JN. Making sense of microarrays. *Genome Biol.*, 2001; 2: 4003.1–4003.2.
6. Wildsmith ES, Archer GE, Winkley AJ, Lane PW, and Bugeski PJ. Maximization of signal derived from cDNA microarrays. *Biotechniques*, 2001; 30: 202–208.
7. Kume H, ed. Photomultiplier Tube. (1994). *Principle to Application*. Hamamatsu Photonics K.K.

6 The Role of Substrates in Microarray Experimentation and How to Choose the Correct Coating for Microarraying

Ruediger Dietrich, Daniel Haines, Rajendra Redkar, and Alistair Rees

CONTENTS

6.1 INTRODUCTION

Microarrays are a widely accepted and powerful analytical tool that allows fast, sensitive, and cost-effective detection of large sets of macromolecules simultaneously. Originally developed to analyze gene expression [1], the value and power of this technology was quickly realized. Several other applications such as single nucleotide polymorphism (SNP) detection [2], comparative genomic hybridization (CGH) [3], tissue arrays [4], microribonucleic acid (miRNA) [5], carbohydrate [6], protein analysis [7], and so on have become increasingly popular.

One of the key factors for achieving successful microarray experimental results is the careful selection of the solid substrate from which the data are generated. Parameters such as appropriate binding chemistry, coating uniformity, batch-to-batch reproducibility, surface energy (hydrophobicity) and roughness, background intensity, detection method sensitivity, and so on are important factors that need to be considered when the user intends to use open microarray systems. Open microarray systems are defined as those microarrays manufactured using robotic or manual deposition of probe molecules on coated substrates, followed by immobilization of probes, which allow users to design arrays exactly to their own needs. Closed microarray systems are defined as those microarrays manufactured using combinatorially synthesized probe molecules on coated substrates, which are fixed in terms of array design. Both open and closed systems have their own advantages and disadvantages. Although relatively expensive, closed microarray systems do not require probe processing. However, for a custom array tailored to specific requirements or when no prefabricated arrays are available, open systems can be a preferred and cost-effective alternative.

The goal of this chapter is to provide an overview of the factors relevant to high-quality microarray substrate production. Furthermore, practical considerations that are important for the selection of the best microarray substrate for a desired application are discussed here.

6.2   SLIDE PROPERTIES

The technical requirements of a substrate for a fluorescence-based detection method dictate that the substrate has high transmission for the wavelengths of interest, exhibits chemical and temperature stability, is resistant to photodegradation, and has minimal auto-fluorescence. A commercial requirement is that the substrate be affordable in the overall context of the experimental cost. The first microarray experiment [1] used standard microscope slides that adequately met all of aforementioned requirements. Low-fluorescence borosilicate glass formulations have been utilized specifically for microarraying by SCHOTT Nexterion™ [8,9]. Corning has selected glass compositions that reduce hazing caused by the leaching of sodium from widely available borosilicate glass [10]. There are materials that have superior properties to glass such as quartz, fused silica, and Zerodur™, but they have significantly higher costs. Lower cost polymeric materials such as polystyrene [11], polymethylmethacrylate [12], and cyclic olefin copolymers [13] have been investigated, but are not preferred due to higher auto-fluorescence and potential outgassing of components used in their manufacture. Components such as plasticizers might interfere with the biological content, which results in lower experimental sensitivity [12].

Nylon and nitrocellulose membranes attached to glass slides have been developed as alternative microarray substrates. Their three-dimensional porous structure compared to glass allows higher probe density captured per spot and compatibility with nonfluorescence detection methods such as isotopic radiolabeling [14] and chemiluminescence. Nylon and nitrocellulose, established materials for standard biological assays such as western blotting, are typically used for specialized microarray experiments (protein arrays), where the cost of biomolecular probes/targets is not a detriment, and lower density arrays (total number of spots) are acceptable. The three-dimensional fibrous structure of the substrates results in decreased spatial resolution (≈20 μm). These substrates also have higher fluorescent backgrounds than glass when using fluorescence-based tests.

Since 1995, many other surfaces have been investigated for microarray applications, but glass remains the substrate of choice for general microarray experimentation. Even as the number of microarray applications, and thus substrate specifications, has expanded, glass continues to be preferred [15] over inorganic substrates such as silicon nitride [16], silicon [17], fused silica [18], mica [19], and organic substrates such as plastics [20] and nylon [21]. In summary, the primary reasons for the preference of glass as a substrate are (1) ease and breadth of chemical group functionalization via coatings; (2) low intrinsic fluorescence; (3) commercially available from numerous suppliers at reasonable cost; (4) consistent and reproducible quality; (5) nonporous and ultra-flat surface with extremely low thickness variation; (6) high spotting densities with maximum resolution; (7) compatible with standard analytical equipment and methods; and (8) chemical resistance to reagents used in microarray protocols. As glass is the most important and commonly used microarray substrate, the emphasis for the remainder of the discussion shall focus on glass.

In all cases glass needs to be cleaned before application of the chemical coating to achieve optimal bonding. Any particulates that are present on the surface prior to coating will become incorporated into the coating and are difficult to remove without damaging the coating. There can be a particular problem with particles of an organic origin as these can be highly fluorescent at the excitation wavelengths used. For good adhesion of the coating the substrate is cleaned and/or activated to provide the necessary functional groups for reaction with the coating moiety. High-quality coated microarray substrates are typically cleaned, coated, and packaged under clean room conditions.

There are many different cleaning procedures available for microarray substrates [22]. Liquid-based cleaning protocols are typically conducted via dipping procedures utilizing water to remove particulates, surfactants and/or bases, and/or nonaqueous solvents such as alcohols and halogenated hydrocarbons to remove hydrocarbon residues, acids for ion exchange and oxidation, and bases for surface dissolution. The effectiveness of a liquid-based cleaning procedure may be increased through the use of ultra-sonication, which uses sound waves generated at frequencies between 10 and 50 kHz. Nonliquid chemical cleaning methods are usually used in addition to or in place of one or more steps in the liquid cleaning procedures. These methods include exposure to ozone and plasma processes.

Many users require a method to validate sample traceability. The most commonly used method is the use of barcodes. Barcodes can be applied as laser-etched foil, ceramic labels or in the form of plastic labels. Label barcodes are a source of possible experimental contamination due to the label-adhesive degradation with the conditions and chemicals used in microarray protocols, whereas etched barcodes can be difficult to read by some scanner designs due to their lower contrast. The quality and effectiveness of the sample traceability method can vary from the commercial substrate vendors, as there is no currently recognized standard.

6.3   CHEMICAL COATINGS

A wide variety of coatings [23] can be applied on glass by different deposition techniques [24] to provide functional groups for the subsequent attachment of biomolecules. There are three types of coatings for microarrays: single, double, and multiple layers. Each layer may have one or more components with one or more properties. The use of the term "layer" is subjective—in this context a layer is defined as being deposited from single precursor source, the precursor source having one or more components. There are many alternative ways to classify these coatings. One major alternative classification addresses the important consideration of the preference of the researcher's choice of the binding mechanism of the coating to the substrate and probe (covalent versus non-covalent), or by the character of the binding moiety itself (amino-, epoxy-, aldehyde-, and NHS-ester coatings; see Table 6.1).

Single-layer coatings can be mono-, bi-, or multifunctional. Monofunctional coatings are the simplest and use the same functional group(s) for attachment to the substrate and biomolecules. Representative examples are nitrocellulose [25], poly-L-lysine [26], and polyhystidine [27], which bind to the substrate surface through a combination of adsorptive [28], electrostatic, and hydrogen bonding interactions [29]. Poly-L-lysine was the coating used in the first microarray experiment [1] primarily due to its low cost, availability, and ease of attachment to glass. With further experience,

TABLE 6.1
List of Common Surface Chemistries

Surface Chemistry	Reacts with/Binds to	Recommended for Probe Molecules	Vendors
Aldehyde chemistry	NH_2 moieties	DNA, proteins, cells, and tissue	SCHOTT Nexterion™, Thermo Scientific
Amino chemistry	Negatively charged moieties such as $-COO^-$, $-PO_4^{3-}$, and so on	Longer DNA as cDNA, BACs, and PACs	Corning, SCHOTT Nexterion™, ArrayIt
Epoxy chemistry	NH_2, $-SH$, and $-OH$ moieties	First choice for all oligo types, cDNA good, even BACs and PACs work, and antibodies	SCHOTT Nexterion™, Thermo Scientific, Corning, ArrayIt
NHS-ester chemistry	NH_2 moieties	DNA and proteins, aminomodified glycans, cells, and tissue	SurModics, SCHOTT Nexterion™
Nitrocellulose	NO_2 moieties	Proteins	GE Healthcare, SCHOTT Nexterion™, and Grace

the use of poly-L-lysine has declined due to the noncovalent binding to substrate (uncertainty of strength) and the low efficiency of deoxyribonucleic acid (DNA) binding [30]. Variability observed in DNA binding as a function of the age of the poly-L-lysine coating has been reported [31]. This led to the development and widespread acceptance of bifunctional single-layer coatings.

Bifunctional single-layer coatings contain one functional group type for anchoring to the substrate and a second functional group for the attachment of biomolecular probes. Organosilanes such as aminosilanes, epoxysilanes, aldehydesilanes, and hydroxyalkylaminosilanes are the predominant precursors, containing four ligands surrounding a silicon atom ($SiL_aL_bL_cL_d$), although dipodal and silsesquioxane precursors can also be used. The ligand structure of organosilanes is relatively straightforward. Three of the ligand positions (L_{a-c}) are normally utilized for surface attachment and/or crosslinking with neighboring organosilanes. These ligands predominantly contain alkoxide ($L = R-OH; R = alkyl$) or chloride moieties to provide a range of reactivity for hydrolysis and condensation reactions to form Si—O—substrate bonds [30]. Alkoxide and chloride ligands also provide convenient manufacturing routes and sites for further substitution. Further substitution of one or two [32] L_{a-c} by alkyl groups has been used to control the level of crosslinking [33]. The fourth ligand (L_d) is designed to bind with probes, although it can be reacted further for cost and reactivity/stability reasons (one of which is binding to other coating layers vide infra). The typical composition of L_d is a hydrocarbon spacer group attached to Si and terminated by the reactive functional group of choice. The functional group binds probes covalently (e.g., epoxy, aldehyde, and NHS-ester) or through hydrogen bonding/electrostatic interactions (e.g., amino and poly-L-Lys). Taken together, a wide variety of functional groups have been used for microarray applications, including aldehyde [34], amine [35], carboxylic acid [34], epoxy [36,37], isothiocyanate [38], mercapto [39], and semicarbazidopropyl [40]. Alternatively, the organosilane can first be attached to probe oligonucleotides and then deposited onto glass substrates [41].

Multifunctional single-layer coatings have one or more components for attaching to the substrate, one or more components for attaching to the biomolecules, and one or more components for additional purposes such as reducing nonspecific target adsorption. Mixed organosilane [42] and hydrogel [43] coatings are representative examples of these types of coatings. Hydrogel coatings are formulated to deter nonspecific protein adsorption on protein microarrays to reduce increased fluorescent background attributable to target adsorption to nonprobe bound areas. Hydrogel coatings have the following characteristics—hydrogen bonding accepting moieties, nonionic moieties, and steric shielding of the substrate surface, while also having functional groups for the probe immobilization (i.e., NHS-esters, epoxides, and so forth). Polyethylene glycol functionalized silanes have been reported to reduce protein adsorption when included as one component of a mixed silane coating [44,45].

Two-layer coatings have a first layer for the attachment to substrate and a second layer for the attachment of biomolecules. A typical multilayer coating is created by depositing a bifunctional coating layer onto the substrate, followed by adding a different bifunctional coating [46]. This strategy has the following advantages over single-layer coatings: control coating thickness to provide more steric shielding of the substrate, introduction of different/additional functional groups [47], easier access to tethered probes through use of more rigid molecules [33], increasing the density of functional groups [48], and reducing nonspecific adsorption [17]. An example of a two-layer coating is a first aminosilane layer with a second polycarbodiimide layer [49,50]. Another example is a streptavidin layer put on top of a hydrogel or polymer layer (H-S slides, SCHOTT Nexterion™), which allows the directed immobilization of biotin-labeled proteins.

Mono- and bifunctional coatings are typically deposited through dip-coating [51], spin-coating, and chemical vapor deposition [52,53] methods. Reproducible, uniform, high-quality coatings in high volumes are nontrivial to manufacture. Microarray coatings are typically produced under clean room conditions to avoid surface contamination by particulates—elimination of >5 μm-sized particles is essential for microarray experimentation. Dip-coating (poly-L-lysine, organosilanes, and alkanethiols) is a highly complex process that produces variable quality coatings [51] from batch-to-batch, unless stringent control is exercised over the manufacturing process. This includes substrate cleaning, control of temperature and humidity, raw material quality, bath conditions (volume, concentration, and pH), coating time, solvent removal, coating drying/curing, and packaging. One of the most important aspects of manufacturing is developing a reliable cleaning procedure [22]. Glass and oxidic surfaces are typically prepared to have a high hydroxyl group content for reaction with chloro- or alkoxide-organosilanes.

There are several routine quality control methods available for assessing the uniformity, reactivity, and cleanliness of the coated substrates. Assessing the uniformity of the functional coating can be accomplished using contact angle and fluorescent staining methods. Macrouniformity of a coating is assessed by determining the surface energy (hydrophobicity/hydrophilicity) as a function of the coated surface area [54]. Microuniformity (<100 µm) of a coating can be assessed by reacting a fluorescent dye [55–57] with surface functional groups to estimate the density of the functional groups by comparing the fluorescence to calibration curves. Alternatively, ultraviolet-absorbing compounds can be reacted with surface functional groups and detected [58,59]. Functional group reactivity to biomolecules is normally assessed by conducting a hybridization control experiment [60] on a batch-to-batch basis and comparing against a database of historical control to determine if the coating meets reactivity specifications. Fluorescence staining can also be used to test for the presence/absence of functional groups. Cleanliness (particulate levels) is normally assessed by normal glass inspection methods such as high-intensity backlighting (http://www.schott.com/optics_devices/english/download/tie-28_bubbles_and_inclusions.pdf).

One major problem with the uniformity and reactivity measurements is the destructive nature of the test methods. Nondestructive spectroscopic test methods such as ultraviolet-visible (UV-VIS) and Fourier-Transformation-Infrared (FTIR) cannot be used for organosilane microarray coatings on glass due to the nanometer thickness [61] of these coatings. SCHOTT Nexterion™ has recently developed a nondestructive uniformity test [62] that utilizes single or multicomponent coatings and assesses coating uniformity through the inherent coating fluorescence. This does not interfere with subsequent microarray performance.

Nonroutine measurements have been used to develop and understand the specific composition of microarray coatings and their effects on probe/target binding. Surface figure has been probed using atomic force microscopy [63] and white light interferometry. Coating composition has been assessed using mass spectroscopy [64], photoelectron spectroscopy [65], time-of-flight secondary-ion mass spectroscopy [66], and so on. Coating thickness has been elucidated using ellipsometry [67]. Monolayer orientation has been assessed using reflection absorption infrared spectroscopy [68]. Coating formation kinetics has been investigated using nuclear magnetic resonance [69]. Additional techniques such as piezoelectric quartz crystal microbalance [70], scanning probe microscopy [67], and zeta potential [71,72] have been used to answer specific questions regarding the structure, reactivity, and stability of deposited coatings.

6.4   OPTICAL AND ANALYTICAL COATINGS

Optical interference and analytical coatings (e.g., gold) can be applied on glass to improve signal intensities (Figure 6.1) and/or sensitivity of the slides or to serve as a basis for label-free on-chip detection methods, such as surface plasmon resonance (SPR) [73,74]. In most cases, the chemical coating required for the immobilization of the probe molecules is placed on top of the optical coating. This allows for processing of the microarrays without any change in the protocol or use of specific equipment.

One such example is a multilayer dielectric optical interference coating designed to enhance the efficiency of the fluorescent excitation for Cy3 and Cy5 dyes. This is achieved by the addition of multiple dielectric layers, each of a precise thickness. Upon irradiation a standing wave is generated creating a high electric field in the vicinity of hybridized fluorescently labeled targets [73,75] (Figure 6.2). Experimental analysis has demonstrated an eightfold signal enhancement with the optical interference slides (HiSens slides, SCHOTT Nexterion™) compared to regular slides with the identical

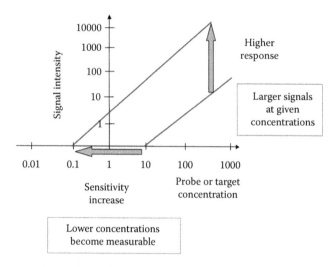

FIGURE 6.1   Beneficial effects of optical coatings include higher signal intensities and/or higher sensitivity.

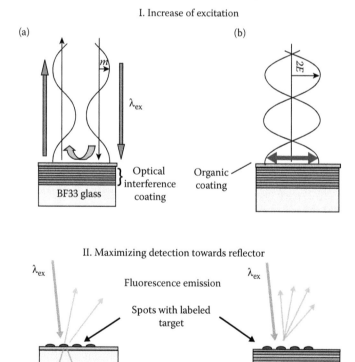

FIGURE 6.2   (See color insert following page 138.) Schematic illustration of the principle of the optical interference coating. (a) Reflecting wave overlaps incoming wave; (b) constructive interference results in amplification of the excitation energy.

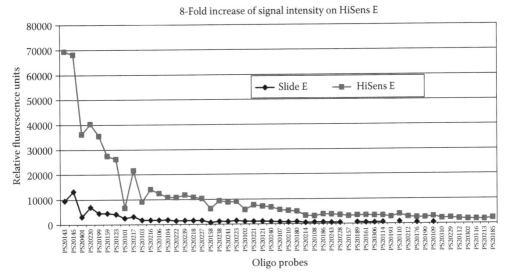

FIGURE 6.3 HiSens slides result in on an average 8-time signal increase compared to standard slides.

chemical coating (for epoxysilane comparison, see Figure 6.3) [75]. This increased detection sensitivity is especially critical for applications in which researchers have only a limited amount of target material, or in which amplification of material is difficult or problematic (Figure 6.4). The latter includes measurement of biopsy samples as well as gene expression monitoring and molecular diagnostics involving limited samples. Even gene expression studies can benefit from higher signals, as weak signals cannot be reliably differentiated from the noise with confidence. This results in up to 20–40% more measurable signals on HiSens E (epoxy) slides compared to the regular E slides. The HiSens surface utilizes both increased absorption of dyes and reflection of hybridized signal into the detector of the scanner and differentiates itself from metallic coatings, which mainly rely on reflection for increasing the signal (Figure 6.3).

A different optical interference coating formerly available from Amersham consisted of a metallic dielectric aluminum coating with a spacer component to put the fluorescently labeled target at the highest electric field maximum [76]. This coating exclusively enhances Cy3 and Cy5 values, as it is

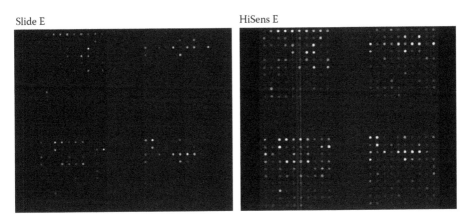

FIGURE 6.4 (See color insert following page 138.) The same amount of miRNA (bladder versus lung samples) target was hybridized to oligo probes printed on HiSens E and Slide E. More signals (low expressor) and higher signals are visible on HiSens E.

known that this dye is especially susceptible to bleaching. Related coatings based on the use of nano-particles [77] or fluorescent signal enhancement through evanescent waves [78] have been reported.

A second type of added functionality is also sometimes required for certain analytical methods. An example would be a gold coating on glass for microarray analysis through SPR experiments [74], used to detect binding of targets to probes without the need for target labeling but requiring a noble metal (Au, Ag)-coated substrate. However, as these approaches require the use of different detection equipments, their usage so far has been limited.

6.5 PRACTICAL CONSIDERATIONS FOR CHOOSING THE RIGHT SUBSTRATE

Firstly, decision should be made regarding the use of "active" or "nonactive" surface chemistries. In the first case, the probe molecules can be covalently attached to the microarray surface via the related functional groups (Table 6.1). As described in Section 6.3, the nonactive surfaces such as aminosilane rely on ionic interactions of probe molecules to the surface. Immobilization of the adhering molecules can be done by either UV-crosslinking or baking steps and produces nondirected or random attachment. On the contrary, active surfaces chemistries such as epoxysilane can be used for both directed and random immobilization. An example for a directed immobilization is the printing of aminolinker-modified oligos onto epoxysilane coating. Here the linker-based ε-NH_2 group reacts with the surface epoxy moiety and forms a covalent bond. Oligos or complementary DNA (cDNA) without an amine linker will also be immobilized on some active surfaces as for example epoxysilane, via noncovalent forces and subsequent immobilization steps. However, it should be realized that this random attachment of probe molecules can occur at different regions of the sequence. This has an impact on hybridization with the target when the related complementary sequence is involved in the attachment to the surface. Because of this reason, a directed immobilization is generally recommended for immobilization of short oligos (<30 nt). A second advantage of active surfaces is that the unused areas on these surfaces can be efficiently inactivated during blocking process, resulting in lower backgrounds after hybridization with target. On the contrary, the nonactive surfaces are generally blocked with bovine serum albumin (BSA) or chemicals to control hybridized background [29].

The second decision is determining which binding chemistry is most appropriate for the probe type. Table 6.1 summarizes the most relevant surfaces for probe molecules. There is not one "universal" surface that meets the needs of all applications and probe types. However epoxysilane-coated slides come closest to being a universal surface, with a high binding capacity, active chemistry that can react with many functional groups, low nonspecific binding, and ease of use. This is the coating of choice for arraying oligo probes [79], and has been used successfully with cDNA [80] or bacteria artificial chromosomes (BACs)/P1-derived artificial chromosomes (PACs) probes for array CGH applications [81]. A number of applications in the protein array field have been demonstrated to work on epoxy coatings [82]. Another widely used active slide surface is the NHS-ester functional group, which reacts with the primary NH_2 groups already present in biomolecules (proteins) or those introduced into DNA by polymerase chain reactions (PCR) or chemical synthesis (such as amino-modified glycans). A number of probe types can be immobilized on this coating, including oligo and cDNA [83], proteins [84], and carbohydrates [6]. Aminosilane chemistry is currently the most widely used surface, probably as a result of being the first commercially available high-quality coating. It is a passive chemistry, and is widely used for printing long DNA molecules such as cDNA [85], BACs, or PACs [86]. Slides coated with nitrocellulose are predominantly used for protein microarrays [87].

After selecting surface chemistry for specific probe sets, the third consideration is the conditions for robotic deposition of probes on the substrate. Control of environmental factors such as temperature, humidity during printing as well as types of pins are critical for obtaining desired spot diameters and spot morphologies. In addition, the printing buffer is very critical for optimal immobilization of probes on a coated substrate. For example, pH of the print buffer drives the opening of the ring structure of epoxysilane chemistry and covalent bond formation. Ideally, the print buffer should not contain compounds that compete for binding places such as tris(hydroxymethyl)aminoethane (TRIS) on epoxysilane or N-hydroxy succinimide NHS-ester surfaces. In some cases probes are already dissolved in a specific print buffer and this dictates the substrate to be selected. For example, DMSO-based printing buffers are not compatible with epoxysilane chemistry and should not be used. Great care should be taken while selecting a printing buffer due to high cost of probes and how the dissolved probes can be stored among different print runs. Some researches prefer to dry down the probes, whereas others will freeze the probes. Both approaches have pros and cons in terms of contamination, change in concentration, and so on, and can have direct effect on the performance of microarrays. An optimal and robust print buffer results in consistent spot size, shape, and uniformity under specific conditions. A round and highly uniform spot shape is desirable beside aesthetic aspects for automated data analysis (see Chapter 15), as it facilitates precise spot/grid recognition and does not require manual intervention. In most cases, a good spot morphology can be accomplished by the use of buffers that are recommended by the slide suppliers. However, in certain cases addition of additives such as betaine [26,88], glycerol [89], or detergents [90] might help to improve spot morphology. On the contrary, printing buffers can contain additives as detergents to increase and adjust the spot size (Figure 6.5) or compounds that are specifically used to reduce the evaporation rate of the buffer especially during long print runs (Figure 6.5), or to obtain significantly more spots per fill. For printing buffer evaluation tests factors such as hydrophobicity of the slide surface, humidity, probe concentration, pin size, pin quality, and stamp (dwell) time are known to affect spot size and uniformity (see also Chapter 2).

(a) Spot diameter on slide E depending on detergent concentration in LEB

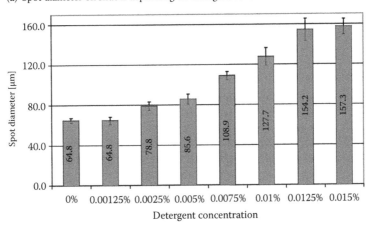

(b) Evaporation rate of LEB versus standard print buffer

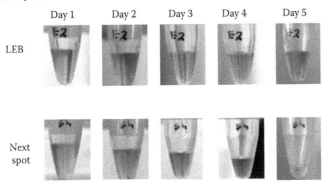

FIGURE 6.5 (See color insert following page 138.) (a) Effect of detergent concentration on spot size in low evaporation buffer (LEB); (b) evaporation features of LEB versus standard spot buffer.

The printing performance can be checked by using fluorescent dyes (such as SYBR 555) or hybridization with small labeled oligomers (Panomer, Invitrogen), or by simply printing a few fluorescently labeled probes on the array. Such spot-checking is especially advisable when new substrates are evaluated, as both printing and immobilization can be verified independently from downstream processing.

The fourth consideration is to evaluate the impact of the blocking and hybridization process in terms of selection of an appropriate surface is relevant to obtain low background values. Parameters that affect background values are (i) nonspecific binding characteristics of the surface; (ii) method used for blocking (BSA blocking versus chemical blocking versus chemical inactivation); and (iii) hybridization process (composition of hybridization buffer, hybridization temperature, and time, ionic strength of the buffer, agitation rate, avoiding of drying during hybridization, and so forth). The applicability of a surface regarding low backgrounds needs to be checked experimentally and often significant differences can be seen among different surfaces. Removal of unincorporated label from labeled target, the presence of nonspecific DNA molecules (e.g., cot 1 DNA), and polyA sequences are also critical for lowering the background. However, the complexity of the target mixture can be altered to reduce the background by exchanging probes and targets, as is done for reverse protein microarrays (RPMAs) in which serum samples are printed and antibody solutions are the target samples [91].

It is important to ensure a proper packaging and storage of the printed microarrays, especially when they are to be used for days, weeks, or even months after the printing. Accurate storage of printed slides is crucial to minimize exposure to ozone [92], volatile plasticizers, CO_2, and H_2O vapor [65], trace air-born contaminants [93], particulate matter, ultraviolet radiation, and oxygen [94], or taken together to ensure optimal performance. In most cases arrays should be stored directly after printing and immobilization but before blocking steps, to keep background values as low as possible. Printed microarrays should be stored typically in the same mailer box they have been shipped, under a dry, dark, and inert atmosphere.

The fifth consideration is to quantify the amount of variation through microarray experimentation. Hence, a sufficient number of replicate microarrays should be envisaged, which on the one hand allow to determine the median signal values; standard deviation, on the other, gives a rough estimation about the variation in general. However, it could be beneficial to specify the amount and source of variation. Technical variation covers printing variability (due to different pins, contamination via pins, and so forth), labeling variation (different RNA concentrations used, different labeling efficiencies, different dye bleaching rates due to protocol variation, and so forth), hybridization variability (due to low target concentration, improper working of hybridization stations, manual hybridization without mixing, contaminated solutions, and so forth), analysis variability (due to nonaccurate spot recognition of the analysis software, comparing different scans by not considering different bleaching rates of the used dyes, and so forth), and sequence-specific variability that affect signal intensities and performance of the slide (nonuniformity of the coating). A possible check of the coating uniformity can be done by printing labeled oligos, which can be included as controls for every printing design. Such an approach separates labeling and hybridization variation. Moreover, statistic science offers help for experimental design, to be able to determine in principle every source and amount of variation [95,96].

For all the considerations taken together, the selected surfaces should be experimentally evaluated to determine optimal print conditions, buffers, and coating uniformity among vendors to find the best surface for each application.

6.6 HIGH SAMPLE THROUGHPUT MICROARRAY SOLUTIONS

In contrast to single-slide experiments where thousands of probe molecules are tested against one target sample, high sample throughput microarray solutions provide the opportunity to test many target samples against low-density microarrays on one substrate simultaneously. This offers the benefits of parallel processing such as time-saving, cost-savings, and reproducibility, which are valid arguments for many users [97]. Basically two different approaches can be utilized: the positional approach in which different target samples are applied on a particular position on a printed microarray, and the tagging approach in which the target molecules are uniquely tagged which allows the attribution of activity values to the respective target molecules during analysis [98]. In the latter case the target molecules must be accessible for tagging, which can be achieved only for some applications such as drug screening or enzymatic assays. However, commonly used microarrays that allow for parallel processing of several target samples belong to the positional approach. Meanwhile, several microarray platforms in the market are available ranging from microsphere-based suspension array platform (Luminex xMAP), bead-based technology on silicon slides (Sentrix Array Matrix, Illumina), UltraPlex technology (SmartBead Technologies Ltd.), nitrocellulose-based (MTP, microtiter plate), 96-well platform (GenTel Path™ HTS), plastic MTP 96-well HTA™ platform (Greiner), and so on. However, the following text focuses on conventional flat glass microarray surfaces that can be processed using standard microarray protocols and hardware.

For some applications such as enzymatic assays or inhibition studies of prognostic diagnostic markers (cathepsin D), target positions can be encoded without a specific patterning by using multiple spotting technology [99]. However, in most cases, the positional approach requires partitioning of substrates into individual reaction chambers or wells, either by polystyrene structures (Nunc, Erie Scientific), special PTFE patternings (SCHOTT Neterion™), or by applying defined nitrocellulose membranes on the slide (Whatman). These partitions serve as a registration aid for probe deposition, prevent cross-contamination between wells, and also ensure proper fixation of superstructures and thus enable incubation of the various target samples in different wells simultaneously. Superstructures can be applied after the printing process (SCHOTT, Erie, Whatman) or multiplexed microarrays are offered with a fixed upper structure.

Multiplexing on glass substrates is commonly available either on slides (processing 16 or 48 wells per target samples) or in an MTP format (for 96 or 384 wells, see also Figure 6.6). Both slide and plate format allow for the use of multichannel pipettes to apply target samples in a fast and convenient way. Although both formats offer a flexible design for efficient printing of probes, the microtiter plate format requires upgrading of most vendors' printing software and hardware (see Table 6.2 for MTP compatible arrayers). Ideally, the dimensions of microarray MTPs correspond to the Society for Biomolecular Screening (SBS) standard ensuring that the microtiter plate is compatible and can be used with liquid handling systems, robotic systems, shakers, or incubation devices. Continuous automatic handling, which could be applied for diagnostic purposes, should be possible, as the need for cost- and time-efficient high-throughput solutions is apparent [100].

The number of probes that can be printed in each well depends on the size of the well, print buffer, pin size, and surface, and can vary between 10 and 3000 spots per well [101].

Presently, the hybridization on multiplexed slides and plates has to be conducted manually, as automatic hybridization stations are not available. The scanning of multiplexed slides (MPX 16/48) is obviously compatible with every standard scanner; however to scan plate format the scanner has to be capable of reading MTP plates (Table 6.3).

Currently, the most relevant surfaces as epoxysliane, aminosilane, NHS-ester, and aldehydesilane are available for MTP or MPX microarrays (Table 6.4) as well. Hence, already adapted protocols for single-slide/target solutions can be easily transferred to the multiplexed formats. Recent publications demonstrate the usability of high sample throughput microarrays for applications such as detection of plant pathogens [102], carbohydrate analysis [103], protein characterization [104], and transcription factor profiling [87]. These publications further demonstrate that reproducible well-to-well uniformity and cross-contamination-free hybridization, the basic requirements of high sample throughput solutions, are guaranteed.

Taken together several solutions are available to researchers for performing high sample throughput microarrays assays that can select between slide versus plate format, surface coating, and the number of samples to be processed for multiplexing. These solutions have to be checked regarding

(a) Standard slide format

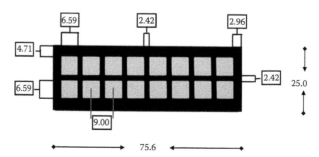

(b) Microtitreplate format

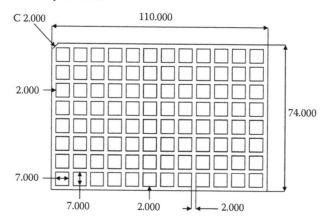

FIGURE 6.6   Dimensions (mm) of SCHOTT's MPX 16-well slide and Microtitreplate formats.

TABLE 6.2

List of Common Arrayers on the Market Capable of Processing MTP Microarrays

Manufacturer	Arrayer	Destination Microplate Capacity	Web Site
Aushon Biosystems	2470	20	http://www.aushon.com/products-and-services/index.shtml.
Bio-Rad Laboratories	BioOdyssey Calli-grapher MiniArrayer	2	http://www.bio-rad.com/
Bio-Rad Laboratories	VersArray ChipWriter Pro	1	http://www.bio-rad.com/
Genetix Ltd.	QArray mini	4	http://www.genetix.com/
Genetix Ltd.	Qarray2	16	http://www.genetix.com/
Genomic solutions	GeneMachines OmniGrid Accent	6	http://www.genomicsolutions.com/
Genomic solutions	MicroGrid	16	http://www.genomicsolutions.com/
Lab Next LLC	THOMAS™ High Capacity Microarray System	9	http://www.labnext.com/
Lab Next LLC	Xact Microplate Arrayer	2	http://www.labnext.com/
PerkinElmer Life Sciences	Piezorray DBC0000	5	http://las.perkinelmer.com/
Telechem	NanoPrint™ Microarrayer	12	http://www.arrayit.com/

TABLE 6.3
List of Common Scanners that are Capable of Reading the MTP Format

Manufacturer	Scanner	Web Site
Alpha Innotech Corporation	NovaRay® detection platform	http://www.alphainnotech.com/
Applied Precision	arrayWoRx®MF multiformat reader	http://www.api.com/lifescience/arrayworxMF.html
Blueshift Biotechnologies, Inc.	IsoCyte™ laser scanning fluorimeter	http://www.blueshiftbiotech.com/isocyte.html
Ditabis	MarS-OEM-MP plate scanner	http://www.ditabis.de/
TECAN	Laser scanner	http://www.tecan.com/

TABLE 6.4
Coatings of Glass-Based High Sample Throughput Microarray Platforms

Product	Substrate	Available Coatings
ArrayIt 96-well microplate	Polymer with glass bottom	Uncoated, amine, epoxy, streptavidin
NUNC ArrayCote™ 16-well slides and 96-well plates	Black polymer well structure with glass bottom	Amino, epoxy, aldehyde
SCHOTT Nexterion™ MTP-96, and MPX 16/48	Kit-based system including glass plate, superstructure, tray, and adhesive foil	Uncoated, amino, epoxy, hydrogel, aldehyde
The Gel Company ArrayPlate	Detachable polymer with glass bottom	Epoxy, aldehyde, lysine
Whatman 16-pad FAST slides	Glass-based slides with nitrocellulose membrane and chip clip system	Nitrocellulose

hardware compatibility, well-to-well reproducibility, and cross-contamination-free hybridization. Beside the diagnostic area, all microarray applications that need to analyze many samples simultaneously benefit from time- and cost-savings provided by multiplexed substrates.

6.7 SUMMARY

The use of an appropriate substrate is an important factor for a successful microarray experiment. The careful selection of the substrate includes consideration of the binding chemistry, microarray hardware, and processing factors affecting spot size and morphology, printing conditions, signal and background intensities as well as coating uniformity and reproducibility tests. Microarray substrates have evolved over a period of time and several formats are available for specific applications. High sample throughput solutions in several formats provide the user the opportunity for customization and parallel processing based on known surfaces and protocols. This will help to expand the role of microarray analysis in the diagnostic field as they offer cost-effective and reproducible alternatives. New optical coatings open the way for increasing the sensitivity of microarray analysis and using lower analyte amounts. Increased detection sensitivity provided by these substrates enable a more detailed analysis elucidating signals from low expressed messenger RNA (mRNAs). Analytical coatings such as gold enable the application of label-free detection methods such as SPR spectroscopy. Furthermore, the development of new coatings for the protein microarray market, that meet the various requirements of the diverse group of proteins is another challenge. Moreover, work around microarray substrates such as protocol optimization, development of new and low evaporation buffers, and so on helps to optimize microarray processing and to provide user with the best possible solution.

REFERENCES

1. Schena M, Shalon D, Davis RW, and Brown PO. Quantitative monitoring of gene expression patterns with a complementary DNA microarray. *Science* 1995; 270: 467–470.
2. Mehrian-Shai R and Reichardt J. A renaissance of "biochemical genetics"? SNPs, haplotypes, function, and complex diseases. *Mol. Genet. Metab.* 2004; 83(1–2): 47–50.
3. Mantripragada KK, Buckley PG, de Ståhl TD, and Dumanski JP. Genomic microarrays in the spotlight. *Trends Genet.* 2004; 20(2): 87–94.
4. Jubb A, Pham T, Frantz G, Peale F, and Hillan K. Quantitative in situ hybridization of tissue microarrays. *Methods Mol. Biol.* 2006; 326: 255–264.
5. Einat P. Methodologies for high-throughput expression profiling of microRNAs. *Methods Mol. Biol.* 2006; 342: 139–157.
6. Paulson J, Blixt O, and Collins B. Sweet spots in functional glycomics. *Nat. Chem. Biol.* 2006; 2(5): 238–248.
7. Becker K, Metzger V, Hipp S, and Hofler H. Clinical proteomics: New trends for protein microarrays. *Curr. Med. Chem.* 2006; 13(15): 1831–1837.
8. Pantano CG, Metwalli E, Conzone S, and Haines D. Modified substrates for the attachment of biomolecules. US Patent 6916541, 2005.
9. Lautenschläger G, Kloss T, von Fintel S, and Schneider K. Method for the production of borosilicate glass with a surface suitable for modification, glass obtained according to said method and the use thereof. European Patent 045862, 2005.
10. Carre ARE. Substrate for array printing. US Patent 6461734, 2002.
11. Willats WGT, Rasmussen SE, Kristensen T, Mikkelsen JD, and Knox JP. Sugar coated microarrays: A novel slide surface for the high-throughput analysis of glycans. *Proteomics* 2002; 2: 1666–1671.
12. Fixe F, Dufva M, Telleman P, and Christensen CB. Functionalizaiton of poly(methylmethacrylate) (PMMA) as a substrate for DNA microarrays. *Nucleic Acids Res.* 2004; 32: e9–e16.
13. Kai J, Sohn YS, and Ahn CS. Protein microarray on cyclic olefin copolymer (COC) for disposable protein lab-on-a-chip. Proceedings of the 7th International Conference of Micro Total Analysis Systems (micro-TAS 2003), California, October 5–9, 2003, pp. 1101–1104.
14. Beier M and Hoheisel JD. Versatile derivatisation of solid support media for covalent bonding on DNA-microchips. *Nucleic Acids Res.* 1999; 27(9): 1970–1977.
15. Southern E, Mir K, and Shchepinov M. Molecular interactions on microarrays. *Nat. Genet.* 1999; 21(Suppl.): 5–9.
16. Manning M and Redmond G. Formation and characterization of DNA microarrays at silicon nitride substrates. *Langmuir* 2005; 21: 395–402.
17. Strother T, Cai W, Zhao X, Hamers RJ, and Smith CL. Synthesis and characterization of DNA-modified silicon (111) surfaces. *J. Am. Chem. Soc.* 2000; 122: 1205–1209.
18. Chrisey LA, Lee GU, and O'Ferrall E. Covalent attachment of synthetic DNA to self-assembled mono-layer films. *Nucleic Acids Res.* 1996; 24(3): 3031–3039.
19. Rouzina I and Bloomfield VA. Macroion attraction due to electrostatic correlation between screening counterions. 1. Mobile surface-adsorbed ions and diffuse ion cloud. *J. Phys. Chem.* 1996; 100: 9977–9989.
20. Goldsmith ZG and Dhanasekaran N. The microrevolution: Applications and impacts of microarray technology on molecular biology and medicine. *Int. J. Mol. Med.* 2004; 13: 483–495.
21. Cheung VG, Morley M, Aguilar F, Massimi A, Kucherlapati R, and Childs G. Making and reading microarrays. *Nat. Genet.* 1999; 21(Suppl.): 15–19.
22. Cras JJ, Rowe-Taitt CA, Nivens DA, and Ligler FS. Comparison of chemical cleaning methods of glass in preparation for silanization. *Biosens Bioelectron* 1999; 14: 683–688.
23. Mittal KL. *Silanes and Other Coupling Agents.* Utrecht, The Netherlands: VSP; 1992.
24. Pulker HK. *Coatings on Glass*, 2nd Ed. Amsterdam, The Netherlands: Elsevier; 1999.
25. Stillman BA and Tonkinson JL. Fast slides: A novel surface for microarrays. *Biotechniques* 2000; 29: 630–635.
26. Diehl F, Grahlmann S, Beier M, and Hoheisel H. Manufacturing DNA microarrays of high spot homogeneity and reduced background signal. *Nucleic Acids Res.* 2001; 29(7): E38.
27. Allemand JF, Bensimon D, Jullien L, Bensimon A, and Croquette V. pH-dependent specific binding and combing of DNA. *Biophys. J.* 1997; 73: 2064–2070.
28. Sambrook J, Fritsch EF, and Maniatis T. *Molecular Cloning: A Laboratory Manual*, 2nd Ed. Cold Springs, New York: Cold Spring Harbor Laboratory Press; 1989.

29. Taylor S, Smith S, Windle B, and Guiseppi-Elie A. Impact of surface chemistry and blocking strategies on DNA microarrays. *Nucleic Acids Res.* 2003; 31(16): e87.

30. Berre VL, Trevisiol E, Dagkessamanskaia A, Sokol S, Caminade AM, Majoral JP, Meunier B, and Francois J. Dendrimeric coating of glass slides for sensitive DNA microarray analysis. *Nucleic Acids Res.* 2003; 31: e88.

31. Hessner MJ, Meyer L, Tackes J, Muheisen S, and Wang X. Immobilized probe and glass surface chemistry as variables in microarray fabrication. *BMC Genomics* 2004; 5: 33–60.

32. Pirrung MC, Davis JD, and Odenbaugh AL. Novel reagents and procedures for immobilization of DNA on glass microchips for primer extension. *Langmuir* 2000; 16: 2185–2191.

33. Li J, Wang H, Zhao Y, Cheng L, He N, and Lu Z. Assembly method fabricating linkers for covalently bonding DNA on glass surface. *Sensors* 2001; 1: 53–59.

34. Zammatteo N, Jeanmart L, Hamels S, Courtois S, Louette P, Hevesi L, and Remacle J. Comparison between different strategies of covalent attachment of DNA to glass surfaces to build DNA microarrays. *Anal. Biochem.* 2000; 280: 143–150.

35. Oh SJ, Cho SJ, Kim CK, and Park JW. Characteristics of DNA microarrays fabricated on various amino-silane layers. *Langmuir* 2002; 18: 1764–1769.

36. Southern E, Case-Green SC, Elder JK, Johnson M, Mir KU, Wang L, and Williams JC. Arrays of complementary oligonucleotides for analysing the hybridisation behaviour of nucleic acids. *Nucleic Acids Res.* 1994; 22(8): 1368–1373.

37. Shi J and Boyce-Jacino MT. Attachment of unmodified nucleic acids to silanized solid phase surfaces. US Patent 5919626, 1999.

38. Guo Z, Guilfoyle RA, Thiel AJ, Wang R, and Smith LM. Direct fluorescence analysis of genetic polymorphisms by hybridization with oligonucleotide arrays on glass supports. *Nucleic Acids Res.* 1994; 22(24): 5456–5465.

39. Rogers YH, Jiang-Baucom P, Huang ZJ, Bogdanov V, Anderson S, and Boyce-Jacino MT. Immobilization of oligonucleotides onto a glass support via disulfide bonds: a method for preparation of DNA microarrays. *Anal. Biochem.* 1999; 266: 23–30.

40. Podyminogin MA, Lukhtanov EA, and Reed RW. Attachment of benzaldehyde modified oligodeoxynucleotide probes to semicarbazide-coated glass. *Nucleic Acids Res.* 2001; 29: 5090–5098.

41. Kumar A, Larsson O, Parodi D, and Liang Z. Silanized nucleic acids: A general platform for DNA immobilization. *Nucleic Acids Res.* 2000; 28(14): E71.

42. Lefkowitz SM, Fulcrand G, Dellinger DJ, and Hotz CZ. Functionalization of substrate surfaces with silane mixtures. US Patent 6258454, 2001.

43. Mao G, Metzger SW, and Lochhead MJ. Functional surface coating. US Patent Application 2004/0115721, 2004.

44. Yang Z, Galloway JA, and Yu H. Protein interactions with poly(ethylene glycol) self-assembled monolayers on glass substrates: Diffusion and adsorption. *Langmuir* 1999; 15: 8405–8411.

45. Lee SW and Laibinis PE. Protein-resistant coatings for glass and metal oxide surfaces derived from oligo(ethylene glycol)-terminated alkyltricholorosilanes. *Biomaterials* 1998; 19: 1669–1675.

46. Okamoto T, Suzuki T, and Yamamoto N. Microarray fabrication with covalent attachment of DNA using bubble Jet technology. *Nat. Biotechnol.* 2000; 18: 438–441.

47. Zhao X, Nampalli S, Serino AJ, and Kumar S. Immobilization of oligodeoxyribonucleotides with multiple anchors to microchips. *Nucleic Acids Res.* 2001; 21: 4955–4959.

48. Kim HJ, Moon JH, and Park JW. A hyperbranched poly(ethyleneimine) grown on surfaces. *J. Colloid Interface Sci.* 2000; 227: 247–249.

49. Kimura N, Oda R, Inaki Y, and Suzuki O. Attachment of oligonucleotide probes to polycarbodiimide-coated glass slides for microarray applications. *Nucleic Acids Res.* 2004; 32: e68–e76.

50. Kimura N, Ichihara T, and Moriya S. Immobilized nucleic acid and method for detecting nucleic acid. US Patent 7037649, 2006.

51. Halliwell CM and Cass AEG. A factorial analysis of silanization conditions for the immobilization of oligonucleotides on glass surfaces. *Anal. Chem.* 2001; 73: 2476–2483.

52. Popat KC, Johnson RW, and Desai TA. Characterization of vapor deposited thin silane films on silicon substrates for biomedical microdevices. *Surf. Coat. Technol.* 2002; 154: 253–261.

53. Haller I. Covalently attached organic monolayers on semiconductor surfaces. *J. Am. Chem. Soc.* 1978; 100(26): 8050–8055.

54. Petri DFS, Wenz G, Schunk P, and Schimmel T. An improved method for the assembly of amino-terminated monolayers on SiO_2 and the vapor deposition of gold layers. *Langmuir* 1999; 15: 4520–4523.

55. Krasnoslobodtsev AV and Smirnov SN. Effect of water on silanization of silica by trimethoxysilanes. *Langmuir* 2002; 18: 3181–3184.

56. Benters R, Niemeyer CM, and Wohrle D. Dendrimer-activated solid supports for nucleic acid and protein microarrays. *ChemBioChem* 2001; 2: 686–694.

57. Blackledge C and McDonald JD. Catalytic activity of silanols on carbamate-functionalized surface assemblies: Monoalkyoxy versus trialkyoxy silanes. *Langmuir* 1999; 15: 8119–8125.

58. Moon JH, Shin JW, Kim SY, and Park JW. Formation of uniform aminosilane thin layers: An imine formation to measure relative surface density of the amine group. *Langmuir* 1996; 12: 4621–4624.

59. Moon JH, Kim HJ, Kim K, Kang T, Kim B, Kim C, Hahn JH, and Park JW. Absolute surface density of the amine group of the aminosilyated thin layers: Ultraviolet-visible spectroscopy, second harmonic generation, and synchrotron-radiation photoelectron spectroscopy study. *Langmuir* 1997; 13: 4305–4310.

60. Jun-Hyeong C, Nam H, Jeo-Young S, Kyeong-Hee K, and Ga-Young P. Quality control method of DNA microarray. US Patent Application 2003/0270672, 2003.

61. Balladur V, Theretz A, and Mandrand B. Determination of the main forces driving DNA oligonucleotide adsorption onto aminated silica wafers. *J. Colloid Interface Sci.* 1997; 194: 408–418.

62. Conzone SD, Burzio L, and Haines D. Non-destructive quality control method for microarray substrate coatings via labeled doping. US Patent Application 2004/0258927, 2004.

63. Reed J, Singer E, Kresbach G, and Schwartz DC. A quantitative study of optical mapping surfaces by atomic force microscopy and restriction endonuclease digestion assays. *Anal. Biochem.* 1998; 259: 80–88.

64. O'Donnell MJ, Tang K, Köster H, Smith CL, and Cantor CR. High-density, covalent attachment of DNA to silicon wafers for analysis by MALDI-TOF mass spectrometry. *Anal. Chem.* 1997; 69: 2438–2443.

65. Vandenberg ET, Bertilsson L, Liedberg B, Uvdal K, Erlandsson R, Elwing H, and Lundström I. Structure of 3-aminopropyl triethoxy silane on silicon oxide. *J. Colloid Interface Sci.* 1991; 147(1): 103–118.

66. Butler JH, Cronin M, Anderson KM, Biddison GM, Chatelain F, Cummer M, Davi DJ, et al. In situ synthesis of oligonucleotide arrays by using surface tension. *J. Am. Chem. Soc.* 2001; 123(37): 8887–8894.

67. Luzinov I, Julthongpiput D, Liebmann-Vinson A, Cregger T, Foster MD, and Tsukruk VV. Epoxy-terminated self-assembled monolayers: molecular glues for polymer layers. *Langmuir* 2000; 16: 504–516.

68. Kurth DG and Bein T. Quantification of the reactivity of 3-aminopropyltriethoxysilane monolayers with the quartz-crystal microbalance. *Angew. Chem. Int. Ed. Engl.* 1992; 31(3): 336–338.

69. Beari F, Brand M, Jenker P, Lehnert R, Metternich HJ, Monkiewicz J, and Siesler HW. Organofunctional alkyoxysilanes in dilute aqueous solution: New accounts on the dynamic structural mutability. *J. Organomet. Chem.* 2001; 625: 208–216.

70. Kurth DG and Bein T. Monomolecular layers and thin films of silane coupling agents by vapor-phase adsorption on oxidized aluminum. *J. Phys. Chem.* 1992; 96: 6707–6712.

71. Watson H, Norström A, Torrkulla A, and Rosenholm J. Aqueous amino silane modification of E-glass surfaces. *J. Colloid Interface Sci.* 2001; 238: 136–146.

72. Metwalli E, Haines D, Becker O, Conzone S, and Pantano C. Surface characterizations of mono-, di-, and tri-aminosilane treated glass substrates. *J. Colloid Interface Sci.* 2006; 298: 825–831.

73. Schultz N, Conzone SD, Becker O, Haines D, Pawlowski E, and Scheumann V. Arrangement for fluorescence reinforcement. US Patent Application 2005/082518, 2005.

74. Kanda V, Kariuki JK, Harrison DJ, and McDermott MT. Label-free reading of microarray-based immunoassays with surface plasmon resonance imaging. *Anal. Chem.* 2004; 76: 7257–7262.

75. Redkar RJ, Schultz NA, Scheumann V, Burzio LA, Haines DE, Metwalli E, Becker O, and Conzone SD. Signal and sensitivity enhancement through optical interference coating for DNA and protein microarray applications. *J. Biomol. Tech.* 2006; 17: 122–130.

76. Kain R, Marason E, and Johnson R. Optical substrate for enhanced detectability of fluorescence, 1998, World Patent Application 9853304.

77. Pacard E, Brook M, Ragheb A, Pichot C, and Chaix C. Elaboration of silica colloid/polymer hybrid support for oligonucleotide synthesis. *Colloids Surf B Biointerfaces* 2006; 47(2): 176–188.

78. Neuschafer D, Budach W, Wanke C, and Chibout SD. Evanescent resonator chips: A universal platform with superior sensitivity for fluorescence-based microarrays. *Biosens. Bioelectron.* 2003; 18: 489–497.

79. Schlingemann J, Thuerigen O, Ittrich C, Toedt G, Kramer H, Hahn M, and Lichter P. Effective transcriptome amplification for expression profiling on sense-oriented oligonucleotide microarrays. *Nucleic Acids Res.* 2005; 33: e29.

80. Wrobel G, Schlingemann J, Hummerich L, Kramer H, Lichter P, and Hahn M. Optimization of high density cDNA-microarray protocols by "design of experiments." *Nucleic Acids Res.* 2003; 31(12): e67.

81. Kok K, Dijkhuizen T, Swart YE, Zorgdrager H, van der Vlies P, Fehrmann R, te Meerman GJ, et al. Application of a comprehensive subtelomere array in clinical diagnosis of mental retardation. *Eur. J. Med. Genet.* 2005; 48: 250–262.

82. Kusnezow W, Jacob A, Walijew A, Diehl F, and Hoheisel JD. Antibody microarrays: An evaluation of production parameters. *Proteomics* 2003; 3: 254–264.

83. Lyne R, Burns G, Mata J, Penkett CJ, Rusticic G, Chen D, Langford C, Vetrie D, and Bühler J. Whole-genome microarrays of fission yeast: Characteristics, accuracy, reproducibility, and processing array data. *BMC Genomics* 2003; 4: 27–42.

84. Sieber SA, Mondala TS, Head SR, and Cravatt BF. Microarray platform for profiling enzyme activities in complex proteomes. *J. Am. Chem. Soc.* 2004; 126: 15640–15641.

85. Rickman DS, Herbert CJ, and Aggerbeck LP. Optimising spotting solution for increased reproducibility of cDNA microarrays. *Nucleic Acids Res.* 2003; 31(18): e109.

86. Vermeesch JR, Melotte C, Froyen G, Vooren Sv, Dutta B, Maas N, Vermeulen S, et al. Molecular karyotyping: Array CGH quality criteria for constitutional genetic diagnosis. *J. Histochem. Cytochem.* 2005; 53(3): 413–422.

87. Ho SW, Jona G, Chen CTL, Johnston M, and Snyder M. Linking DNA-binding proteins to their recognition sequences by using protein microarrays. *Proc. Natl. Acad. Sci. USA* 2006; 103: 9940–9945.

88. McQuain M, Seale K, Peek J, Levy S, and Haselton F. Effects of relative humidity and buffer additives on the contact printing of microarrays by quill pins. *Anal. Biochem.* 2003; 320(2): 281–291.

89. Olle E, Messamore J, Deogracias M, McClintock S, Anderson T, and Johnson K. Comparison of antibody array substrates and the use of glycerol to normalize spot morphology. *Exp. Mol. Pathol.* 2005; 79(3): 206–209.

90. Deng Y, Zhu X-Y, Kienlen T, and Gou A. Transport at the air/water interface is the reason for rings in protein microarrays. *J. Am. Chem. Soc.* 2006; 128(9): 2768–2769.

91. Hultschig C, Kreutzberger J, Seitz H, Konthur Z, Büssow K, and Lehrach H. Recent advances of protein microarrays. *Curr. Opin. Chem. Biol.* 2006; 10: 4–10.

92. Schoenfisch MH and Pemberton JE. Air stability of alkanethiol self-assembled monolayers on silver and gold surfaces. *J. Am. Chem. Soc.* 1998; 120: 4502–4513.

93. Pulker HK. *Coatings on Glass.* Amsterdam, The Netherlands: Elsevier; 1984.

94. Ye T, Wynn D, Dudek R, and Borguet E. Photoreactivity of alkylsiloxane self-assembled monolayers on silicon oxide surfaces. *Langmuir* 2001; 17: 4497–4500.

95. Kane M, Jatkoe T, Stumpf C, Lu J, Thomas J, and Madore S. Assessment of the sensitivity and specificity of oligonucleotide (50mer) microarrays. *Nucleic Acids Res.* 2000; 28: 4552–4557.

96. Zolman J. *Biostatistics: Experimental Design and Statistical Interference.* Oxford: Oxford University Press; 1993.

97. Conzone SD and Redkar RJ. Reproducible, low cost arraying achieved with a multiplexed approach. *Eur. Pharm. Rev.* 2004; (Spring): 52–55.

98. Li Y, Cu Y, and Luo D. Multiplexed detection of pathogen DNA with DNA based fluorescence nanobarcodes. *Nat. Biotechnol.* 2005; 23(7): 885–889.

99. Angenendt P, Lehrach H, Kreutzberger J, and Glokler J. Subnanoliter enzymatic assays on microarrays. *Proteomics* 2005; 5(2): 420–425.

100. Koehne J, Chen H, Cassell A, Ye Q, Han J, Meyyappan M, and Li J. Miniaturized multiplex label-free electronic chip for rapid nucleic acid analysis based on carbon nanotube nanoelectrode arrays. *Clin. Chem.* 2004; 50(10): 1886–9183.

101. Ahmed F. Microarray RNA transcriptional profiling: Part I. Platforms, experimental design and standardization. *Expert Rev. Mol. Diagn.* 2006; 6(4): 535–550.

102. Szemes M, Bonants P, Weerdt M, Baner J, Landegren U, and Schoen C. Diagnostic application of padlock probes—multiplex detection of plant pathogens using universal microarrays. *Nucleic Acids Res.* 2005; 33: e70.

103. Manimala J, Li Z, Jain A, VedBrat S, and Gildersleeve J. Carbohydrate array analysis of anti-Tn antibodies and lectins reveals unexpected specificities: Implications for diagnostic and vaccine development. *ChemBioChem* 2005; 6(12): 2229–2241.

104. Saxinger C, Conrads T, Goldstein D, and Veenstra T. Fully automated synthesis of (phospho)peptide arrays in microtiter plate wells provides efficient access to protein tyrosine kinase characterization. *BMC Immunol.* 2005; 6: 1–15.

7 Toward Highly Efficient Automated Hybridizations

Kris Pappaert, Johan Vanderhoeven, Gert Desmet, and Paul Van Hummelen

CONTENTS

7.1   INTRODUCTION

Hybridization-based approaches to study gene expression, mutation detection, or genome analysis have become a common technology and allow the analysis of hundreds of thousands of genes or complete genomes in parallel. The power and universal appeal of deoxyribonucleic acid (DNA) microarrays as experimental tools are derived from the exquisite specificity and affinity of complementary base-pairing of the nucleic acids. During the past decade, DNA microarrays have become an important biological tool for obtaining high-throughput genetic information in studies of human disease states [1], toxicological research [2], and gene expression profiling [3]. However, regarding the technology a number of questions still need to be addressed and some protocols need further development to circumvent a number of inherent shortcomings such as lack of uniformity, slow hybridization speed, and so on. One of the problems with conventional microarrays is that without active agitation, the number of target molecules available for hybridization is limited by molecular diffusion. For a typical DNA molecule, this distance has been estimated to be <1 mm during an overnight experiment [4,5], as can be seen from Einstein's law of diffusion

$$l^2 = \frac{2D_{mol}t}{4} \tag{7.1}$$

71

using a value of $D_{mol} = 10^{-11} \, m^2/s$ to represent the rate of diffusion of DNA strands in a typical microarray analysis. This suggests that, for any given spot on an 18×654 mm array, <0.3% of the complementary targets present in the sample can reach the given spot during an overnight diffusion assay. If a target is in low abundance, it may become depleted near the complementary probe spot [5].

In general, hybridization signal and efficiency depends on the diffusion rate, the concentration, and time. Systems that can enhance any of these three components will greatly enhance the sensitivity of the microarrays, but also the reproducibility within and between different laboratories. Indeed, incomplete hybridization reactions are frequently found to result in a systematic hybridization bias, which may in part also explain the seemingly bad correlations among microarray platforms or laboratories.

However, current goals with microarray hybridization are centered on three objectives, a reduction in hybridization time from several hours to 1 to 2 h, sample size reduction permitting needle biopsy material or the use of small numbers of cells, and automation. To accomplish these goals, an obvious evolution is miniaturization to increase the concentration of the sample and bypassing the diffusion-driven reactions to decrease the reaction time.

In this chapter, we will discuss the main hybridization systems and how they increase the hybridization rate, decrease sample, or just increase the fluorescent signal strength. In addition, we will also present in more detail an approach that we developed to circumvent diffusion resulting in a considerable reduction in hybridization time and sample volume. This chapter is not intended to present a complete overview of all strategies developed or employed to enhance hybridization efficiency, but to highlight the majority of commercial hybridization systems currently available, and present several interesting experimental strategies and future lab-on-chip developments.

Increasing the efficiency of hybridization via automation can be largely divided into two categories: the systems that start with classical microarrays where the genes are spotted on a support with the dimensions of a standard microscope slide, and those that redesign the arrays into more complex fluidics systems.

7.2 HYBRIDIZATION SYSTEMS COMPATIBLE WITH STANDARD MICROARRAYS

It is difficult to completely change a current microarray design printed in the laboratory or core facility or purchased from a commercial source. Therefore hybridization systems that are compatible with existing glass or plastic supports, which conform to the dimensions of the standard microarray slide (20×60 mm), are the preferred choice. Scanning the literature yielded many publications that present static experimental systems and report five- to tenfold increase in hybridization signal. Wei et al. [6] presented an elegant approach combining microfluidic channels and the classic microarray microscope format. A microscope slide with a printed array was covered and carefully aligned with a poly(methylmethacrylate) (PMMA) microtrench channels in which the labeled sample was introduced and scrambled with discrete plugs. With the combination of the channels and plugs, a continuous flow could be introduced together with chaotic mixing of the sample. This approach required only 1 µL sample and hybridization equilibrium could be reached in 500 s.

Hui Liu et al. [7] described a phenomenon of cavitation microstreaming, providing a mechanism for achieving rapid and homogeneous mixing in a hybridization chamber. The microarray was sealed with a polycarbonate coverslip that contains uniformly distributed air pockets, on which a piezoelectric (PZT) disk was tightly glued. The sound generated by the PZT disk served to vibrate the air in the air pockets creating a microstreaming that was sufficient to mix the sample over the entire surface of the microarray (Figure 7.1). The cavitation microstreaming resulted in up to fivefold enhancement of the hybridization signal compared to results from conventional diffusion-based hybridization, for given time of 2 h.

Finally, Agilent technologies have achieved 10-fold enhancement of hybridization using microfluidic planetary centrifugal mixing [1]. This approach combined the small chamber sizes of coverslips with the uniform mixing capacity of large-volume hybridizers. Planetary centrifugal mixing

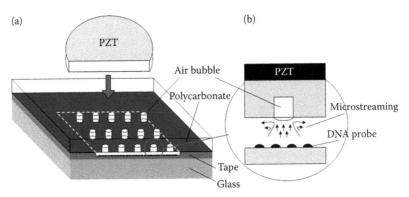

FIGURE 7.1 Schematic showing a number of air pockets in the top layer of the DNA biochip chamber: (a) overview and (b) side view [7].

was achieved by placing four microarray slides in four separate mixing chambers mounted on a centrifugal plate but with each rotated on its own axis separately. Chamber rotation caused the fluids to redistribute and maintain their menisci relative to the central rotor axis and not the chambers themselves. The presence of an air bubble and the high-g centrifugal forces enabled constant and thorough mixing. Each chamber stacked up to three microarray slides allowing as many as 12 slides to be hybridized at the same time. The minimum hybridization volume was ~60 μL and according to the authors 10-fold increases in signal were achieved as compared to conventional coverslip hybridizations using the same probe concentration.

7.3 COMMERCIALLY AVAILABLE SYSTEMS

The Lucidea Automated Slide Processor or SlidePro (GE Healthcare, Piscataway, New Jersey) allows environmentally controlled hybridization and washing of up to 30 slides at a time. Hybridizations are performed in independently controlled chambers, allowing customized process parameters to be used for each slide. Mixing is accomplished by pumping the 210-μL hybridization volume with an adjustable flow rate of 1150 μL/s. This toggles flow back and forth through the chamber (volume 1050 μL). Slides can be washed and air-dried so that the slides are ready for scanning. An advantage of the SlidePro is that each module can be programmed differently and run simultaneously or at different starting times.

Tecan's HS4800 Pro hybridization stations are designed for full automation, from prehybridization and on-board denaturation to automatic slide drying. The HS4800 has an active bubble suppression system that minimizes the risk of bubble formation during a hybridization step, which helps to eliminate artifacts and enhances consistency of results. The hybridization volume depends on the slide chambers but is ~100 μL for a full slide. Actively mixing is achieved by pumping a small amount of liquid back and forth, similar to the SlidePro. The biggest advantage of the latest Pro-version is its novel dual chambers enabling the processing of two independent subarrays per slide with no risk of carry-over. This part was developed in collaboration with Agilent technologies and is therefore also compatible with their dual array slides. The number of slides that can be hybridized ranges from 4 to 12 per module, and four modules can be placed in series. The HS4800 Pro cannot run different protocols in parallel, except for differences in hybridization or wash temperature.

The MicroArray User Interface (MAUI) hybridization device from BioMicro Systems (www.biomicro.com) is comprised of consumable hybridization chambers and a base unit, which provides the mixing action and maintains a constant incubation temperature. The MAUI Mixer adheres to the microarray slide via an adhesive gasket forming a uniform, low volume, sealed hybridization chamber. This chamber is clamped into one of the heated slide bays in the base unit, where hybridization

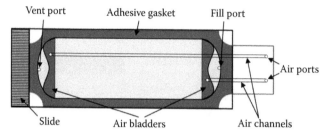

FIGURE 7.2 Consumable hybridization chamber or mixer. Mixing is established by placing sequentially pressure on the left and right air bladders. (From www.biomicro.com.)

takes place. The mixing is accomplished by sequentially placing pressure on two air-driven bladders that are part of the hybridization chamber (Figure 7.2). This pressure pumping will rock the hybridization mix back and forth over the slide. The MAUI system can hold 4 or 12 slides that can be hybridized with different protocols in parallel [2,3]. The hybridization signal can be increased from three- to fivefold compared to diffusion-driven hybridization under coverslip. The advantage of the MAUI is that many different types of consumable chambers can be purchased, containing one to 16 different chambers, and for even the largest chamber only 40 μL of hybridization, volume is needed. Unfortunately, the system cannot wash and dry the slides automatically.

The GeneMachines HybArray 12 (http://las.perkinelmer.com or www.genomicsolutions.com) is an automated hybridization system for both hybridization and post-hybridization washing. A single instrument can process up to 12 slides at a time in pairs and has multiprotocol software, enabling the user to run all six modules with different protocols or all arrays under the same conditions. The hybridization volume ranges from 100 to 160 μL depending on the chambers. Sample mixing is accomplished by a similar mixing as for the SlidePro and HS4800.

The a-Hyb hybridization station of Macs Molecular (www.miltenyibiotec.com) is a fully automated system from the initial hybridization step to the dried microarray. This device has the only

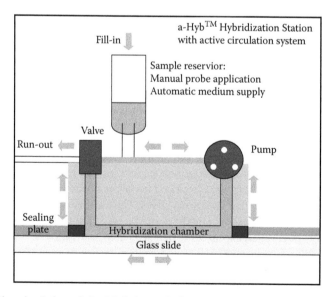

FIGURE 7.3 Active circulation of the labeled sample by a complete pump around system. (From www.miltenyibiotec.com.)

real pump around system that continuously circulates the hybridization mix over the slide from a reservoir (Figure 7.3). This leads to an even distribution of liquid over the surface of the slides and accelerates diffusion-controlled processes. The a-Hyb can only handle four standard slides in parallel and can be combined with standard laboratory robot systems enabling its integration into fully automated microarray processing. The total sample volume is 200 μL.

Advalytix is selling two devices, ArrayBooster and SlideBooster (www.advalytix.de, [4]), that mix hybridization solutions using surface acoustic waves (SAW) to effectively agitate the sample solution during the incubation of DNA and protein microarrays. The incubation chambers accept all standard slide formats and are for the SlideBooster, compatible with regular coverslips or lifterslips. Sample volumes can be as low as 15 μL. The heart of the ArrayBooster is the AdvaCard—a microagitation chip card available in three different sizes adapted to different spotting areas. To ensure optimal agitation efficiency AdvaCards contain one, two, or three agitation chips. A radio frequency voltage fed to the chips induces quasi-chaotic streaming patterns in the hybridization solution, which is sandwiched between the AdvaCard and the microarray. Special fluidics prevents bubble formation while loading the sample. The fluidic system has no moving parts making it very reliable. Unfortunately, only four slides can be hybridized in parallel.

7.4 NEXT GENERATION MICROFLUIDICS DESIGNS

Parallel with the growth of microarray technology, the field of microfluidics has grown considerably during the last decade and both technologies can be seen as competitive in some aspects and complementary in others. The concept of "Micro Total Analysis Systems" or μ-TAS was first introduced by Manz et al. [8] in 1990. The original vision was to develop one single automated miniaturized device where all the necessary parts and methods to perform a chemical analysis are integrated (from sample treatment through detection). An inherent aspect of the μ-TAS concept is the elimination of the dependence on external laboratory analysis making real-time diagnostic control of chemical production processes feasible [9]. This is not simply shrinking down traditional benchtop techniques; it requires innovative research in many fields, from chemistry and biology to optics, materials, fluid mechanics, and microfabrication. Microfluidic systems are fabricated with standard photolithographic methods on silicon or glass substrates to produce channel networks in two dimensions for sample transport, mixing, separation, and detection systems on a monolithic chip.

Developments in microfluidic systems are currently the furthest along. Recently several reports have been published where microfluidic microarray devices are discussed. In most cases the microfluidic system can be fully integrated onto a standard microarray slide; others described were more complex multifunctional devices.

Peytavi et al. [5] described a removable microfluidic structure that can be attached to a normal glass microarray slide. The fluid transport is created by fixing the devices in a rotational device and applying a centrifugal force to drive the sample and buffers onto the microarray slide. A volume of 2 μL was flown through the hybridization chamber containing 16 probe spots at a continuous velocity of 0.4 nL/min (corresponding to a 2-μL sample volume). This device can accommodate up to 150 spots in the hybridization chamber.

Another system described by Wei et al. [6] (cf. *Hybridization systems compatible with standard microarrays*) also integrated microfluidic channels with the microarray slide layout. Plugs of the sample solution are shuttled over the microarray spots inducing the hybridization. Only 1 μL of the sample and about 10 min are needed for the hybridization reaction. A total number of 5000 spots can be accommodated in this microchannel.

Bowden et al. [10] presented a more integrated platform that combined fiber-optic microarrays with a microfluidics system. 50 μL of the target DNA samples are flown through a PMMA microfluidic chip at a flow rate of 1 μL/min. The use of an integrated optical fiber array enabled the detection of 10 aM in only 15 min.

Recently, Hashimoto et al. [11] delivered a proof of concept of a microfluidic device that enables a polymerase chain reaction (PCR), a ligation detection reaction (LDR), and a hybridization reaction in the same microfluidic device.

Three other microfluidic systems were developed further and were successfully commercialized. Pamgene (www.pamgene.com) and Metrigenix (www.metrigenix.com) both have developed a flow-through microarray platform. Instead of using the two-dimensional surface of a microarray slide, a three-dimensional matrix of microchannels that traverse the thickness of the substrate is used to immobilize the probe molecules. The target fluid sample flows through the chip and the hybridization occurs in the microchannels. The sample can be pumped back and forth several times (twice per minute). Up to 400 spots can be hybridized in such devices where a total sample volume of 25 µL is necessary.

The third platform is the Geniom platform (www.geniom.de). The technology integrates microarray production (through *in situ* synthesis using a photolithographic method) with the hybridization and detection steps in one unit. This technology also uses three-dimensional microchannels to immobilize the probe molecules and a small reaction volume (10–15 µL) is transported. Unfortunately, although the oligos are synthesized in capillaries, the system does not create a flow-through mixing for the hybridization. Indeed, hybridization occurs statically and is governed completely by diffusion. This is a missed opportunity to also increase the hybridization efficiency to its maximum.

Common in most of these reports were the very low sample volumes needed and the short analysis time. It was difficult to compare these systems as several important parameters varied, such as the nature of the probes and targets. Moreover, the pumping method, the fluid velocity, and fluid layer height above the probe molecules depended on the platform. Therefore, there is an increasing interest and demand in the need for a more profound view on the molecular basics of the hybridization reaction and the influence of a velocity field near the microarray spots. Profound theoretical [12–14] and experimental studies [15–18] are crucial for the further development of microfluidic microarray hybridization systems.

7.5 STUDY OF HYBRIDIZATION PROCESSES USING SHEAR-DRIVEN HYBRIDIZATION

7.5.1 INTRODUCTION TO HYBRIDIZATION PROCESSES

To study aspects of hybridization kinetics and the influence of diffusion, we have developed and tested a working prototype that enabled combination of a small microfluidic system (small sample volume) with an active form of mixing using a shear-driven flow approach [19,20] (Figure 7.4). In a shear-driven system, the fluid flow is generated by moving two surfaces (one flat carrier and the other

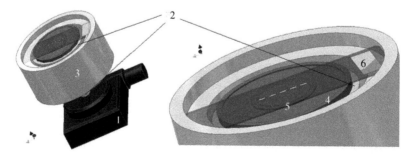

FIGURE 7.4 Bird's eye view schematic of the rotating microchamber hybridization setup, showing the rotation stage (1) and the stainless holding shaft (2). The PVC-cup (3) is used for temperature control. The etched glass consists of an outer etched ring (4) made to limit the contact area between the rotating bottom substrate and the microarray slide surface, and the microchamber (5). Finally, the microarray slide (6) seals the entire system.

containing the channel structure) relative to each other. The fluid in-between the two surfaces will move due to the dragging effect of the moving channel. The generated flow has exactly half the velocity of the moving wall and does not exhibit any pressure-drop or double-layer-overlap limitation as in the traditionally employed pressure and electrically driven transport enhancement methods [21,22].

7.5.2 Effects of Velocity, Mode of Convection, and Channel Depth

This setup has been used for various types of experiments. A first set of experiments were performed with a simple microarray layout consisting of a small number of probe spots and target sequences (cf. the methods section). This setup was primarily used to deliver a proof of concept and to investigate several specific aspects of this hybridization technique. The hybridization process consists of two major steps: the formation of a nucleation site between the target and the probe strands and in case of matching sequences, a fast zippering reaction. The fluid flow will not have a great effect on the zippering reaction, but will especially influence the formation of the nucleation site. Therefore certain aspects including the influence of the channel depths, fluid velocities, and the fluid flow method (discontinuous versus continuous flow) were investigated. Figure 7.5 shows a large number of hybridization experiments with the rotating microchamber setup, conducted in four different microchambers (1.6, 3.7, 10, and 70 μm), different types of fluid flow (discontinuous versus continuous), different types of discontinuous movement, and different analysis times. All results are compared to the results of overnight experiments under coverslip and 12-h experiments in a commercial Slide Processor (SlidePro, GE Healthcare).

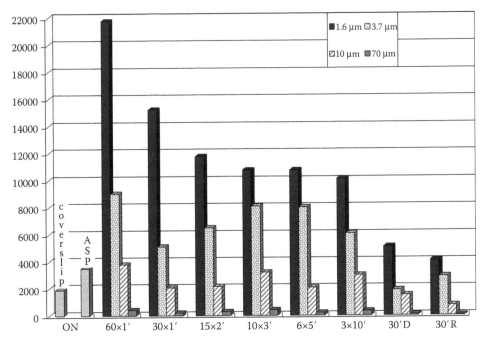

FIGURE 7.5 Comparison between the hybridization intensities obtained after 30 min discontinuous rotation hybridization for the four considered microchamber depths: 1.6 μm (black), 3.7 μm (dotted), 10.0 μm (striped), and 70.0 μm (dark gray), with stop periods of 1(30 × 1′), 2(15 × 2′), 3(10 × 3′), 5(6 × 5′), and 10(3 × 10′) min. These results can be compared to the data obtained after 30 min diffusion (30′D), 30 min continuous rotation (30′R), and a 60-min discontinuous rotation hybridization (60 × 1′) for the four considered microchambers. The values after overnight (ON) diffusion-driven hybridization under coverslip and hybridization in the SlidePro system (light gray) are also displayed.

The results revealed that the currently proposed rotating microchamber system allowed the combination of a miniaturization of the hybridization chamber with the creation of a strong lateral convective transport. In only 30 min, the system yielded hybridization intensities about 4 times greater than a conventional overnight (16 h) experiment under a coverslip and 5 times greater than in an experiment using the same amount of DNA conducted in the SlidePro. One of the essential keys to this gain seemed to be that the convective displacement was performed in a discontinuous mode (a rotation of 5 s at a rate of 3 rpm followed by a stationary interval), because the continuous rotation mode only yielded intensities which were of the same order as those obtained in the conventional diffusion-driven mode. It was found that the optimal condition for hybridization in the microchamber system was to perform the hybridization in the shallowest microchamber in a discontinuous contact mode with a stop period of 1 min. The optimal rotation conditions changed depending on the microchamber used. For a 3.7-μm chamber a stop period of 3 min was determined to be optimal. The hybridization time varied between 30 min and a few hours. As for the influence of the velocity on the hybridization rate experiments were performed in a continuous mode at different velocities. Here again the microchamber depth was found to be important. In a 3.7-μm chamber the hybridization rate increased to a small optimum around $\omega_{opt} = 1$ rpm followed by a sharp decrease with increasing rotation rate [16].

7.5.3 EFFECTS OF CONVECTION AND CONCENTRATION

The unique combination of small sample volumes combined with active mixing was responsible for the efficiency of the shear-driven hybridization (SDH) system. As most of the volume gain was due to the small chamber depths that were used (<5 μm), convection was absolutely necessary in order to enhance the hybridization rate. As the sample volume was reduced the concentration of the sample could be enhanced. Figure 7.6 (taken from data in Pappaert and Desmet [14]) shows the importance of convection and the enhancement of the concentration. With low concentrations

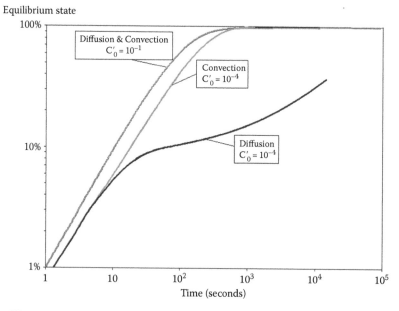

FIGURE 7.6 These curves, obtained from simulated data [14], show the influence of a dimensionless initial concentration parameter (C_0') on the hybridization rate in the case of diffusion-driven (black curves) and the corresponding convection-driven systems (gray curves). The values for C_0' are 10^{-1} and 10^{-4} corresponding to target concentrations of 40 nmol/L (abundant strands) and 40 pmol/L (rare strands), respectively.

(dimensionless initial concentration $C_0' = 10^{-4}$), the difference between ideal fluid convection (gray) and relying on pure diffusion was enormous. For a concentration ($C_0' = 10^{-1}$) with a factor 1000 higher than this difference was almost nonexistent. With such high concentrations however, diffusion-driven systems would consume enormous amounts of sample, while a convection-driven system in very small micro- or nanochannels would reduce sample consumption.

7.5.4 COMPARISON OF SDH WITH DIFFUSION UNDER COVERSLIP AND AUTOMATED HYBRIDIZATION

To further evaluate the efficiency of the SDH system, a second series of hybridizations with complex samples (mouse spleen and heart) were performed on slides containing >21,000 complementary DNA (cDNA) fragments. To compare SDH with other types of hybridization approaches all experiments were carried out with the same amount of labeled nucleic acid (i.e., 40 pmol Cy3/Cy5-labeled cDNA, Table 7.1).

By keeping the amount of labeled sample the same, SDH had an immediate advantage in that the sample concentration could be increased from 3- to 21-fold as compared to the conventional coverslip and SlidePro, respectively. An experimental comparison of these three types of hybridization systems (Table 7.1) revealed that, compared to the results obtained with an automated slide processor (SlidePro, GE Healthcare), the relative intensity of the spots increased by a factor 2 for the coverslip hybridization and by a factor 4 for the SDH. Furthermore, the total number of spots that could be flagged as being hybridized rose from 67% with the SlidePro system and 80% under the coverslip to 87% with the SDH system. This significant rise in the number of spots points to the great potential of SDH as it allows detection of a larger number of hybridized spots with the same amount of sample, or detection of the same amount of spots with lesser sample consumption. An even better gain in performance could be obtained by making the SDH last longer than 2 h. All experiments with the SDH system lasted 2 h, whereas the SlidePro and coverslip hybridizations lasted for 16 h.

Figure 7.7 shows images of experiments identical to the ones listed in Table 7.1. It was clear from these images that the overall intensity of the spots was much greater for the SDH system than for the coverslip and SlidePro hybridization. For this series of experiments, the SlidePro hybridization was conducted with a total amount of 40 pmol incorporated Cy3/Cy5 dyes, whereas the SDH and coverslip hybridizations were performed with only 20 pmol incorporated Cy3/Cy5 dyes (Figure 7.7a). It was even possible to drop the concentration even further to 5 pmol incorporated Cy3/Cy5 dyes in the sample (Figure 7.7b) and still achieve good results with the SDH system compared to SlidePro, where

---

TABLE 7.1
Comparison of Hybridization Conditions on the Shear-Driven Rotating Microchamber System, the Diffusion-Driven Coverslip System, and the SlidePro (GE Healthcare) for a Mouse Heart and Spleen Total RNA Sample on the VIB_Mouse_5KI Microarray Chip

	SDH System	Diffusion Under Coverslip	SlidePro-Automated Hybridization
Amount incorporated Cy3/Cy5 (pmol)	40	40	40
Sample volume (μL)	10	30	210
Sample concentration (pmol Cy3/μL)	3.00	1.00	0.14
Hybridization time (h)	2	16	16
Surface microchamber (mm²)	314	1000	1000
Hybridized spots (%)	87	80	67
Relative intensity (%)	396	189	100

(a)

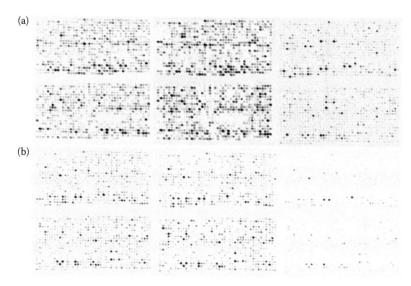

(b)

FIGURE 7.7 A series of different hybridization experiments done with 20 pmol (coverslip and SDH) and 40 pmol (SlidePro) of incorporated Cy3/Cy5 dyes (a) and with 5 pmol of incorporated Cy3/Cy5 dyes (b). All experiments were done with mouse heart (Cy3) versus spleen (Cy5) total RNA sample on the VIB_Mouse_21K_II microarray chip.

almost no hybridization occurred. For the experiments with 5 pmol incorporated dyes, the relative intensity compared with the coverslip experiment was 137% and 21%, respectively, for the SDH and SlidePro systems. Furthermore, the number of hybridized spots increased from 23.5% with the SlidePro system to 45.9% for the coverslip experiment and 66.5% for the SDH system.

7.5.5 SUMMARY AND DISCUSSION

To study the different aspects of the hybridization process and to investigate the influence of the presence of a convective fluid flow near the microarray surface, we have developed and tested a working prototype for microarray hybridizations based on the shear-driven technology. The early experiments with the SDH system were performed on a simple microarray layout to develop a better understanding of the different aspects of the hybridization reaction. In the presence of a convective fluid flow, a compromise had to be found between the refreshment rate and the fluid velocity. High refreshment rates mean high velocity fluid flows that might disturb the hybridization reaction. Low velocity flows result in lower refreshment rates. The most ideal convective flow seemed to be a discontinuous flow system where short periods of fluid flow were alternated with longer periods where the fluid was kept stationary. For smaller depths of the used microchamber, the ideal stop period was reduced accordingly.

The key factor for the success of the SDH method was the combination of the mixing that circumvented the slow molecular diffusion, and the reduction of sample volume that generated more concentrated samples and higher final intensity values of the microarray spots. The concentration was indeed found a major factor on the effect of convection to reach higher signals intensities (Figure 7.7).

A further comparison of the SDH system with diffusion under coverslip and automated hybridization was done on microarray slides containing >21,000 cDNA fragments with complex heart and spleen total ribonucleic acid (RNA) target samples. A comparison of the SDH system with conventional hybridization methods such as diffusion under coverslip and an automated hybridization system (SlidePro) demonstrated the value of the system. Higher raw signal intensities were obtained and a larger number of probe spots could be flagged as being hybridized. Hence the SDH system enhances the detection limit, achieving identical results as conventional systems with lower sample amounts or obtaining more data with the same amount of sample.

The drawback of the current SDH prototype is that the rotating design only covered 30% of the microarray slide. To use shear-driven convection on standard microarray slides, we need to redesign the prototype from a rotating to a lateral movement or change the standard microarray slides to microarrays on round disks.

7.6 CONCLUSIONS

In this chapter, we provided an overview of several experimental hybridization systems, reviewed current efforts in microfluidics or lab-on-chip developments, and critiqued the most prominent automated hybridization devices in the market. In addition, we tried to touch upon some aspects of the hybridization process using a new hybridization system based on shear-driven convective flow.

Concerning the commercial automated hybridization systems, we have tested several devices and found that the hybridization uniformity and signal strength were comparable if they were used to their maximum potential. As all the systems have active mixing of the labeled sample, hybridization signals were higher than the diffusion-driven static hybridization.

From the experiments with the SDH, we concluded that convection or mixing of the labeled sample is the key to increase the hybridization efficiency and signal strength. However, the mode (continuous/discontinuous) and the velocity of mixing/convection and concentration of the sample determined this effect to a great extent. These aspects should be taken into consideration not only in the design of new hybridization devices, but also in optimization and improvement of existing protocols. Optimizing protocols for new slide chemistries, labeling strategies, or novel microarray applications is a painstaking process. Knowledge of the theoretical aspects of hybridization process and fluid dynamics will greatly help these optimizations. For example, a pause in the mixing can help to establish nucleation of the target on the probe or slowing down the velocity of wash buffers, flushed into the hybridization chamber, can increase the efficiency of mixing and the clearing-out of the labeled sample.

However, it is interesting to note that all major microarray producers, including Affymetrix, Agilent, Applied Biosystems, and Codelink still did not develop a specific hybridization device for their microarrays. All of them still use a diffusion-driven approach where the labeled sample is placed in a chamber or under a sealed coverslip and mixing is achieved by rotation or brisk shaking. At the same time, many researchers who have purchased hybridization stations in the past are moving away from them and are going back to chambers usually in combination with coverslips or lifterslips (Erie Scientific), such as Hybex (SciGene), Hybriwell sealing system (Invitrogen), Hybridization chambers from Corning Life Sciences, Takara Bio or Scienon; Versarray hybridization chambers from Bio-Rad. The main reason for moving away from stations is that they are still expensive as regards purchase and maintenance, and that the performance of the devices may drop after a couple of years when the fluidic parts clog due to continuous salt deposition. In addition, new microarray developments are rapidly emerging, such as multiple high-density arrays on one slide, making it very difficult for the companies, developing the hybridization devices, to keep abreast.

Regardless, automated hybridization systems are indispensable for medium or large core facilities to insure the turnaround time and consistency over several months or years.

To choose the right one for your laboratory settings, one has to compare the devices with the specific requirements. These can be

- What is the throughput, 10 or 50 slides a week? At least 12 chambers need to be processed in parallel.
- Are you handling small sample sizes like needle biopsies? Choose a small volume device.
- Do you want to process slides that contain single or multiple arrays?

- Do you want to process several protocols in parallel or always the same one?
- Do you have a lot of bench space? Some devices take much more space than others.
- What is your budget? Although list prices are comparable, some companies are more willing to negotiate than others.

Finally, the ultimate dream is to change the current microarray systems completely and to fabricate, through micro- and nanotechnology, either single or multiple, integrated, self-containing laboratories of a size equivalent to that of a computer chip. It is hoped that these systems will eventually be capable of executing analysis in which reactions, separations, and various forms of detection are integrated into a single method. In time, this will lead the way to point-of-care diagnostics and field analysis.

In the drive towards miniaturization, it is not only the reduction of volume and mass transfer distances that are involved in the analysis process. There are numerous other benefits that come with miniaturization:

- Reduction of reagent and sample consumption as well as waste production
- An increased level of automation
- Low cost (mass production and/or the use of inexpensive materials)
- Shorter analysis times (diffusion scales to the second power of the distance)
- High number of assays that can be run in parallel (high-throughput)
- An increased level of integration of systems components

As both microarray and microfluidic technologies have grown in parallel with each other during the past decade, it is evident that a combination of both would lead to a system that combines the advantages of both technologies. A portable, low sample consuming fully integrated and automated system that is able to deliver the accurate, high-throughput data of a microarray.

7.7 MATERIALS AND METHODS

The two main components of the employed microchamber/microarray slide system are schematically shown in Figure 7.4. The circular microchamber (diameter = 2 cm) was etched in the center of a round, flatly polished (flatness = $\lambda/4$ at $\lambda = 512$ nm) borosilicate glass wafer (Radiometer Nederland, The Netherlands) with thickness 6 mm and diameter 5 cm. The etching was carried out using a 50% HF solution (Fluka Chemie GmbH, Buchs, Switzerland) for different time periods depending on the desired depth of the chamber. After the etching, the channel depth was measured using a Talystep™ step profiling apparatus (Rank Taylor Hobson Ltd., UK). The wafer was subsequently clamped in a home-built, hybridization device (Figure 7.4). The wafer was rotated using an automated rotation stage equipped with a stepper motor (M-060, Physik Instrumente, Karlsruhe, Germany), and controlled with NetMove420-software (Physik Instrumente, Karlsruhe, Germany) running on an ME Windows controlled PC. With this automated rotation stage, the rotation rate can be varied between $\omega = 0$ rpm and $\omega = 60$ rpm. Different rotation regimes were programmed for this study (intermittent and continuous rotation). The circular polyvinyl chloride (PVC) holding cup surrounding the glass wafer makes it possible to control the temperature using a Julabo® F32-MD Heating (and Refrigerated) Circulator (JD Instruments Inc., Houston, Texas, USA), set at $T = 42°C$ during all the experiments.

To carry out the hybridization experiments, a freshly spotted microarray slide was gently deposited (spotted surface facing downwards) on the top surface of the glass wafer. Since in the shear-driven rotating disk flow-system more sample volume is needed than only the microchamber volume, for example a volume of 10 µL was applied in the 3.7-µm deep microchamber with diameter 2.0 cm. This was needed because of the possible formation of air bubbles caused by the nonperfect flatness of the microarray and substrate surface, leading to a higher effective microchamber volume than the calculated one, and due to a mechanical friction problem, caused by glass-on-glass contact, resulting in a possible evaporation of the sample. This problem would most probably be solved if microarray slides with a smaller degree of large-scale waviness would be used. For the time being, the problem is solved by adding a small volume excess, forming a lubrication layer between the microarray and the substrate surface.

During the rotation of the glass wafer, the microarray slide was kept stationary and a pressure of 1 bar was used to push the microarray substrate and the microchamber against each other.

For the diffusion-driven hybridization under coverslip, the standard overnight hybridization procedure of the MicroArray Facility Lab [23] was followed. The procedure simply consisted of pipetting a 30-μL sample and putting it on the microarray slide, which was then topped by a thin glass coverslip (2 by 5 cm^2) and sealed off with rubber glue. The slides were subsequently kept overnight (16 h) in an incubator at 42°C.

Hybridizations using the SlidePro system (GE Healthcare) are performed in independently temperature-controlled chambers, allowing customized process parameters to be used for each slide. The system consists of up to five modules, each containing six chambers that clamp to slide surfaces. Labeled sample is injected via a septum injection port and wash solutions are circulated from wash storage bottles via tubing to each chamber. Up to 200 μL can be injected into each chamber. Chambers are constructed with chemically resistant polyetheretherketone (PEEK) and have a patented o-ring design. Wash solutions are pumped in and out of each chamber using a 1000-μL syringe pump. Flow rate can be adjusted with the ability to toggle flow back and forth through the chamber (10–50 μL). Clean, dry air is used to dry slides as well as actively pump wash solutions back and forth across the slide surface [24].

All hybridization experiments were conducted using conventional microarray procedures and made use of the VIB_Mouse_21K_II chip (MicroArray Facility, Leuven, Belgium). As array elements, mouse spleen and heart total RNA was used (BD Biosciences Clontech, Palo Alto, California, USA). All different cDNA strands were obtained by PCR amplification and were arrayed on Amersham type 7 star slides (Amersham Biosciences, Buckinghamshire, UK) using a commercial Generation III Array Spotter (Amersham Biosciences, Buckinghamshire, UK). The array elements were all spotted at a concentration of 200 ng/μL. The spots (diameter 100 μm and spaced 20–30 μm apart), situated in the circular spotted area were grouped in 2×5 different blocks of 21×44 spots.

Prior to each experiment, the concentration of the samples and the amount of incorporated Cy3 molecules was measured with a NanoDrop™ spectrophotometer (NanoDrop Technologies, Montchanin, Delaware, USA). To achieve maximal hybridization, the arrays were slightly dampened and UV-crosslinked at 50 mJ. The slides were then prehybridized in $2 \times$ SSPE/0.2%SDS at 26°C for 30 min, rinsed with MilliQ water and dried by centrifugation. After heat denaturation of the samples, the hybridization was carried out at 42°C, either on the microchamber system, under coverslip, or using the SlidePro system.

After hybridization using the rotating disk flow-system and under coverslip, the slides were washed 3 times for 10 min in different sodium dodecyl sulphate (SDS)/sodium chloride/sodium citrate (SSC) solutions ($1 \times$ SSC/0.2%SDS, $0.1 \times$ SSC/0.2%SDS, and $0.1 \times$ SSC/0.2%SDS, respectively) at 56°C and 1 time for 1 min in $0.1 \times$ SSC at room temperature. After washing, all slides were rinsed with MilliQ water and finally dried by centrifugation. The washing procedure after hybridization in the SlidePro system is a fully automated process: after hybridization, the microarray slides are washed 2 times for 10 min, 1 time for 4 min, and another time for 10 min in $1 \times$ SSC/0.2%SDS, $0.1 \times$ SSC/0.2%SDS, $0.1 \times$ SSC, and MilliQ water, respectively. The slides were then scanned at 532 nm using a commercial Array Scanner Generation III (Amersham Biosciences, Buckinghamshire, UK). Image analysis was done using Array Vision 7.0 software (Imaging Research, Ontario, Canada) and Spotfire 8-software (Spotfire Inc., Somerville, Massachusetts, USA).

All experiments were started from 5 μg spleen total RNA and 5 μg heart total RNA per sample. This was first amplified by T7 *in vitro* transcription and 5 μg amplified RNA (aRNA) was used for every labeling reaction. After labeling with Cy3/Cy5 dye, both samples were mixed and for every experiment, identical amounts of sample, i.e., 5, 20, or 40 pmol Cy3/Cy5 dye, were applied. The cDNA was further dissolved in a solution containing 50% formamide, 25% buffer, 15% Mouse COT, and 7.5% polydT35. This cDNA mixture was then respectively denatured at 96°C for 3 min, set on ice for 2 min, heated to 42°C for 5 min, centrifuged at $\omega = 12,000$ rpm for another 5 min, and subsequently held on ice before applying it on the microarray slide.

ACKNOWLEDGMENTS

The authors want to thank Ruth Maes, Kirsten Deschouwer, Kizi Coeck, Kathleen Coddens, and Tom Bogaert from the MicroArray Facility, for their excellent technical help. The shear-driven work was financially supported by a GBOU-grant from the Institute for the Promotion of Innovation through Science and Technology in Flanders (IWT-Vlaanderen). Kris Pappaert is supported through a specialization grant from the same institute (grant no. SB/01/11324).

REFERENCES

1. Bynum, M.A. and Gordon, G.B. 2004. Hybridization enhancement using microfluidic planetary centrifugal mixing. *Anal. Chem.*, **76**, 7039–7044.
2. Adey, N.B., Lei, M., Howard, M.T., Jensen, J.D., Mayo, D.A., Butel, D.L., Coffin, S.C., et al. 2002. Gains in sensitivity with a device that mixes microarray hybridization solution in a 25-μm-thick chamber. *Anal. Chem.*, **74**, 6413–6417.
3. Schaupp, C.J., Jiang, G., Myers, T.G., and Wilson, M.A. 2005. Active mixing during hybridization improves the accuracy and reproducibility of microarray results. *Biotechniques*, **38**, 117–119.
4. Toegl, A., Kirchner, R., Gauer, C., and Wixforth, A. 2003. Enhancing results of microarray hybridizations through microagitation. *J. Biomol. Tech.*, **14**, 197–204.
5. Peytavi, R., Raymond, F.R., Gagne, D., Picard, F.J., Jia, G., Zoval, J., Madou, M., et al. 2005. Microfluidic device for rapid (<15 min) automated microarray hybridization. *Clin. Chem.*, **51**, 1836–1844.
6. Wei, C.W., Cheng, J.Y., Huang, C.T., Yen, M.H., and Young, T.H. 2005. Using a microfluidic device for 1 μl DNA microarray hybridization in 500 s. *Nucleic Acids Res.*, **33**, e78.
7. Liu, R.H., Lenigk, R., Druyor-Sanchez, R.L., Yang, J., and Grodzinski, P. 2003. Hybridization enhancement using cavitation microstreaming. *Anal. Chem.*, **75**, 1911–1917.
8. Manz, A., Miyahara, Y., Miura, J., Watanaba, Y., Miyagi, H., and Sato, K. 1990. Design of an open-tubular column liquid chromatograph using silicon chip technology. *Sens. Actuators B*, **1**, 249–255.
9. Manz, A., Verpoorte, E., and Raymond, D.E. 1994. μTAS: Miniaturized total chemical analysis systems. In: *Proceedings from Micro Total Analysis System*, pp. 5–27.
10. Bowden, M., Song, L., and Walt, D.R. 2005. Development of a microfluidic platform with an optical imaging microarray capable of attomolar target DNA detection. *Anal. Chem.*, **77**, 5583–5588.
11. Hashimoto, M., Barany, F., and Soper, S.A. 2006. Polymerase chain reaction/ligase detection reaction/hybridization assays using flow-through microfluidic devices for the detection of low-abundant DNA point mutations. *Biosens. Bioelectron.*, **21**, 1915–1923.
12. Pappaert, K., Van Hummelen, P., Vanderhoeven, J., Baron, G.V., and Desmet, G. 2003. Diffusion-reaction modeling of DNA hybridization kinetics on biochips. *Chem. Eng. Sci.*, **58**(21), 4921–4930.
13. Gadgil, C., Yeckel, A., Derby, J.J., and Hu, W.S. 2004. A diffusion-reaction model for DNA microarray assays. *J. Biotechnol.*, **114**, 31–45.
14. Pappaert, K. and Desmet, G. 2006. A dimensionless number analysis of the hybridization process in diffusion- and convection-driven DNA microarray systems. *J. Biotechnol.*, **123**(4), 381–396.
15. Vanderhoeven, J., Pappaert, K., Dutta, B., Van Hummelen, P., Baron, G.V., and Desmet, G. 2004. Exploiting the benefits of miniaturization for the enhancement of DNA microarrays. *Electrophoresis*, **21–22**, 3677–3686.
16. Vanderhoeven, J., Pappaert, K., Dutta, B., Van Hummelen, P., and Desmet, G. 2005. DNA microarray enhancement using a continuously and discontinuously rotating microchamber. *Anal. Chem.*, **77**(14), 4474–4480.
17. Chung, Y.C., Lin, Y.C., Shiu, M.Z., and Chang, W.N. 2003. Microfluidic chip for fast nucleic acid hybridization. *Lab Chip*, **4**, 228–233.
18. Kim, J., Marafie, A., Zoval, J., and Madou, M. 2006. Characterization of DNA hybridization kinetics in a microfluidic flow channel. *Sens. Actuators B*, **113**, 281–289.

19. Pappaert, K., Vanderhoeven, J., Van Hummelen, P., Dutta, B., Clicq, D., Baron, G.V., and Desmet, G. 2003. Enhancement of DNA micro-array analysis using a shear-driven micro-channel flow system. *J. Chromatogr. A*, **1014**, 1–9.
20. Vanderhoeven, J., Pappaert, K., Dutta, B., Van Hummelen, P., and Desmet, G. 2005. Comparison of a pump-around, a diffusion-driven, and a shear-driven system for the hybridization of mouse lung and testis total RNA on microarrays. *Electrophoresis*, **26**(19), 3773–3779.
21. Desmet, G. and Baron, G.V. 2002. Chromatographic explanation for the side-wall induced band broadening in pressure-driven and shear-driven flows through channels with a high aspect-ratio rectangular cross-section. *J. Chromatogr. A*, **946**(1–2), 51–58.
22. Desmet, G., Vervoort, N., and Clicq, D. 2002. Shear-flow-based chromatographic separations as an alternative to pressure-driven liquid chromatography. *J. Chromatogr. A*, **948**(1–2), 19–34.

8 Characterization of Microarray Hybridization Stoichiometry*

Richard J.D. Rouse and Gary Hardiman

CONTENTS

8.1 INTRODUCTION

Stoichiometry or reaction stoichiometry is defined as the calculation of quantitative relationships of reactants in a given reaction. Over the last decade microarray-based assays have evolved from a nascent technology into a key molecular tool widely used in basic, translational, and clinical research [1]. Their primary utility has been gene expression analyses, facilitating high-throughput transcriptome profiling [2]. Microarrays have also found utility in genotyping and re-sequencing applications, comparative genomic hybridization, and genome-wide (epigenetic) localization studies [3–6]. Regardless of the platform, a microarray is comprised of a library of nucleic acid sequences immobilized in a grid on a solid surface. The nucleotide sequence contained in each unique microarray feature is termed a "probe" and is derived from a specific gene. The probe detects a specific labeled target sequence derived from cellular messenger ribonucleic acid (mRNA) via complementary base pair hybridization [7].

There are two general microarray formats: one-color systems that have historically been associated with Affymetrix Gene Chips, and two-color schemes used with complementary DNA (cDNA) and oligonucleotide arrays [8,9]. The one-color approach involves a one-sample per-chip hybridization. However, two-color experiments are competitive hybridizations utilizing two samples, one labeled with Cy3 and the other with Cy5. Gene expression levels are commonly reported in

* Rouse RJ, Espinoza CR, Niedner RH, and Hardiman G. Development of a microarray assay that measures hybridization stoichiometry in moles. *Biotechniques* 2004; 36: 464–470. With permission.

ratiometric terms, as this assay compares transcript abundance between the two different biological samples on the same microarray. One fluorescent target is prepared from a control mRNA and the second from mRNA isolated from the treated cells or a diseased tissue under investigation. The control is generated from "normal" cells or tissues or a standardized mRNA mix, sometimes termed a "universal control," collected from the transcriptomes of a variety of cells or tissues to ensure a reference signal for the majority of features. Although this method can accurately compare expression levels between the cy3- and cy5-labeled samples, the relative nature of the results requires elaborate normalization methods (discussed in Chapters 9 and 11) to enable the comparison of data from different chips.

When data from new chips are introduced into an experiment, the entire dataset needs to be re-normalized to include the additional data. Two-color assays also suffer from labeling bias between the fluorescent dyes and uneven hybridization [10,11]. Despite extensive research and improvements in this field, a microarray based assay that addresses these problems by recording an absolute value for each feature would be very useful. Such a measure would directly correlate with the copy number of the complementary transcript in the biological sample. These per-chip datasets could potentially simplify cross-laboratory and cross-platforms comparisons.

As a means of achieving this aim we developed a hybridization stoichiometry assay [12]. In this chapter, we expand on our initial approach and describe in greater detail the constituent components of this noncompetitive assay. Furthermore, we describe the advantages of utilizing hybridization stoichiometry as a per-chip quantitative measure.

8.2 RESULTS

8.2.1 ASSAY DESIGN

The molar relationship between a given array probe and its corresponding target is very useful for a variety of applications. Our approach for determining this molar relationship is two-fold. Firstly, we measure the amount of deoxyribonucleic acid (DNA) present in every array feature post-printing and post-hybridization. Secondly, a separate series of dye printing experiments are performed in parallel to determine the amount of cyanine dye present in the hybridized cDNA sample.

The schema of how array features are measured in moles is outlined in Figure 8.1a. A dilution series of a known amount of DNA probes (or standards) is printed. A universal reference sequence labeled with Cy3 is subsequently hybridized to these standards. The signal intensities derived from this hybridization are plotted against the molar amount of DNA printed. A standard curve is generated in which the formula that characterizes the slope of the line is used to transform signal intensities into moles of DNA present in specific array features.

Dye printing experiments are performed to determine the amount of cyanine dye present in the hybridized cDNA sample. Figure 8.1b illustrates how this measurement is performed. A dilution series of a known amount of cyanine dye is printed on a blank microarray slide comparable to the one used in the assay, and the slide is scanned using scanner settings identical to those used for the array hybridization experiments. A standard curve is plotted from the data which correlate signal intensity with molar amounts of cyanine dye printed. The formula characterizing the slope of this line is used to transform the signal intensities derived by the hybridized cDNAs to the amount of Cy5 dye detected in moles (i.e., the amount of Cy5-labeled target hybridized). For this assay to work, a labeling system that incorporates a constant defined amount of dye molecules per DNA molecule must be used. To accomplish this we have used exclusively the 3DNA Array Detection from Genisphere (Hatfield, Pennsylvania). This labeling technique employs branched dendrimers containing cyanine dyes [13].

In practice, we have utilized spotted cDNA arrays in concert with a Cy3-labeled synthetic universal amplicon (SUA) target, which hybridizes to every feature (Figures 8.1c and 8.2). The array probes are polymerase chain reaction (PCR) amplicons engineered to possess a universal sequence.

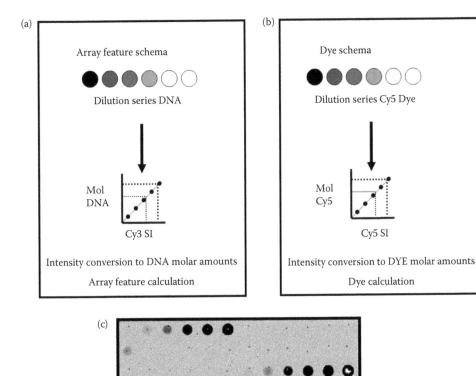

FIGURE 8.1 The process of quantifying hybridization stoichiometry in moles. (a) Schematic of the method for measuring array features in molar quantities. A dilution series of known molar amounts of DNA is spotted on an array. A Cy3-labeled SUA reference is then hybridized to these standards. The derived signal intensities are plotted against the molar amount of DNA printed and a curve fit is calculated. This enables the transformation of signal intensity into molar amounts. (b) Illustration of how the assay determines the amount of dye present in a hybridized cDNA sample. The plot correlates signal intensity to the molar amounts of cyanine dye printed. (c) Dilution series as it appears on a hybridized array.

This is achieved via amplification with a single conserved primer set. As the Cy3-SUA hybridizes to the universal sequence tag present in every feature, it can be used to measure precisely the amount of DNA present in a given feature (Figure 8.2). This approach can also be adapted for oligonucleotide arrays.

This system generates approximately 360 ± 20 molecules of dye per labeled DNA by hybridizing DNA dendrimers containing cyanine dyes to the modified cDNA sequences [13]. Using this labeling system, the amount of DNA hybridized can be measured on a molar basis. A stoichiometric unit that represents the relationship between known amounts of a hybridized target, and its homologous array probe can be calculated.

8.2.2 Measuring Hybridization Stoichiometry of a Reference RNA Sample

We performed an experiment to determine the hybridization stoichiometry of the Stratagene Universal Mouse Reference RNA (UMRR). The hybridization mixture consisted of three different labeled targets: (a) Cy5-labeled cDNA derived from the UMRR; (b) Cy5 individually labeled dynamic range spikes; and (c) a Cy3-SUA that recognizes every array probe. In transforming the derived dataset from this hybridization into units whereby stoichiometry can be measured, we compared the signal intensities of the hybridized array probes to signal intensities that were

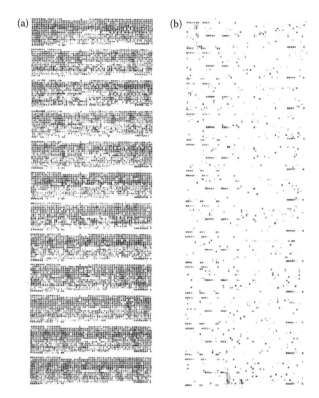

FIGURE 8.2 Comparison of a microarray hybridized with an SUA and biological sample hybridized to a microarray. Approximately 9K cDNA clones derived from the Mouse UniGene set and controls were printed on Type 7* reflective slides and hybridized with the Cy3-labeled SUA control sequence, the Cy5-labeled Clontech mouse spleen polyA⁺ mRNA-derived cDNA, and the Cy5-labeled xenogenic spikes. (a) Image from the Cy3 channel representing hybridization with the Cy3-labeled SUA. This sequence hybridizes to a conserved sequence present in every array feature. Thus, the intensity of the spots is directly related to the amount of DNA present in every feature. Empty spots indicate the absence of probe DNA. (b) Image from the Cy5 channel representing hybridization with the UMRR and dynamic range controls.

generated upon printing Cy5 dye. This dye printing experiment was performed on a microarray slide similar to that used in the assay and scanning was carried out using the same scanner settings. Figure 8.3 shows the plot of the amount of dye printed versus the corresponding signal intensity. The curve-fit regression derived from this assay generated the following formula: $[y = (7 \times 10^{-10}) \times 1.09]$. This was used to calculate the picomole amount of Cy5 dye that was hybridized to a specific array probe.

The Cy3-SUA is a DNA sequence that hybridizes to conserved regions on every array probe. Therefore, hybridization of this sequence to a dilution series of array probes printed on the slide can be used as a reference to calculate the amount of DNA molecules printed (Figure 8.1c). Figure 8.4 shows the plot of the amount of picomoles of probe DNA printed versus the signal intensity of the Cy3-SUA. A regression formula can be derived from this dataset and is used to calculate the amount of DNA printed for every array probe. For this experiment the regression formula was $[y = (1 \times 10^{-7})x + (2 \times 10^{-5})]$. Based on measurements of the sample cDNA sequence and microarray features, the stoichiometric relationship between labeled target samples and array probes can be measured. Table 8.1 presents the data for the most highly expressed sequences, measured for the Cy5-dendrimer labeled UMRR. The data revealed that the probe was in excess even for the highest molar amount of Cy5-labeled sequence bound.

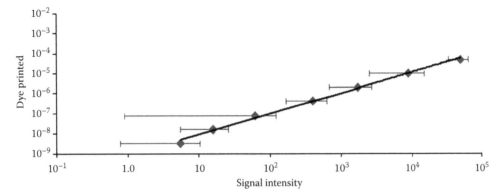

FIGURE 8.3 Correlation of Cy5 dye molecules to signal intensity. Printing was carried out using a titrated dilution series of Cy5 dye. The plot represents the average signal intensity of 48 printed replicates. The Cy5 signal intensity is plotted on the x-axis, and Cy5 dye in picomoles is plotted on the y-axis. After plotting, a curve fit was calculated, and the formula $[y = (7 \times 10^{-10}) \times 1.09]$ was used to transform signal intensity derived from microarray hybridization experiments to the amount of dye molecules present in each hybridized spot.

Table 8.2 displays data generated using the dynamic range control spikes. By using known concentrations in the hybridization mixture, this approach serves as a guide for determining the upper and lower limits of sensitivity. The table contains data for the percentage of array probe saturation as well as raw signal intensity values. As the concentration of the target DNA in the hybridization increased, so did the percentage of array probe saturation with the dynamic range controls.

TABLE 8.1
Measurement of the Stoichiometry of cDNA Target Sequences Bound to Array Features

Accession no.	Cy5	Cy3	Cy5 (pmol) DNA	Cy3 (pmol) DNA	Probe Saturation (%)
AA166336	50,430	654	2.58×10^{-7}	8.54×10^{-5}	3.02×10^{-1}
W18585	42,809	1580	2.16×10^{-7}	1.78×10^{-4}	1.21×10^{-1}
AA030527	40,085	4315	2.01×10^{-7}	4.52×10^{-4}	4.45×10^{-2}
AA276835	35,617	896	1.77×10^{-7}	1.10×10^{-4}	1.61×10^{-1}
AA271223	26,201	1717	1.27×10^{-7}	1.92×10^{-4}	6.60×10^{-2}
AA271588	24,782	991	1.19×10^{-7}	1.19×10^{-4}	1.00×10^{-1}
W14332	22,172	1503	1.05×10^{-7}	1.70×10^{-4}	6.19×10^{-2}
AA277159	18,676	55	8.75×10^{-8}	2.55×10^{-5}	3.43×10^{-1}
AA000655	18,454	3827	8.64×10^{-8}	4.03×10^{-4}	2.14×10^{-2}
AA221886	17,232	19938	8.01×10^{-8}	2.01×10^{-3}	3.98×10^{-3}
AA058055	16,334	2043	7.56×10^{-8}	2.24×10^{-4}	3.37×10^{-2}

Notes: Calculated picomolar abundance of labeled targets and array features. The most abundant sequences derived from the Stratagene mouse universal reference total RNA sample that were hybridized to the array as represented in the Cy5 (pmol) DNA column. The molar amount of array probe is represented in the Cy3 (pmol) DNA column. After transforming the hybridization data into picomoles, the percentage of array probe saturation was subsequently calculated (% probe saturation).

TABLE 8.2
Array Signal Intensities and Transformed Datasets of the *Bacillus subtilis* Dynamic Range Control Spikes

Clone	Hyb (ng)	Hyb (pmol)	Cy5	Cye5 SD	Cy3	Cy3 SD	Target DNA (pmol)	Probe DNA (pmol)	Probe Saturation (%)
ybhR	2500	1×10^{-2}	62,317	5117	776	208	3.25×10^{-7}	9.76×10^{-5}	3.33×10^{-1}
ybaQ	500	3×10^{-3}	22,491	4987	961	347	1.07×10^{-7}	1.16×10^{-4}	9.25×10^{-2}
ycxA	100	5×10^{-4}	9045	1139	612	217	3.97×10^{-8}	8.12×10^{-5}	4.89×10^{-2}
ybaS	40	2×10^{-4}	5290	1242	1218	449	2.22×10^{-8}	1.42×10^{-4}	1.57×10^{-2}
ybaF	20	1×10^{-4}	4683	848	1048	383	1.94×10^{-8}	1.25×10^{-4}	1.56×10^{-2}
ybdO	4	2×10^{-5}	828	319	465	231	2.96×10^{-9}	6.65×10^{-5}	4.44×10^{-3}
ybaC	0.2	1×10^{-6}	1133	501	1280	541	4.16×10^{-9}	1.48×10^{-4}	2.81×10^{-3}
yacK	0	0	879	591	2329	939	3.19×10^{-9}	2.53×10^{-4}	1.26×10^{-3}

Notes: These data are derived from the dynamic range control spike hybridizations. The clone, hyb (ng), and hyb (pmol) columns, respectively, represent the *B. subtilis* clone names and nanograms and picomoles of the spike DNA present in the hybridization mixture, respectively. The Cy5 and Cy3 signal intensities and the corresponding standard deviations of twenty-four replicates are presented. The target DNA column lists the mean value of the target DNA hybridized, and the probe DNA column lists the mean value of printed corresponding microarray probe that was measured; picomole measurements are shown. The probe saturation column lists the percentage of saturation observed for every spike sequence that was hybridized.

8.3 DISCUSSION

Comprehensive data collection, annotation, and interpretation will facilitate the comparison of data from many laboratories and improve the utility of relational gene expression databases, particularly in integrating data derived from various transcriptional profiling and massively parallel signature sequencing methods. The approach to microarray analysis described in this study utilizes data thoroughly analyzed for target nucleic acid labeling, array feature quality, signal dynamic range, and sensitivity. This approach examines the accuracy and precision of the array manufacturing and assay process, and addresses potential sources of variation across different experiments.

Ramakrishnan et al. [14] demonstrated that array probes need to be present in molar excess compared to the targets for the signal intensities to remain within linear range. To construct high-quality arrays it is important to determine at which concentration the probes are in molar excess. Our system accurately measures this parameter, because the amount of array probe is measured in the assay. In designing quantitative microarrays, consideration is required for the actual amount of target DNA hybridized as well as the amount of probe DNA available to the target.

Many groups have contributed to improving the overall microarray manufacturing process. To improve data quality, microarrays fabrication needs to be consistent [15]. Close attention to the spotting process, and particularly the operation of array printing pins, is necessary to ensure that spotted arrays are of high quality. Capillary pins frequently become clogged during high-density printing, which often causes uneven dispensing of DNA or in a worst-case scenario results in a spot drop out phenomenon. Spot reproducibility is a measure of the variation that occurs during the printing process. Variation exists from slide to slide during a printing run even when the pins are not clogged. This is attributed to mechanical differences between pins, and minor variation in slide surface properties. The system described in this report is very useful in characterizing array quality by measuring the DNA content for every array feature and thereby developing a level of confidence in every signal generated.

The use of a common reference nucleic acid as one of the labeled targets has been shown to improve data consistency [10,11]. Data derived from multiple microarray experiments can be compared to a reference sample. When comparing two samples in a two-color experiment it may be difficult to discern the difference between actual signal and dye-specific bias, as the Cy3 and Cy5 dyes possess different spectral properties [16–18]. Using a homologous control reference such as the SUA rather than a biological reference improves the consistency of the data, because it eliminates batch variation that may exist in reference samples. The SUA hybridizes to every array feature and generates data points with all probes.

Our method generates a set of values that represent the absolute amount of cDNA present in the hybridization milieu across the dynamic range of expected target concentrations. This approach also permits quality control since it allows the determination of the actual amount of probe DNA for every printed feature. The uniform signal improves the accuracy of spot-finding, identifying, and flagging unusable array features.

Microarray data are most useful when it can be compared to other gene expression measurement systems, such as sequencing, real-time quantitative PCR, northern blots, and other array formats [14,19,20]. Ideally, it should be possible to reliably compare published array datasets. Translating microarray data into a natural unit representing absolute mRNA abundance as opposed to the widely used ratiometric format could help improve the accuracy of cross platform and laboratory comparisons. Creating a dataset that correlates the absolute abundance of single transcripts in a given biological sample would simplify the comparison of multiple datasets. Moreover, it would alleviate the need for inter-chip normalization between experiments and allow new experimental data points to be easily introduced.

8.4 METHODS

8.4.1 MOUSE 9K ARRAYING AND POSTPROCESSING

The mouse UniGEM Clone List sequence-verified clones were amplified as described previously [12,21]. PCR products corresponding to the 9182 mouse cDNAs and a series of *B. subtilis* control sequences (see later) were printed on Amersham Type 7 slides using the Molecular Dynamics GenIII spotter (GE Healthcare, Piscataway, New Jersey). The GenIII spotter contains 12 capillary printing pins, with each depositing 0.8 nl per slide contact. Postprinting, DNA probe adhesion to the slide was achieved by crosslinking in a Stratalinker Model 2400 ultraviolet (UV) illuminator (Stratagene, La Jolla, California) with UV light at 254 nm and the energy setting at 500 mJ.

8.4.2 CONSTRUCTION OF THE SUA

To fabricate the mouse array, the cDNAs were amplified with the following primers (forward primer 5′-ctgcaaggcgattaagttgggtaac-3′ and reverse primer 5′-gtgagcggataacaatttcacacaggaaacagc-3′). This ensured that in addition to the complete cDNA insert being amplified, some vector sequence was also amplified. These sequences from the various vector plasmids were aligned using the CLUSTALW program. All the plasmids contained homologous sequences 5′ and 3′ of the polylinker-cloning sites, 79 and 29 bp, respectively, in length. This conserved sequence was ultimately used to validate the amount of bound probe on a microarray.

An SUA was first constructed. The initial step was an extension reaction with oligonucleotides representing the homologous sequences. The oligonucleotides were as follows:

- 56-mer sense primer
- 5′-gtgctgcaaggcgattaagttgggtaacgccagggttttcccagtcacgacgttg-3′
- 71-mer anti-sense primer with a complementary sequence overhang and
- 5-ggaattgtgagcggataacaatttcacacaggaaacagctatgaccatgattacgcaacgtcgtgactggg-3′.

The extension reaction was carried out using PCR conditions described previously [12] on a Tetrad thermocycler (MJ Research, Inc., Waltham, Massachusetts). The extension cycle was as follows: 94°C for 5min, 60°C for 2 min, and 68°C for 5 min. A double-stranded DNA product was created and 1 μL was used as a template for PCR amplification using forward primer 5′-ctgcaaggc-gattaagttgggtaac-3′ and reverse primer 5′-gtgagcggataacaatttcacacaggaaacagc-3′. The PCR program consisted of a 94°C hot start for 5 min; 25 cycles of 94°C for 30 s, 58°C for 30 s, 68°C for 30 s, and a final extension for 5 min at 68°C. 1 μL of the resultant 101bp amplicon was used as a template for a final round of PCR to attach a T7 promoter sequence. PCR was performed with the following sense primer 5′-taatacgactcactataggggagggtttcccagtcacgacgttg-3′ and the reverse primer utilized as described above. The resultant 127-bp amplicon was transcribed using the MegaShortScript *in vitro* transcription kit (Ambion, Austin, Texas), and the resultant RNA product was subsequently reverse transcribed into Cy3-labeled cDNA target using the Cyscribe indirect labeling kit (GE Healthcare, Piscataway, NJ).

8.4.3 Organization of Array Controls

The controls used to measure the amount of probe DNA deposited and to measure signal dynamic range were derived from *B. subtilis*. Seven *B. subtilis* sequences were used. The annotations for these sequences are ybbR|Z99104|CAB11952.1, yabQ|Z99104|CAB11837.1, ycxA|Z99106|CAB12161.1, ybaS|Z99104|CAB11935.1, ybaF|Z99104|CAB11923.1, ybdO|Z99105.1|CAB11999.1, and ybaC |Z99104|CAB11890.

The *B. subtilis* sequences were positioned in a printing plate so that each clone was printed once with each of the 12 printing pins. Each of the clones was diluted at the following concentrations in the printing plate: 200, 20, and 2 ng/μL. For the *ycxA* clone the following dilutions were made: 200, 50, 20, 5, 2, 0.5, and 0.05 ng/μL. The printing plate was used twice during the printing run ensuring twenty-four replicates for every clone at each dilution point.

8.4.4 Target Labeling

8.4.4.1 Genisphere Labeling Method

The Array 350RP™ Kit (Genisphere, Hatfield, PA, USA) was used to label 2 μg of UMRR (Stratagene). The cDNA samples were quantified by optical density at A260 and A280 nm.

8.4.4.2 Incorporation Labeling Methods

In this method, 5 μg of SUA RNA were labeled by reverse transcription and incorporation of amino-allyl-labeled 2′-deoxyuridine 5′-triphosphate (dUTP), using the CyScribe Indirect labeling kit (GE Healthcare, Piscataway, NJ).

After labeling the Cy3- or Cy5-labeled DNA, targets were assayed at A260 nm to measure DNA, at A550 nm to measure Cy3, and at A650 nm to measure Cy5. To determine how much Cyanine dye was incorporated into the fluorescent target, the following equations were used:

Cy5 pmol dye/target = (OD 650 × dilution × volume)/0.25
Cy3 pmol dye/target = (OD 650 × dilution × volume)/0.15

8.4.5 Hybridization, Scanning, and Image Analysis

Four hundred nanograms Cy5-UMRR, cy5-*B. subtilis* spikes (concentrations outlined in detail in Table 8.2) and 10 pmol (250 ng) of Cy3-labeled SUA were mixed in 50% formamide, 8 × standard saline citrate (SSC), 1% sodium dodecyl sulfate (SDS), and 4 × Denhardt's solution. To avoid nonspecific hybridization, 2 μL of Array 350RP dT Blocker (from the Genisphere kit) were added to the hybridization mixture. All hybridizations were performed in 1 × hybridization buffer

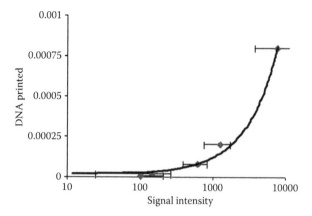

FIGURE 8.4 Calculation of printed array probe utilizing a Cy3-labeled homologous reference labeled target. This plot represents the correlation of known amounts of spotted DNA probe with signal intensity after hybridization with the Cy3-SUA. The average signal intensities of 24 replicates are plotted. The Cy3 signal intensity is plotted on the *x*-axis, and the picomole amount of feature DNA on the *y*-axis. The curve fit was calculated $[y = (1 \times 10^{-7}) x + (2 \times 10^{-5})]$, and was used to determine the picomole amount of every microarray feature.

(GE Healthcare, Piscataway, NJ), 50% formamide, and 50 ng of A80 oligonucleotide. The hybridizations were conducted under a cover slip at 42°C for 12 h as described previously [12]. After hybridization the slides were washed using the following concentrations: $1 \times$ SSC 0.2%SDS for 10 min, $0.1 \times$ SSC 0.2%SDS for 10 min, $0.1 \times$ SSC for 10 min, $0.1 \times$ SSC for 1 min, isopropanol for 1 min, and a 1-min air flush. Slides were scanned using a 4000-A scanner (Axon Instruments Inc., Union City, California) at the photomultiplier tube (PMT) settings of 500 and 600 for Cy3 and Cy5, respectively. After scanning, spot-finding and image quantification were done using Imagene (Biodiscovery, El Segundo, California). The signal intensities for each spot were calculated by summation of the pixel intensities for each spot.

8.4.6 Dye Printing Experiment

A dye printing experiment was performed to measure the correlation between the cyanine dye amount and signal intensity. The Cy5-labeled deoxycytosinetriphosphate (dCTP) was purchased from GE Healthcare (Piscataway, NJ), and a dye dilution series was printed as 24 replicates. The concentration of dye in the source printing plate ranged from 1×10^{-6} to 10 pmol/μl. As the printing pins deposited 0.8 nl per contact, an accurate calculation of the amount of dye deposited per slide was feasible. Postprinting, the slides were scanned on an Axon Scanner at the Cy5 setting at 600 PMT. The average intensity values for each concentration range were determined and a plot of the signal intensity versus the molar dye amount was generated (Figure 8.3).

8.4.7 Calculation of Array Features

The *B. subtilis ycxA* sequences were positioned in a printing plate such that each clone was printed with each of the 12 printing pins. The clones were diluted at the following concentrations (ng/μl) in the source printing plate: 200, 50, 20, 5, 2, 0.5, and 0.05. As the plate was used twice during the printing run, 24 replicate spots were created for every clone at each dilution point. As the printing pins deposited 0.8 nl of material per contact, the picomoles of DNA deposited on the slide could be determined. These values were plotted against the Cy3 signal intensity values following hybridization of the Cy3-labeled SUA (Figure 8.4).

REFERENCES

1. Hardiman G and Carmen A. (2006). DNA biochips—past, present and future: an overview. In: *Biochips as Pathways to Discovery* (A. Carmen and G. Hardiman, eds). Taylor & Francis, New York, pp. 1–13.

2. Brown PO and Botstein D. Exploring the new world of the genome with DNA microarrays. *Nat Genet* 1999; 21: 33–37.

3. Waring JF, Ciurlionis R, Jolly RA, Heindel M, and Ulrich RG. Microarray analysis of hepatotoxins *in vitro* reveals a correlation between gene expression profiles and mechanisms of toxicity. *Toxicol Lett* 2001; 120: 359–368.

4. Hamadeh HK, Amin RP, Paules RS, and Afshari CA. An overview of toxicogenomics. *Curr Issues Mol Biol* 2002; 4(2): 45–56.

5. Johnson JA. Drug target pharmacogenomics: An overview. *Am J Pharmacogenomics* 2001; 1(4): 271–281.

6. Kruglyak L and Nickerson DA. Variation is the spice of life. *Nat Genet* 2001; 27: 234–236.

7. Hardiman G. Microarray platforms—comparisons and contrasts. *Pharmacogenomics* 2004; 5: 487–502.

8. Lipshutz RJ, Fodor SP, Gingeras TR, and Lockhart DJ. High density synthetic oligonucleotide arrays. *Nat Genet* 1999; 21(Suppl. 1): 20–24.

9. Schena M, Shalon D, Davis RW, and Brown PO. Quantitative monitoring of gene expression patterns with a complementary DNA microarray. *Science* 1995; 270(5235): 467–470.

10. Dudley AM, Aach J, Steffen MA, and Church GM. Measuring absolute expression with microarrays with a calibrated reference sample and an extended signal intensity range. *Proc Natl Acad Sci USA* 2002; 99(11): 7554–7559.

11. Yang YH, Dudoit S, Luu P, Lin DM, Peng V, Ngai J, and Speed TP. Normalization for cDNA microarray data: A robust composite method addressing single and multiple slide systematic variation. *Nucleic Acids Res* 2002; 30(4): e15.

12. Rouse RJ, Espinoza CR, Niedner RH, and Hardiman G. Development of a microarray assay that measures hybridization stoichiometry in moles. *Biotechniques* 2004; 36: 464–470.

13. Stears RL, Getts RC, and Gullans SR. A novel, sensitive detection system for high-density microarray using dendrimer technology. *Physiol Genomics* 2000; 3: 93–99.

14. Ramakrishnan R, Dorris D, Lublinsky A, Nguyen A, Domanus M, Prokhorova A, Gieser L, Touma E, Lockner R, Tata M, Zhu X, Patterson M, Shippy R, Sendera TJ, Mazumder A. An assessment of Motorola CodeLink™ microarray performance for gene expression profiling applications. *Nucleic Acids Res* 2002; 30(7): E30.

15. Hegde P, Qi R, Abernathy K, Gay C, Dharap S, Gaspard R, Hughes JE, Snesrud E, Lee N, and Quackenbush J. A concise guide to cDNA microarray analysis. *Biotechniques* 2000; 29(3): 548–550, 552–554, 556 passim.

16. Chen Y, Dougherty ER, and Bittner ML. Ratio-based decisions and the quantitative analysis of cDNA microarray images. *J Biomed Opt* 1997; 2: 364–374.

17. Tseng GC, Oh MK, Rohlin L, Liao JC, and Wong WH. Issues in cDNA microarray analysis: Quality filtering, channel normalization, models of variations and assessment of gene effects. *Nucleic Acids Res* 2001; 29(12): 2549–2557.

18. Goryachev AB, Macgregor PF, and Edwards AM. Unfolding of microarray data. *J Comput Biol* 2001; 8(4): 443–461.

19. Taniguchi M, Miura K, Iwao H, and Yamanaka S. Quantitative assessment of DNA microarrays—comparison with Northern blot analyses. *Genomics* 2001; 71(1): 34–39.

20. Stoeckert C, Pizarro A, Manduchi E, Gibson M, Brunk B, Crabtree J, Schug J, Shen-Orr S, and Overton GC. A relational schema for both array-based and SAGE gene expression experiments. *Bioinformatics* 2001; 17(4): 300–308.

21 Rouse R, Verdun K, and Hardiman G. (2003). DNA microarrays and the core facility. In: *Microarray Methods and Applications*, G. Hardiman, (ed.), vol. 3. DNA Press Inc., Eagleville, PA, pp. 37–66.

9 Data Normalization Selection

Phillip Stafford

CONTENTS

9.1 INTRODUCTION

Deoxyribonucleic acid (DNA) microarray manufacturers have been through a rigorous market compression yielding only a few robust and trusted commercial platforms, each with significant market share. Agilent, Affymetrix, and Illumina remain the most ubiquitous and durable expression arrays, and have continued increasing density and quality. Many comparative analyses have shown the strengths and weaknesses of each platform, and have proposed methods to obtain the best cross-platform comparisons [1–12]. Several recent papers from the Microarray Quality Consortium (MAQC) [13–18] have shown that care in ribonucleic acid (RNA) preparation yields substantial benefits in high correlation across different laboratories, platforms, and sample preparation methods. High-quality RNA remains one of the most important sources of variance, and although some platforms are more susceptible than others, the Achilles' heel of commercial expression platforms is RNA integrity. Little can be done to rescue data from samples derived from degraded RNA, however normalization techniques help control for platform-specific biases that are exacerbated by degraded samples. Unfortunately normalization methods are the second largest source of variance. An improper normalization will alter expression data even more than degraded RNA. Of the many normalization methods for expression data, each has its pros and cons. These can be leveraged in order to obtain the most benefit from a normalization technique based on the type of array, the size of the experiment, and the hypotheses proposed by the experimenter.

Expression arrays continue to be used for classification [19–21] and identification of gene function, drug target prediction [22], transcription factor targeting [23], toxicology evaluations [24], and pathway analysis [25]. Classification has been highly useful in grouping patients who may or may not respond to a certain drug, which patient population would benefit from adjuvant chemotherapy, or which patients should report to their physician for follow-up, more frequently based on their tumor expression profile. Thus, classification has taken the expression array from the laboratory to the clinic, in so doing creating a medical device that requires mathematical manipulation in order to begin treating humans. Precision and accuracy are now not just buzzwords, but are terms relevant to human health. Classification error is largely controlled by the appropriate selection of features that avoid overfitting or bias [26], and are robust for data variability. Good features are not always those that represent the most significant difference between sampled means, or that have good cluster separation. However, there are some characteristics of good features that implicitly define their performance within classification algorithms. To examine this phenomenon, we examined how normalization and platform affect classification error. Normalization methods have an effect on data specificity and sensitivity, as well as accuracy. A perfectly normalized dataset would correct for platform-specific bias, but might not fix a lack of cross-platform correlation. Ideally, we would want two platforms such that a classifier could be trained on one and tested on another with low error. Here I tested data across three healthy human tissues, two commercial platforms, and nine different normalization methods using a simple sequential forward search feature selection. I present the results to propose a normalization method that minimizes error and maximizes features with the highest predictive power.

9.2   MICROARRAY EXPRESSION PLATFORMS

Commercial expression arrays are by far the most developed of the wide assortment of high-throughput molecular profiling devices that measure biomolecules at the whole-genome level. Physically these devices most often consist of a high-quality glass slide or other nonreactive substrate, made suitable for gene transcript hybridization by eliminating surface imperfections, reactive groups, trace oxidants, fluorescent residuals, and microscopic fractures [25,27–29]. Some alternate technologies rely on fluorescent beads [30,31], gel-based separation, electronic detection [32], sequencing-based methods such as serial analysis of gene expression (SAGE) [33–35] and massively parallel signature sequencing (MPSS) [35,36] molecular calipers [37], even a unique complementary metal-oxide-semiconductor (CMOS)-like device that directs microelectrodes to alter local pH and detritylation during on-chip probe synthesis [38]. In this chapter, we will focus on the typical glass-surface oligomer-based expression microarray. The need for precision manufacturing is necessary when the output of a single observation requires a precise count of the actual number of photons from single physical feature—a feature that averages 15–150 mm in diameter, with 50,000 to millions of features per slide [39]. As technology moves the field forward, these arrays will accommodate more features and the corresponding feature size will continue to shrink. The new Affymetrix Exon Array will exceed 5,000,000 features per slide at 5 mm per feature [40]. To obtain meaningful data, this requires highly precise and sensitive optics and chemistry. Slight variations in the distance from the feature to the confocal scanner caused by minor warps or imperfections will drastically change the readings, especially when measuring slight changes in quantity. Auto-fluorescence will change the local and global background, adding random and systematic noise to the overall signal. The surface of the slide is also exposed to myriad reactive chemicals during processing; the coating on the slide surface must not interfere with the hybridization or binding of probe and target yet it must provide sufficiently strong binding to the printed probes to prevent inadvertent removal of the spot from the slide during wash steps. All these factors play a role in the resulting technical variance of the array—the goal is to

create an environment where noise is mitigated, where the device will provide the highest reproducibility and sensitivity when exposed to labeled messenger RNA (mRNA), even at low concentrations. The ideal situation is where biological variance overwhelms all nuisance factors. This is
arguably the current state of microarray expression technology; however, commercial arrays are
still highly dependent on the quality of the starting mRNA.

In 2006, the major expression platforms are manufactured by Agilent (Santa Clara, California)
and Affymetrix (Palo Alto, California); both have made substantial gains in performance, price,
and market share, enabling them to dominate the field. Custom complementary DNA (cDNA) arrays
are still the most prevalent expression technology in academia due to price, increased spotting precision [41,42], and data quality, but are not standardized in any way. Public data repositories have
paralleled the market prevalence, reflecting a near 50:50 split between custom cDNA arrays and the
two major commercial arrays. Owing to this valuable resource of historical data, the impetus for
better normalization never ends. Older data are often supplied with the original Tagged Image File
Format (TIFF) image, so the best normalization algorithm from the date of the experiment may not
be the best, available today.

9.3   EXPRESSION PLATFORM TYPES

Agilent and Affymetrix arrays have always used some sort of mathematical normalization to correct for systematic biases and error [7,9–11,43–46], however it was the prevalence of cDNA arrays
that have forced the development of highly advanced normalization methods that improve precision
in self-spotted long cDNA arrays. These arrays are ubiquitous due to their relatively low cost.
However comparison of cDNA data between labs or across arrays is not an easy task. Even commercial cDNA arrays require sophisticated normalization routines that short oligo arrays do not.
Agilent leveraged their expertise in ink-jet technology to build a robust cDNA array with more than
22,000 features. These arrays were very reproducible but presented Agilent with a normalization
challenge, which they met by adopting two-channel Loess normalization. Within a few years Agilent
introduced a method of applying nucleosides via ink-jet applicators to enable *in situ* synthesis of
60-mer oligonucleotides on the surface of the array. The synthesis is >99.5% complete through 60
rounds of synthesis, yielding ~75% full-length 60-mer probes but a significant proportion much
longer than 50 bp. The physical characteristics of the array are shown in Table 9.1 and represent the
current product line. Agilent offers single- and dual-color modes for their array, but for the two-
channel mode they require two dyes, CY3 and CY5, be used as labels in a competitive hybridization.
For any two-channel arrays, several physical characteristics play a role in causing an intensity-based
deviation from perfect correlation between the channels. The differential dye incorporation and
energy output cause severe biases. Although hybridizing two species simultaneously in one solution
is the most accurate way to measure the relative differences between two samples, the uncorrected
color bias completely eliminates this benefit. Fortunately, scatter plot smoothing, or Loess, a first-
order smoothing using local quadratic regression, or Lowess, a second-order smoothing using local
linear regression, mitigates the effect of differential dye effects by using a locally weighted linear
regression to accommodate this known effect. The cause of dye bias is predictable and therefore
well suited for a regression approach. A first-degree polynomial least squares regression (Loess) or
a second-degree polynomial fit (Lowess) closely approximates the highly correlated nature of the
relationship between fluorescence intensity and the deviation of the color channels from one another.
Often the channels are CY5 and CY3, but other fluorescent dyes are also available. Currently, the
Agilent expression products use Loess and a designer error model within their Feature Extraction
software to produce a robust set of measurements; Agilent also recommends going without background subtraction because of the stringency of the wash steps and the correspondingly low background fluorescence. The software does provide a measure of the raw, mean, median, subtracted,
processed, and error estimates for each calculation, so the normalization effect can be calculated
per spot if desired.

TABLE 9.1
Sample Size, Normalization Methods, Platform, and Tissues Used

Platform	Normalization Methods	Probes (Gene$_i$)	Overlap	Tissues (Sample$_j$)	N$_j$
Agilent Human 1Av2	BSUB (*gBGSubSignal col62 and rBGSubSignal col63 Feature Extraction 8.1*)	18,703	11,504	Liver, lung, spleen	6
Agilent Human 1Av2	MEAN (*gMeanSignal col33 and rProcessed col34 Feature Extraction 8.1*)	18,703	11,504	Liver, lung, spleen	6
Agilent Human 1Av2	PROC (*gProcessed col23, 80 and rProcessed col 24, 81 Feature Extraction 8.1*)	18,703	11,504	Liver, lung, spleen	6
Affymetrix U133Av2	MAS5 (*GCOS 1.2*)	22,215	11,504	Liver, lung, spleen	6
Affymetrix U133Av2	GC-RMA (*GeneSpring 7.2*)	22,215	11,504	Liver, lung, spleen	6
Affymetrix U133Av2	RAW (*Bioconductor Affy package, mean PM*)	22,215	11,504	Liver, lung, spleen	6
Affymetrix U133Av2	PM (*dChip 2006 Perfect Match only model*)	22,215	11,504	Liver, lung, spleen	6
Affymetrix U133Av2	PM–MM (*dChip 2006 Perfect Match–Mismatch difference model*)	22,215	11,504	Liver, lung, spleen	6

Notes: Agilent's MEAN value is the signal intensity per channel plus local and global background. BSUB is MEAN minus local background. Local background is calculated using negative controls, mean local background, and a spatial detrending calculation based on scanner-induced low-frequency multiplicative noise. PROC is background subtracted, spatially detrended, Lowess normalized, and error-modeled data. The error model separates the lower additive components error for low intensity, the multiplicative components for high intensity, and adds the squared results of all error terms plus the error from the simple background subtracted signal. Affy Microarray Suite 5.0 (MAS5) is the mismatch-subtracted data from GCOS. GC-RMA is the GeneChip-modified robust multi-array variance stabilizing method. The dChip PM and PM–MM (perfect match–mismatch) methods are iterative model-based outlier identification and isolation methods that automatically exclude high error data points.

Affymetrix has been the leader in shadow-masking technology. Affymetrix' technology relies on photolithography to create a 25 bases long *in situ*-synthesized probe. Current shadow-masking technology is ~90% efficient at each step, yielding ~7% full-length probe. New technologies have increased this to ~95% efficiency, yielding 27% full-length probe. This synthesis efficiency has led to some precision problems that Affymetrix has accommodated by including mismatch (MM) probes. The MM probe is designed to provide data about ectopic hybridization. Perfect match minus mismatch was an early method of removing background signal from an ideal match. Advancements in normalization have occurred since, and one of the most popular methods actually discards mismatch data [29]. As Affymetrix pushes the envelope in shrinking feature size to enable high-density exon and single nucleotide polymorphism (SNP) arrays, the issue of normalizing data from hybridization to a population of 25-mer and shorter probes becomes vital in order to provide sufficient precision. Physical characteristics and performance metrics are listed in Table 9.1.

The two major sources of expression data tend to fall into a dichotomy of cDNA arrays that make up the majority of expression data both within public databases and in the academic scientific community, and synthetic oligo arrays, made by commercial manufacturers. cDNA arrays were first introduced and remain very economical; as such they will continue to dominate public databases. Commercial expression arrays have fallen into two highly competitive categories: Affymetrix GeneChip® array, a single-color technology using 25-mer probes constructed using photolithography, and Agilent's dual-mode expression arrays that can use a one- or two-channel protocol, depending upon the experimental design.

Agilent's arrays utilize proprietary ink-jet technology that produces *in situ*-synthesized 60-mer probes. The ink-jet technology uses an electronic charge to direct individual phosphoramadite bases onto a prepared glass surface. Each of the 60 total rounds of synthesis is >99% efficient, and adds a single base to the growing chain, resulting in a uniform population of 60-mer oligonucleotides in a spot that is currently 130 mm in diameter; new products will see a 50 mm or smaller probe to enable higher density and content. Each probe's nucleotide content is computer-controlled; consequently, each array is effectively completely customized. Affymetrix markets a single-channel platform using shadow-masking to synthesize 25-mer probes. Each round of synthesis is ~90–95% efficient; incomplete probes are physically capped, preserving the correct probe sequence but yielding a mixture of full- and partial-length probes. Affymetrix relies on mismatch probes to accommodate ectopic hybridization and other noise. The debate continues whether 25 bases provide sufficient thermodynamic stability to prevent mishybridization, and conversely, whether 60 bases allow sufficient selectivity to distinguish highly related genes such as those in the cytochrome P450 family. Both technologies utilize fixed probe lengths, which create a distribution of nonoptimal hybridization temperatures because the T_m of the probes is rarely matched to the actual temperature of hybridization. Only NimbleGen creates an isothermal probe set by modifying the total probe length [47]. Agilent and Affymetrix accommodate the nonoptimal hybridization conditions by using mismatch probes, replication, positive and negative control probes, and proprietary normalization techniques to measure and adjust for thermodynamic biases. These techniques are quite mature and have proved sufficient to the task of exploring the dynamic nature of the transcriptome in a variety of biological contexts. Precision is so good for currently available arrays that reasonably accurate predict of cancer outcome, recurrence, and drug resistance is commonplace and even commercially available [21,48–50]. The two major technologies have leveraged their vastly different technologies to improve the noise reduction methods to an extent that technical imprecision is sufficiently low to ensure highly precise and sensitive measurement in the context of the same array type. Issues are still present that prevent a perfect concordance between measurements of the same gene from two or more commercial expression platforms.

9.4 SEVERAL ALTERNATIVE GLASS-SUBSTRATE EXPRESSION TECHNOLOGIES

Several technologies that differ in technology from the industry-standard Affymetrix and Agilent arrays are attracting renewed attention. NimbleGen's (Madison, Wisconsin) maskless array synthesis (MAS) technology utilizes a DLP-like digital micromirror device developed by Texas Instruments at a resolution of 1024×768 (~15 mM feature size). The light projection enables much tighter control over the *in situ* probe synthesis, and is not subject to the problem of repeatedly manufacturing a disposable shadow mask that is subject to photonic degradation [21]. NimbleGen's technology allows probe lengths up to 70 bases, allowing for varying length probes and an isothermal hybridization if desired (Figure 9.1).

Combimatrix (Seattle, Washington) utilizes CMOS-like technology to create a dense surface of microelectrodes that are independently addressable. Each electrode directs the chemical synthesis of biomolecules in a contained microenvironment. The customization of this design makes it possible to create a catalog of timely focused arrays such as the cancer genome, or inflammation or stress. Combimatrix also makes the array manufacturing equipment available as a desktop device for researchers who require proprietary sequences to be generated.

MPSS, from Lynx Technologies, used a large number of identifiable tabs and fluorescent beads coupled with fluorescently activated cell sorting (FACS) to quantitatively identify all available mRNA species. This technique was remarkably precise and accurate; nearly all misidentifications are caused by problems in the RNA sample preparation.

Illumina (San Diego, California) leads the way in bead-based molecular profiling. They offer their flexible BeadChip devices, which profile not only mRNA but also SNPs. The gene expression bead array uses presynthesized 50-mer oligo probes that are chemically bonded to the bead and

FIGURE 9.1 Images of several different expression array technologies. Top left to right: NimbleGen's DLP digital micromirror device uses MAS; Combimatrix technology directs microelectrodes to alter chemistry during probe synthesis at individually addressable locations; Illumina's BeadArray utilizes probes bound to Dynal beads yielding high precision through high replication and probe quality. Bottom left to right: cDNA pin spotters can attach many different types of chemical compounds from nucleic acids to proteins, peptides, siRNA probes, microRNA probes, and antibodies; Agilent arrays are available in one- and two-color modes; and the Affymetrix GeneChip is available in a variety of densities.

affixed to the substrate with preetched wells at a 3-mM feature size. Within a relatively small size, it offers more than 30× redundancy providing high confidence and a good estimate of precision. The BeadChip profiling devices contain 6–8 sets of 23–46K Refseq genes, but only requires 50–100 ng total RNA. Owing to the small feature size, a special scanner must be used to scan and extract submicron features.

9.5   RNA DEGRADATION NORMALIZATION

These alternate technologies often benefit from the mistakes and knowledge gained during the decade of expression array development. The precision of the Illumina platform is known industry-wide as a gold standard, although more expensive than competing technologies. For newer technologies including bead-based and MPSS, the raw data are very close to the actual molecule count, and needs very little normalization. Little technical bias is created with these technologies, so normalization methods are very minimal, yielding a dataset that is reliable, accurate, and essentially the ground truth. However, a different type of normalization is often done with these technologies: probe modeling in these cases is done on the amount of RNA degradation. RNA degradation can happen in many ways, but physical and enzymatic degradation are the most important methods that affect expression profiling. Using a highly precise expression technology means that the effect of degradation can be measured and modeled, so intact, slightly degraded, and highly cross-linked formalin-fixed paraffin-embedded (FFPE) samples have the potential to provide highly reproducible, accurate RNA profiles. The 5′ degradation can be determined by sequencing or tiled probes and a repeatable labeling method, such as 3′ oligo-dT-based labeling. Differences in the amount of intact transcript would tend to drop precisely along the length of the target. Spikes in abundance can be attributed to the type of degradation that is the most prominent. This is only possible using precise and absolute quantitative expression profiling technologies such as MPSS or bead-based multi-probe detection. The type of normalization that is likely to prevail in the future will be concerned with self-measurements of the level and type of RNA degradation.

9.6   NORMALIZATION METHODS: HISTORY

Benefiting from the past 10 years of research efforts, a set of rigorous measurement, quality control, and normalization steps have been developed for expression profiling. Numerous methods account for and correct the intrinsic uncertainty or technical biases in the signals that make up expression array data. Much of the credit must be given to those researchers who constantly battled to improve the quality of self-spotted cDNA data during the time when high imprecision was routine. Variance continues to stem from sources that range from differences in dye incorporation, differential energy output, scanner variation, various types of background fluorescence from different array surfaces and substrates, probe design, and increasingly important, ectopic/failed hybridization, differential mRNA degradation, RNA integrity, and biological variability. For both cDNA and oligo-based expression arrays, quality control is a major issue, although the details differ. Normalization will correct for technical biases, but in some cases, the adjustment is so extreme that the normalized data bear little resemblance to raw data, and interpretation of the results relies on the quality of the normalization technique [7,9–11,43–45,51–55] peculiarities and performance issues of each array platform. Ratio measurements are commonly used in two-color cDNA arrays [56] as well as the Agilent expression system. Ratio measurements have been shown to be very precise and accurate because of the kinetics of competitive hybridization. Measuring two mRNA species at the same time in the same hybridization solution reduces inter-array variability and provides a robust comparison of treatment to control. Expression intensities from single-color data provide a semi-direct assay of transcript abundance, which is common practice for experimental designs that look at hundreds or thousands of conditions. Even the traditional Agilent two-channel oligo array, which has benefited from the increased accuracy that results from competitive hybridization, now offers protocols for both to accommodate the new paradigm of highly parallel large-scale experiments.

During the development of an expression platform containing the whole human transcriptome, both Affymetrix and Agilent have fallen prey to a situation where certain probes are precise when measuring a defined set of tissues, but highly unpredictable when exposed to a different set of samples. Routinely, probes from otherwise highly precise cDNA and oligo arrays may show unexpected variance when measuring some tissues but not others. To date, this behavior is highly anecdotal, but is not unexpected. Both Affymetrix' U133A and Agilent's Human 1A are on revision two; major revisions to these companies' expression products are rare but are often the result of an accumulation of probes that warrant replacement based on supporting evidence that they do not identify either a real gene or a gene's mRNA abundance correctly. The reasons for the unusual thermodynamic behavior of these problematic probes are rarely discussed, principally because there is rarely an identifiable cause. However, given sufficient data drawn from a panel of normal healthy tissues, one could examine the discrepancies between probes and tissues on a probe-by-probe basis and make (albeit tenuous) connections between probe performance and chemical composition.

A compelling method to correcting for expression bias and probe variability would be the construction of a full library of human transcripts. Given this resource, one can spike a series of dilutions into a series of complex mixtures of mRNA. This method, which is being developed by the External RNA Advisory Committee, would allow one to measure the endpoint sensitivity of everyone of 40,000+ probes, and would help identify the factors that might cause an increase in imprecision.

Several different combinations of mRNA can serve as the background into which dilutions are spiked. Varied sources such as diseased and health tissues, developmental tissue, highly specific or multi-purpose tissue, cell lines, pooled samples, and so on generate experimentally useful biological milieus where a probe can lose some of its precision due to competition and solution complexity. In turn, these experimental data may help generate a probe–response curve for individual probes. Non-linearity in probe–response can be modeled precisely, rather than applying a global error model to the entire set of probes. Model-based probe-by-probe normalization would be the most efficient way to compensate for variant probes, and would reward the expression platform that is the most precise. In this scenario, the most precise expression array device would become the most accurate.

In a more realistic scenario, genomic DNA can be used as a reference to measure the fluorescence that results from the binding of a known concentration of target—in this case the known concentration of target would be from a single pair of molecules, the genomic DNA for a particular gene. This has in fact been attempted and works moderately well for cDNA arrays [57]. However, several problems arise: the complexity of the genome is several orders of magnitude greater than the complexity of the transcriptome; 25-mers or even 60-mers designed to bind mRNA sequences, may lack the specificity to uniquely bind single-copy genomic DNA without high imprecision.

9.7 NORMALIZATION METHODS

9.7.1 MAS5

Affymetrix uses a short 25-mer probe designed for maximum specificity. However, short probes are prone to mishybridization; in an attempt to rectify this Affymetrix added a centered mismatch probe that in theory accounts for ectopic hybridization. However, typically 20–40% of mismatch probes per array show a signal higher than the matched PM probes, leading to the development of the ideal match (IM). The IM calculation now used in MAS5 no longer yields negative values, and has been shown to provide higher accuracy in spike-in studies. MAS5 normalization with its present, marginal, and absent flags is the most commonly reported data normalization in public databases. MAS5 does suffer anecdotally from a relatively high degree of false positives and there is a noticeable exaggeration, or expansion, of ratios and a concomitant compression of signal which makes it difficult to estimate true copy number.

9.7.2 Robust Multi-Array Averaging and GeneChip-RMA

Robust multi-array averaging (RMA) is a variance stabilization method [46,54] that normalizes data toward a common distribution under the assumption that very few genes are differentially expressed across two or more conditions. As we will see, this method works best for large numbers of experiments with small changes between conditions. The dataset examined in this chapter tends to put RMA back on its heels a bit since the tissues used are highly differential at the transcriptome, causing some problems for RMA. This method tends toward false negatives, again anecdotally, but has among the highest precision of any of the most popular GeneChip normalizations. RMA differs from GC-RMA in that the latter utilizes the sequence information of the MM probes to calculate probe affinity values during background correction. In this way, although a little general, at least some of the thermodynamic qualities of the probe can be incorporated into the normalization.

9.7.3 dChip

Li and Wang developed a model-based approach for estimating unusual but predictable behavior in Affymetrix probe set data. This method greatly reduces the effect of outlier probes, and the model for the most part predicts the majority of the variance from individual probe sets. The model uses an "explained energy" term that predicts deviance from an expected norm, and much like the GC modification to RMA, is probe-specific. This method can use or ignore mismatch data, but does a much better job in predicting expression values from probes that are close to background, probes that would be called "absent" in the MAS5 system. dChip also automatically identifies and corrects for slide imperfections, image artifacts, and a general class of probe outliers. dChip has been shown to be quite robust and compares favorable to the other common Affy normalization methods [9–11,52,58,59].

9.7.4 PLIER

Probe logarithmic intensity error (PLIER) is a relatively new method developed at Affymetrix that again uses background correction but in a less damaging way than MAS5. PLIER assumes that the variance of PM probes is substantially better than MM probes, which yields estimates of error that are far below other methods. However, this makes the data appear much more precise than it actually is, and PLIER tends to make other nonbiological assumptions that create a situation that is not very useful to biologists, regardless of the apparent precision [11].

9.7.5 QUANTILE NORMALIZATION

Quantile normalization was introduced by Bolstad et al. [46] and specifically targeted for early two-channel cDNA arrays; this method assumes that the distribution of gene abundance is almost identical for all samples and takes the pooled distribution of probes on all arrays (the so-called target distribution). The algorithm subsequently normalizes each array by transforming the original value to that of the target distribution, quantile-by-quantile. The target distribution is obtained by averaging the sorted expression values across all arrays, and reassigning a transformed value to each array, then resorting back to the original order. This method has been shown to work best for samples from similar tissues or conditions with little large-scale transcriptional changes. As with RMA, this method suffers from false negatives and false positives that are caused by inappropriately low intra-sample variance and compressed differentially expressed ratios.

9.7.6 LOESS AND CYCLIC LOESS NORMALIZATION

The most common normalization for two-channel cDNA arrays is Loess in which a piece-wise polynomial regression is performed. For two-channel array data, Loess records the changes needed to achieve a convergence of center-weighted means followed by a calculation of the distance of each point to the best-fit curve through the data. The distance is reduced iteratively until no further changes in data will affect the magnitude of the residuals. Loess is highly dependent on the span of points that will be used for calculating each local estimate, and selecting a span is not automatic, but must be carefully chosen to prevent overcorrection. Cyclic Loess is based on the concept of the M/A plot, where M is the difference in log expression and A the average of log expression values. Although often used for two-channel cDNA arrays, this method can also be used for pairs of single-channel data, or by using all possible pairwise combinations, for multiple arrays at once. With each iteration, the curve from the M/A plot is redrawn, and adjustments are made to flatten this curve. Overfitting with this method simply results in pairs of arrays being highly similar while still allowing extreme outliers to remain far from the majority of the data. Fastlo by Therneau and colleagues [59] is one of the fastest and most controllable of the iterative smoothing algorithms.

9.7.7 OTHERS

Although dChip has been shown to possess qualities that tend to balance false-positive and false-negative errors when making estimates for the amount of correction necessary to account for systematic biases in probe-level expression values, like RMA it requires several arrays to perform well. The position-dependent nearest-neighbor has been shown to have similar performance to dChip PM but achieves this level of performance by doing a chip-by-chip normalization. In this case, Zhang et al. use the tendency of probe to have similar sequence or to physically overlap the same sequence when probing the same gene [50].

9.8   CAUSES AND CONSEQUENCES OF IMPRECISE MEASUREMENTS

Both obvious and subtle problems arise because of improper microarray normalization, beyond the simple lack of precision or correlation to raw values. Each individual spot represents a probe designed to detect some minimum number of mRNA molecules, the minimum detectable fold change. There are physical constraints that are imposed on the probe design process; sufficient sensitivity to detect the dynamic range of mRNA molecules in a cell and adequate selectivity to distinguish related or spliced versions of the same gene or gene family. The dynamic range within a cell may be quite large, from <1 mRNA to thousands per cell. Typical dynamic range for commercial arrays must therefore exceed three orders of magnitude minimally; most commercial arrays meet or exceed this but the problems lie in the nonlinearity of probe response to mRNA copy number and intensity-dependent bias. In addition, a small percentage of probes on an array will randomly fail or perform below specifications. Some probes, however, will fail for complex thermodynamic reasons. Even Universal RNA [60], a broad collection of pooled RNA from cell lines, may not contain the mRNA concentration that can identify poorly behaving probes—one might only see that unique expression profile in cancer or other disease. Normalization methods usually do not help in situations like this—no amount of coercing data into a hypothetical normal data distribution would work for nonresponsive probes, however the knowledge we obtain from cross-platform and real time polymerase chain reaction (RT-PCR) experiments brings us closer to creating the ideal probe for each gene in the human genome. Other failure modes include mishybridization, excessive competition for a target, crossing splice sites, or missing splice variants. In these cases, a biological analysis of the resulting data would lead the investigator to an incorrect conclusion due to a false negative or a false positive, and in a clinical setting, this might mean a faulty diagnosis or drug treatment, which can be immediately damaging to a patient. Classification of disease phenotypes requires precise and accurate observations in order to minimize classification error. Precision is necessary to ensure low error across both training and validation datasets. Accuracy is important when the classifier is tested on data generated outside of the training set, such as an alternate expression platform.

9.9   CROSS-PLATFORM COMPARISONS: EXPERIMENTAL DESIGN

To fully understand the complex interactions among expression technology, normalization methods, and the results of expression analysis in a biological context, I feel that an actual controlled biological example would speak volumes. I present data from pooled commercial RNA extracted from several healthy human tissues (Table 9.1). Many publications have presented cross-platform or cross-normalization comparisons [1,8,13,14,42,50,61–69]. Several approaches have been used; comparisons of healthy tissues or cell lines or diseased versus healthy paradigms have been the predominant categories. I favor experiments that utilize healthy tissues so that one can examine the cause-and-effect relationship between normalization, array, tissue type, and biological interpretation without having to deal with the underlying alteration in the cell's regulatory mechanisms that results from, or is caused by, the diseased phenotype. These complex interactions are worth of study without any additional nuisance factors. Most importantly, specialized tissues have such carefully regulated processes that the expression profile tends to be very stable, especially when examining pooled samples.

Although the Affymetrix and Agilent platforms are free from the most egregious technical errors, there are issues in the design and implementation of their full complement of probes that are still somehow responsible for discordance between the platforms, between normalization methods, and between array and quantitative real time polymerase chain reaction (qRT-PCR) data. It is well known that an array company can produce excellent correlation between their own and qRT-PCR data simply by using the same probe sequences for both—this trick is been well documented, although precious little definitive work has been published on the effect of probe position, nucleotide content, kinetics, and thermodynamics. I have identified several human tissues (Table 9.1) that show diverse expression patterns and the effect of normalization on the differential expression results.

Both Agilent and Affymetrix expression arrays were run using commercially available (Stratagene, La Jolla, California) human spleen, liver, and lung RNA using the Agilent Human 1Av2 and the Affymetrix U133Av2 arrays. Although Agilent data are usually presented as ratios, I extracted the CY5 and CY3 channels separately in order to compare with single-channel Affymetrix data, which may tend to underestimate the Agilent precision, but have made the comparisons as simple as possible. The Agilent data are examined in the context of three normalizations: MEAN, PROCESSED, and BSUB corresponding to columns X, Y, and Z from Agilent's Feature Extraction software. Affymetrix data are examined in the context of MAS5, GC-RMA, and RAW signals (plus some sets of dChipPM and dChipPM–MM) generated from the .CHP and .CEL files output from GCOS. RAW data were generated by averaging all PM probes per probe set; GC-RMA, dChip, and MAS5 data were generated using the .CEL files and their respective software or algorithms. Figure 9.1 shows the diversity of comparisons we can make. The scatter plots of single-channel and ratio data are plotted in order to convey a feel for reproducibility for both platforms. To make impartial comparisons, I generated ratios from the Affymetrix data, and extracted each of the CY5 and CY3 channels independently from Agilent. Samples were run at two different laboratories for both the Agilent and Affymetrix data in order to introduce a background of moderate technical variation, the impact of which is encompassed in the complete analysis. For the Agilent experimental design, tissues were assigned CY5 or CY3 labels randomly. The associated dye-flips were then selected to complete the design. This experimental design allows single-channel analysis of the effect of channel crossover on reproducibility, since some of the same tissues had different tissues on the opposite channel.

9.10 STATISTICAL MEASURE OF MICROARRAY VARIABILITY

With a wide array of platforms available, the diversity of detection accuracy is as great as the number of platforms. Microarray data can suffer from a variety of potential problems, some specific to the technology on which they are based and some attributable to the experimental design [2,15,70,71]. Possible sources of inaccuracy include high false positives (results show genes are expressed when they are not), high false negatives (results show genes are not expressed when they are), poor sensitivity (inability to detect rare transcripts), high background (low signal-to-noise), or poor spot-to-spot and array-to-array reproducibility. Each of these sources of variation is measurable and can be accommodated statistically in a formal experimental design. To understand how microarray variance is measured, a brief overview of statistical measurements is required.

Statistics provide a description of a population of data and the variance inherent in that population. In microarray technology, variance derives from repeated measurements of the relative level of mRNA transcripts in a biological sample. Two terms can be used to define the source of variance: technical variance describes the difference between repeated measurements that can be attributed to the performance of the measuring system. This can be thought of as var_m, or variance due to the imperfections in the detection device, and is integral to the platform. Other sources of technical variance that enter the system subsequent to the manufacture of the array include sample preparation, RNA storage or extraction techniques, differences in laboratory location or environment, or even sample handling by different laboratory technicians. This variance can be defined as var_t, or variance due to technical factors physically separate from the array. Biological variance, on the contrary, is due to the heterogeneity of the transcriptional response of the organism being studied, or var_b. It is incumbent upon the experimenter to determine the variance of the biological system a priori. Although each individual component of the total variance in a microarray experiment can be identified and measured, it is often more expedient to measure all technical variance in one trial (var_t) and all biological variance in another trial (var_b), thereby obtaining a clear and distinct delineation of the factors that need to be accounted for in the eventual design of the experiment. Any array manufacturer should make var_t values available to the end user, thus total variance would be accounted for by the terms: $var_{tot} = var_b + var_t + var_m$.

Note that technical variance is often well below the biological variance seen in many biological systems. Bakay et al. [72] found that the greatest source of variance in microarray analysis of human muscle biopsies was the physical location within the biopsy where cells were extracted. RNA extraction, cRNA generation, and other aspects of technical variance inherent in the construction of the microarray platform were all swamped by the biological variability. As a confirmation, results of a power analysis with CodeLink microarrays showed a 1.28-fold slide-to-slide detection limit using the same cRNA; two subsequent cRNA preps of the same total RNA resulted in a 1.38-fold slide-to-slide detection limit (data not shown) suggesting that the cRNA prep inherently adds roughly 0.10-fold of uncertainty to the detection limit. Other experiments with data from the CodeLink platform indicate that the difference between total RNA and poly-A RNA adds 0.5-fold, and the difference between starting with 200 ng or 5 µg of total RNA adds 0.9-fold, but an analysis of RNA from two growth forms of the pathogenic yeast *Candida albicans* run on a separate array platform indicates that the biological variance alone can add three- to fourfold worth of "noise" making the eventual selection of differentially expressed genes very challenging.

9.11 DISCUSSION

Since classification has become a much-used method in clinical disease prognosis and diagnosis [50,51,68,73], it is important to understand the sources of variability in gene expression data, both biological and technical, that can cause misclassification. Microarray normalization methods, especially Loess and model-based, often cause large nonlinear changes that make the data difficult to model and sometimes difficult to correlate to qRT-PCR. In general, ratios tend to expand, false positives are higher, and total intensity is lower when comparing expression data to RT-PCR data. The normalization methods sometimes greatly affect the outcome and interpretation of array analysis, especially when looking for features that work independently of the platform. Techniques such as feature selection and classification are highly susceptible to these changes and biases, often resulting in nonoverlapping sets of genes being selected as the best classifiers without a robust way of accounting for cross-platform and cross-normalization differences, or incorporating prior knowledge as selection criteria. In a parallel study, we examined every probe that showed the best and worst correlation across the different platforms and normalizations. As expected, the most highly correlated probes either overlap or are spaced near one another along the gene [8]. Because feature selection is so highly empirical, and rarely uses biological priors, this tends to result in features that are best for a specific platform (training data) but do not work as well for another platform (test data). Without adding additional criteria, such as robustness to platform variance, biological context, physical probe location, cluster quality, and others, feature selection methods will continue to generate predictors that are susceptible to the fragile nature of expression data where probe hybridization biases are still mostly mysterious. No normalization method is able to independently model the response of single probes because we do not have a complete transcriptome of spike-in controls. Without being able to model the response of each probe on an array, we are left with global normalizations that are susceptible to complex effects, such as probe thermodynamics and secondary structure. GC-RMA takes into account the difference in $G ::: C$ and $A :: T$ hydrogen bonds. This small bit of thermodynamic information at least increases the concordance between GC-RMA normalized data from 25-mer probes and Agilent's 60-mers.

Without array normalization, accurate data interpretation is almost impossible. I examined the GO and pathway analysis results from the Affymetrix RAW dataset using both MetaCore from

GeneGo (St. Joseph, Michigan) and IPA from Ingenuity Pathway Analysis (Santa Clara, California). The RAW data produced results that deviated substantially from both MAS5 and GC-RMA, and from the Agilent results. Pathways tended to match well when using GC-RMA and the three Agilent normalizations, suggesting that at least some network analyses can ignore some missing or incorrect genes within a gene list as long as the majority of those genes are part of a unique physiological pathway. Too many incorrect genes will cause pathway analysis to fail to detect the correct metabolic pathways, and results would be highly misleading (Table 9.2). The process of normalization is clearly necessary, but judging which method is best for highest accuracy, precision, and lowest classification error takes some prior knowledge of the comparative performance, precision, and accuracy. I propose that using probe position information, biologically relevant genes, and probes that have been previously shown to correlate well with RT-PCR and across several platforms should produce better tie-break algorithms in feature selection and more robust classifications across current and newly developed expression platforms.

A direct correlation does exist between normalization of microarray expression products and precision of the resulting data. I have discussed expression technology in both short and long oligo technologies, two-color and single-color, and have shown the biological interpretations that can be made from simple differential gene lists. In so doing, I wish to emphasize several generalizations:

1. cDNA arrays have made significant progress in the last few years in terms of accuracy and precision. Most new pin spotters are capable of producing a commercial-grade expression array that can easily cover the annotated human genome on one slide. Precision of cDNA arrays can be exceptional depending on the quality of the starting genomic library.
2. Oligo expression products are available in a wide assortment of technologies, from NimbleGen, Combimatrix, ABI, Illumina, Luminex, Clontech, Solexa/Lynxgen, and many others. Some technologies use fluorescent microbeads, some use microsequencing, gel filtration, microfluidics, or e-detection.
3. Commercial expression arrays will differ in expression values on a gene-by-gene basis, but the biological pathways they identify will be very similar.

Generalization about Affymetrix normalization methods is much more important than Agilent or Illumina normalizations, due to the large number of approaches each normalization method takes. Some ignore mismatch data, some include it, some normalize variance, and others include slide-by-slide normalizations. Each method has strengths and weaknesses, and each has a drastic effect on data (Table 9.3, Figures 9.2 through 9.5). In general, MAS5 is quite accurate at the cost of high false-positives. Given enough replicates, MAS5 is recommended when looking at historical samples, as this is a slide-by-slide method. MAS5 has been the standard at GEO and other databases when the original .CEL files are unavailable. dChip PM–MM is another background subtraction method that is very accurate but has a moderate amount of false positives (less than MAS5). Classification error rates are typically low with dChip PM–MM, and pathway analysis usually matches other platforms when other Affymetrix normalization methods do not. The dChip PM-only approach has many more biases imposed by the cross-hybridization seen on Affymetrix arrays. However the precision of dChip PM-only is second only to GC-RMA, and considering this method does not depend as highly on the other samples in an experiment, is less biased by high differential expression across samples. GC-RMA is best for experiments with many slides that have subtle changes between samples. Large changes such as tissue-to-tissue are hard for GC-RMA and other variance stabilizing methods to cope with, since these methods assume a small change from one sample to another. The data shown in this chapter are not entirely fair for RMA or GC-RMA due to the large number of replicates and drastic expression changes across the few conditions (only three tissues). For a sample like the Expression Oncology set (www.intgen.org), a set of 1200+ arrays of different tumor samples with high heterogeneity and high sample number, GC-RMA would be recommended. However, it should be noted that this size experiment takes over 64G of RAM to

TABLE 9.2

GeneMapp, BioCarta, and KeGG Metabolic Pathways that Resulted from a List of 100 Most Significant Genes per Tissue Ratio and per Normalization

Dataset	Database	Liver : Spleen	Liver : Lung	Spleen : Lung
Agilent BSUB	BioCarta	Intrinsic prothrombin activation	Intrinsic prothrombin activation	NFAT and hypertrophy
	GenMapp	Blood clotting cascade	Blood clotting cascade	Inflammatory response
	KeGG	Complement and coagulation	Complement and coagulation	Cytokine cytokine receptor
Agilent MEAN	BioCarta	Complement pathway	Intrinsic prothrombin activation	Nuclear receptors in lipid metabolism and toxicity
	GenMapp	Ribosomal proteins	Blood clotting cascade	GPCRDB rhodopsin-like
	KeGG	Complement and coagulation	Complement and coagulation	Cell communication
Agilent PROCESSED	BioCarta	Fibrinolysis	Complement pathway	NFAT and hypertrophy
	GenMapp	Blood clotting	Complement activation classical	Inflammatory response
	KeGG	Complement and coagulation cascade	Complement and coagulation cascade	Cytokine cytokine receptor
Affymetrix MAS5	BioCarta	Intrinsic prothrombin pathway	Intrinsic prothrombin pathway	Oxidative stress-induced gene expression
	GenMapp	Ironotecan pathway	Ironotecan pathway	Inflammation response
	KeGG	Complement and coagulation cascade	Complement and coagulation cascade	Cell communication
Affymetrix GC-RMA	BioCarta	Intrinsic prothrombin activation	T helper cell surface molecules	Role of Src kinases in GPCR signaling
	GenMapp	Irinbotecan pathway	GPCRDB rhodopsin like	GPCRDB class A rhodopsin-like
	KeGG	Complement and coagulation cascade	Neuroactive ligand receptor interaction	Cytokine–cytokine receptor interaction
Affymetrix RAW	BioCarta	TSP1-induced apoptosis	Toll-like receptor pathway	Regulation of splicing
	GenMapp	Smooth muscle contraction	Apoptosis	Smooth muscle contraction
	KeGG	MAPK signaling	MAPK signaling	MAPK signaling
Affymetrix PM–MM	BioCarta	Intrinsic prothrombin activation pathway	Intrinsic prothrombin activation pathway	B lymphocyte surface molecules
	GenMapp	Blood clotting cascade	Blood clotting cascade	GPCRDB class A rhodopsin-like
	KeGG	Complement and coagulation cascade	Complement and coagulation cascade	Cell communication
Affymetrix PM	BioCarta	METS effect on macrophage differentiation	Fc epsilon receptor I signaling in mast cells	T-cell receptor signaling pathway
	GenMapp	Apoptosis	GPCRDB class A rhodopsin-like	GPCRDB class A rhodopsin-like
	KeGG	Cell cycle	Leukocyte transendothelial migration	Insulin signaling pathway

NFAT, nuclear factor of activated T-cells.

Note: Each list was used to test for the most likely gene regulatory pathway using www.biorag.org.

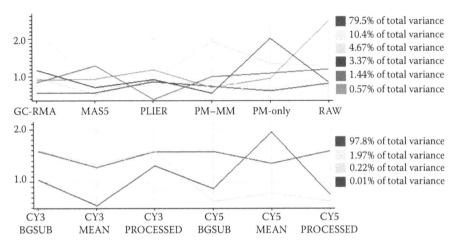

FIGURE 9.2   (See color insert following page 138.) Principal components analysis of 2215 probes for Affymetrix and 18,513 probes for Agilent. X-axis is the mean for each tissue and normalization, Y-axis is the underweighted dot products for each condition's expression profile. Top panel (Affymetrix) shows the six most probably components of variance. The fourth component shown in blue accounts for ~3% of the total variance and indicates that GC-RMA and dChip PM are the furthest away from the other normalizations, as borne out by analysis of variance (ANOVA) overlap and clustering. Bottom panel (Agilent) shows that the second and fourth components in yellow and blue, respectively, account for the differences between raw (MEAN) and processed signal for both channels, and only amount to ~2% of the total variance. The cyan line is the variance component caused by the difference between channels, and is remarkably low.

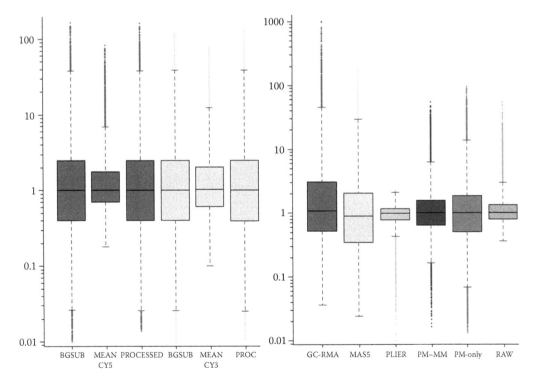

FIGURE 9.3   (See color insert following page 138.) Boxplot showing the range of values for all samples per normalization. Left panel (Agilent) shows the compression and right-shift skew of near-raw data. Normalization creates a Gaussian distribution that is significantly different from raw data. Left panel (Affymetrix) shows the compression of PLIER (left-skewed) and RAW (right-skewed) data. Only dChip methods have a near-Gaussian distribution.

TABLE 9.3
Sensitivity Results

Dataset	Average $\Delta_{\iota\kappa}$	Average MDFC (95th Percentile Ratio)	Median MDFC (95th Percentile Ratio)	N_j
Agilent BSUB	1.13 ± 0.03	1.37 ± 0.08	1.34	3
Agilent MEAN	1.14 ± 0.08	1.30 ± 0.07	1.15	3
Agilent PROCESSED	1.28 ± 0.07	1.61 ± 0.13	1.37	3
Affymetrix MAS5	1.99 ± 0.69	2.38 ± 0.52	2.16	3
Affymetrix GC-RMA	1.32 ± 0.21	1.31 ± 0.26	1.43	3
Affymetrix RAW	1.56 ± 0.21	1.58 ± 0.14	1.19	3
Affymetrix PM	1.85 ± 0.19	2.3 ± 0.14	2.16	3
Affymetrix PM–MM	1.65 ± 0.11	2.01 ± 0.25	1.99	3

Notes: Delta is the minimum detectable difference at $\alpha = 0.05$, $\beta = 0.20$, $N = 3$, in fold-change units. Delta was averaged per probe, per case, and per tissue with the standard deviation shown. The minimum detectable fold-change (MDFC) is the ratio of two technical replicates at the 95th percentile probe. The average was taken across all probes, all tissues, and all possible technical replicates. The median MDFC was the middle value across all possible cases k.

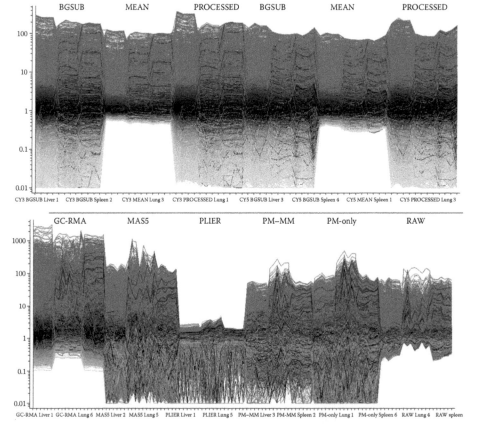

FIGURE 9.4 (See color insert following page 138.) Visualization of signal dynamic range per tissue and normalization. *X*-axis is sample, right panel is $\log_{10}$ signal. Top panel (Agilent) shows the compression of low-signal strength probes that are expanded by the normalization methods. Bottom panel (Affymetrix) again emphasizes the left compression of PLIER, the general compression of RAW signals, and the right skew of GC-RMA.

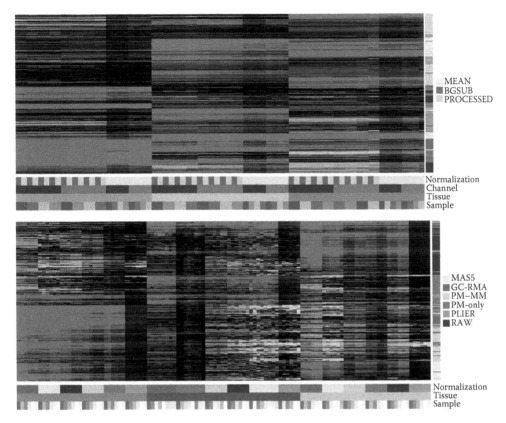

FIGURE 9.5 **(See color insert following page 138.)** Clustering of 1000 ANOVA-selected probes for Agilent (top) and Affymetrix (bottom). Clustering is not tissue-weighted because the Model 1 ANOVA selected only those probes that differed from the total sums of squares per probe. As seen in the top panel the difference between the tissues dominates, and PROCESSED and BSUB normalizations are virtually indistinguishable. Channel has a slight effect in the intensity scale, but little in the clustering results. Bottom panel shows the different alignments to which each normalization approach conforms. Occasionally MAS5 and dChip PM–MM cluster, showing the differential effect of background-subtraction methods versus variance stabilization methods. Note also that groupings are not consistent per tissue.

process in R for RMA, GC-RMA, and MAS5. PLIER takes even more memory, and is not recommended due to the assumption that mismatch data should be less than perfect match data. Several other biologically unsound principles make up PLIER, and it has a tendency to compress data somewhat. PLIER tends to generate many negative values, and adds a left skew to most array data. dChip PM–MM does a much better job of using mismatch data than PLIER, and is recommended for small experiments with moderate noise or marginal RNA quality, such as from laser-capture microdissected tissues or RNA from plants. dChip PM is better than PLIER for low-to-moderate slide number experiments with moderate-to-high quality. dChip PM is a good model-based system but suffers from the bias and inaccuracies that occur when increasing levels of cross-hybridization are not accommodated.

I focused on the two main expression products, the Agilent Human 1A and the Affymetrix U133A. Between these two arrays Affymetrix tends to have a lower inherent precision per probe than Agilent, no matter the normalization method. However, if the single Agilent probe location for a particular gene is in a poor location, Affymetrix has the advantage of using multiple probes spanning more regions in the gene. Most of these problematic genes have been identified over the years, and the large majority of Agilent probes are well-placed. Considering that the Agilent arrays used the Insight Gold sequence database and Affymetrix used public NCBI data, many more probes on the early Agilent arrays were placed in a better position to detect transcript abundance than the corresponding Affymetrix probe. Although there has not been a truly systematic approach to probe-by-probe validation at the whole-genome scale, anecdotal evidence suggests that both platforms correctly identify the expression levels of sufficient numbers of gene transcripts to accurately identify the appropriate metabolic or gene regulatory pathways, and to use classification techniques to train on one platform and test on another. This enables much more robust feature selection, and helps individuals validate their bio-signatures much more quickly and accurately. The ability to correctly identify biological processes is the fundamental aspect of expression microarrays that has driven the amazing progress in genomics research over the last 10 years, and will continue to do so.

REFERENCES

1. Hardiman, G. 2004. Microarray platforms—comparisons and contrasts. *Pharmacogenomics*, 5, 487–502.
2. Stafford, P. and Liu, P. 2003. Microarray technology comparison, statistical analysis, and experimental design. In: G. Hardiman (ed.), *Microarrays: Methods and Applications*, p. 273. Eagleville, PA: DNA Press.
3. Tan, P.K., Downey, T.J., Spitznagel, E.L., Xu, P., Fu, D., Dimitriv, D.S., Lempicki, R.A., Raaka, B.M., and Cam, M.C. 2003. Evaluation of gene expression measurements from commercial microarray platforms. *Nucleic Acids Research*, 31, 5676–5684.
4. Yauk, C.L., Berndt, M.L., Williams, A., and Douglas, G.R. 2004. Comprehensive comparison of six microarray technologies. *Nucleic Acids Research*, 32(15), e124.
5. Irizarry, R.A., Warren, D., Spencer, F., Kim, I.F., Biswal, S., Frank, B.C., Gabrielson, E., et al. 2005. Multiple-laboratory comparison of microarray platforms. *Nature Methods*, 2, 329–330.
6. Larkin, J.E., Frank, B.C., Gavras, H., Sultana, R., and Quackenbush, J. 2005. Independence and reproducibility across microarray platforms. *Nature Methods*, 2, 337–343.
7. Cheadle, C., Becker, K.G., Cho-Chung, Y.S., Nesterova, M., Watkins, T., Wood, W., 3rd, Prabhu, V., and Barnes, K.C. 2006. A rapid method for microarray cross platform comparisons using gene expression signatures. *Molecular and cellular Probes*, 21, 35–46.
8. Kuo, W.P., Liu, F., Trimarchi, J., Punzo, C., Lombardi, M., Sarang, J., Whipple, M.E., et al. 2006. A sequence-oriented comparison of gene expression measurements across different hybridization-based technologies. *Nature Biotechnology*, 24, 832–840.
9. Liu, W.M., Li, R., Sun, J.Z., Wang, J., Tsai, J., Wen, W., Kohlmann, A., and Williams, P.M. 2006. PQN and DQN: Algorithms for expression microarrays. *Journal of Theoretical Biology*, 243, 273–278.
10. Millenaar, F.F., Okyere, J., May, S.T., van Zanten, M., Voesenek, L.A., and Peeters, A.J. 2006. How to decide? Different methods of calculating gene expression from short oligonucleotide array data will give different results. *BMC Bioinformatics*, 7, 137.
11. Seo, J. and Hoffman, E.P. 2006. Probe set algorithms: Is there a rational best bet? *BMC Bioinformatics*, 7, 395.
12. Barnes, M., Freudenberg, J., Thompson, S., Aronow, B., and Pavlidis, P. (2005) Experimental comparison and cross-validation of the Affymetrix and Illumina gene expression analysis platforms. *Nucleic Acids Research*, 33(18), 5914–5923.
13. Guo, L., Lobenhofer, E.K., Wang, C., Shippy, R., Harris, S.C., Zhang, L., Mei, N., et al. 2006. Rat toxicogenomic study reveals analytical consistency across microarray platforms. *Nature Biotechnology*, 24, 1162–1169.
14. Shi, L., Tong, W., Fang, H., Scherf, U., Han, J., Puri, R.K., Frueh, F.W., et al. 2005. Cross-platform comparability of microarray technology: Intra-platform consistency and appropriate data analysis procedures are essential. *BMC Bioinformatics*, 6(Suppl. 2), S12.
15. Shi, L., Reid, L.H., Jones, W.D., Shippy, R., Warrington, J.A., Baker, S.C., Collins, P.J., et al. 2006. The MicroArray Quality Control (MAQC) project shows inter- and intraplatform reproducibility of gene expression measurements. *Nature Biotechnology*, 24, 1151–1161.
16. Patterson, T.A., Lobenhofer, E.K., Fulmer-Smentek, S.B., Collins, P.J., Chu, T.M., Bao, W., Fang, H., et al. 2006. Performance comparison of one-color and two-color platforms within the MicroArray Quality Control (MAQC) project. *Nature Biotechnology*, 24, 1140–1150.

17. Tong, W., Lucas, A.B., Shippy, R., Fan, X., Fang, H., Hong, H., Orr, M.S., et al. 2006. Evaluation of external RNA controls for the assessment of microarray performance. *Nature Biotechnology*, 24, 1132–1139.
18. Canales, R.D., Luo, Y., Willey, J.C., Austermiller, B., Barbacioru, C.C., Boysen, C., Hunkapiller, K., et al. 2006. Evaluation of DNA microarray results with quantitative gene expression platforms. *Nature Biotechnology*, 24, 1115–1122.
19. Golub, T.R., Slonim, D.K., Tamayo, P., Huard, C., Gaasenbeek, M., Mesirov, J.P., Coller, H., et al. 1999. Molecular classification of cancer: Class discovery and class prediction by gene expression monitoring. *Science*, 286, 531–537.
20. Pomeroy, S.L., Tamayo, P., Gaasenbeek, M., Sturla, L.M., Angelo, M., McLaughlin, M.E., Kim, J.Y., et al. 2002. Prediction of central nervous system embryonal tumour outcome based on gene expression. *Nature*, 415, 436–442.
21. van 't Veer, L.J., Dai, H., van de Vijver, M.J., He, Y.D., Hart, A.A., Mao, M., Peterse, H.L., et al. 2002. Gene expression profiling predicts clinical outcome of breast cancer. *Nature*, 415, 530–536.
22. Marton, M.J., DeRisi, J.L., Bennett, H.A., Iyer, V.R., Meyer, M.R., Roberts, C.J., Stoughton, R., et al. 1998. Drug target validation and identification of secondary drug target effects using DNA microarrays. *Nature Medicine*, 4, 1293–1301.
23. Tavazoie, S., Hughes, J.D., Campbell, M.J., Cho, R.J., and Church, G.M. 1999. Systematic determination of genetic network architecture. *Nature Genetics*, 22, 281–285.
24. Waring, J.F., Ciurlionis, R., Jolly, R.A., Heindel, M., and Ulrich, R.G. 2001. Microarray analysis of hepatotoxins in vitro reveals a correlation between gene expression profiles and mechanisms of toxicity. *Toxicology Letters*, 120, 359–368.
25. Theilhaber, J., Connolly, T., Roman-Roman, S., Bushnell, S., Jackson, A., Call, K., Garcia, T., and Baron, R. 2002. Finding genes in the C2C12 osteogenic pathway by *k*-nearest-neighbor classification of expression data. *Genome Research*, 12, 165–176.
26. Raudys, S. 2000. How good are support vector machines? *Neural Networks*, 13, 17–19.
27. DeRisi, J.L., Iyer, V.R., and Brown, P.O. 1997. Exploring the metabolic and genetic control of gene expression on a genomic scale. *Science*, 278, 680–686.
28. Hughes, T.R., Marton, M.J., Jones, A.R., Roberts, C.J., Stoughton, R., Armour, C.D., Bennett, H.A., et al. 2000. Functional discovery via a compendium of expression profiles. *Cell*, 102, 109–126.
29. Iyer, V.R., Eisen, M.B., Ross, D.T., Schuler, G., Moore, T., Lee, J.C., Trent, J.M., et al. 1999. The transcriptional program in the response of human fibroblasts to serum. *Science*, 283, 83–87.
30. Oliphant, A., Barker, D.L., Stuelpnagel, J.R., and Chee, M.S. 2002. BeadArray technology: Enabling an accurate, cost-effective approach to high-throughput genotyping. *BioTechniques*, (Suppl.), 56–58, 60–61.
31. Ye, F., Li, M.S., Taylor, J.D., Nguyen, Q., Colton, H.M., Casey, W.M., Wagner, M., Weiner, M.P., and Chen, J. 2001. Fluorescent microsphere-based readout technology for multiplexed human single nucleotide polymorphism analysis and bacterial identification. *Human Mutation*, 17, 305–317.
32. Sosnowski, R., Heller, M.J., Tu, E., Forster, A.H., and Radtkey, R. 2002. Active microelectronic array system for DNA hybridization, genotyping and pharmacogenomic applications. *Psychatric Genetics*, 12, 181–192.
33. Man, M.Z., Wang, X., and Wang, Y. 2000. POWER_SAGE: Comparing statistical tests for SAGE experiments. *Bioinformatics*, 16, 953–959.
34. Elowitz, M.B., Levine, A.J., Siggia, E.D., and Swain, P.S. 2002. Stochastic gene expression in a single cell. *Science*, 297, 1183–1186.
35. Siddiqui, A.S., Delaney, A.D., Schnerch, A., Griffith, O.L., Jones, S.J., and Marra, M.A. 2006. Sequence biases in large scale gene expression profiling data. *Nucleic Acids Research*, 34, e83.
36. Brenner, S., Johnson, M., Bridgham, J., Golda, G., Lloyd, D.H., Johnson, D., Luo, S., et al. 2000. Gene expression analysis by massively parallel signature sequencing (MPSS) on microbead arrays. *Nature Biotechnology*, 18, 630–634.
37. Mouradian, S. 2002. Lab-on-a-chip: Applications in proteomics. *Current Opinion Chemical Biology*, 6, 51–56.
38. Tesfu, E., Maurer, K., Ragsdale, S.R., and Moeller, K.F. 2004. Building addressable libraries: The use of electrochemistry for generating reactive Pd(II) reagents at preselected sites in a Chip. *Journal of the American Chemical Society*, 126(20), 6212–6213.
39. Fodor, S.P., Read, J.L., Pirrung, M.C., Stryer, L., Lu, A.T., and Solas, D. 1991. Light-directed, spatially addressable parallel chemical synthesis. *Science*, 251, 767–773.

40. Fan, W., Khalid, N., Hallahan, A.R., Olson, J.M., and Zhao, L.P. 2006. A statistical method for predicting splice variants between two groups of samples using GeneChip expression array data. *Theoretical Biology and Medical Modelling*, 3, 19.

41. Stee, J., Wang, J., Coombes, K., Ayers, M., Hoersch, S., Gold, D.L., Ross, J.S., et al. 2005. Comparison of the predictive accuracy of DNA array-based multigene classifiers across cDNA arrays and Affymetrix GeneChips. *Journal of Molecular Diagnostics*, 7, 357–367.

42. Rise, M.L., Douglas, S.E., Sakhrani, D., Williams, J., Ewart, K.V., Rise, M., Davidson, W.S., Koop, B.F., and Devlin, R.H. 2006. Multiple microarray platforms utilized for hepatic gene expression profiling of GH transgenic coho salmon with and without ration restriction. *Journal of Molecular Endocrinology*, 37, 259–282.

43. Harr, B. and Schlotterer, C. 2006. Comparison of algorithms for the analysis of Affymetrix microarray data as evaluated by co-expression of genes in known operons. *Nucleic Acids Research*, 34, e8.

44. Wu, W., Dave, N., Tseng, G.C., Richards, T., Xing, E.P., and Kaminski, N. 2005. Comparison of normalization methods for CodeLink Bioarray data. *BMC Bioinformatics*, 6, 309.

45. Parrish, R.S. and Spencer, H.J. 2004. Effect of normalization on significance testing for oligonucleotide microarrays. *Journal of Biopharmaceutical Statistics*, 14, 575–589.

46. Bolstad, B.M., Irizarry, R.A., Astrand, M., and Speed, T.P. 2003. A comparison of normalization methods for high density oligonucleotide array data based on variance and bias. *Bioinformatics*, 19, 185–193.

47. Albert, T.J., Norton, J., Ott, M., Richmond, T., Nuwaysir, K., Nuwaysir, E.F., Stengele, K.-P., and Green, R.D. 2003. Light-directed 5′ → 3′ synthesis of complex oligonucleotide microarrays. *Nucleic Acids Research*, 31, e35.

48. van de Vijver, M.J., He, Y.D., van 't Veer, L.J., Dai, H., Hart, A.A., Voskuil, D.W., Schreiber, G.J., et al. 2002. A gene-expression signature as a predictor of survival in breast cancer. *New England Journal of Medicine*, 347, 1999–2009.

49. Weigelt, B., Hu, Z., He, X., Livasy, C., Carey, L.A., Ewend, M.G., Glas, A.M., Perou, C.M., and van 't Veer, L.J. 2005. Molecular portraits and 70-gene prognosis signature are preserved throughout the metastatic process of breast cancer. *Cancer Research*, 65, 9155–9158.

50. Warnat, P., Eils, R., and Brors, B. 2005. Cross-platform analysis of cancer microarray data improves gene expression based classification of phenotypes. *BMC Bioinformatics*, 6, 265.

51. Wu, W., Xing, E.P., Myers, C., Mian, S., and Bissell, M.J. 2005. Evaluation of normalization methods for cDNA microarray data by k-NN classification. *BMC Bioinformatics*, 6, 191.

52. Li, J., Spletter, M.L., and Johnson, J.A. 2005. Dissecting tBHQ induced ARE-driven gene expression through long and short oligonucleotide arrays. *Physiological Genomics*, 21, 43–58.

53. Naef, F. and Magnasco, M.O. 2003. Solving the riddle of the bright mismatches: Labeling and effective binding in oligonucleotide arrays. *Physical Review. E, Statistical, Nonlinear, and Soft Matter Physics*, 68, 11906.

54. Irizarry, R.A., Bolstad, B.M., Collin, F., Cope, L.M., Hobbs, B., and Speed, T.P. 2003. Summaries of Affymetrix GeneChip probe level data. *Nucleic Acids Research*, 31, e15.

55. Wolber, P.K., Shannon, K.W., Fulmer-Smentek, S.B., Collins, P.J., Lenkov, K., Troup, C.D., Connell, S.D., et al. 2002. Robust local normalization of gene expression microarray data. *Agilent Technical Note*, 1–4.

56. Chen, Y., Dougherty, E.R., and Bittner, M.L. 1997. Ratio-based decisions and the quantitative analysis of cDNA microarray images. *Journal of Biomedical Optics*, 2, 364–374.

57. Shyamsundar, R., Kim, Y.H., Higgins, J.P., Montgomery, K., Jorden, M., Sethuraman, A., van de Rijn, M., Botstein, D., Brown, P.O., and Pollack, J.R. 2005. A DNA microarray survey of gene expression in normal human tissues. *Genome Biology*, 6, r22.21–r22.29.

58. Li, C. and Wong, W.H. 2001. Model-based analysis of oligonucleotide arrays: Expression index computation and outlier detection. *Proceedings of the National Academy of Sciences of the USA*, 98, 31–36.

59. Eckel, J.E., Gennings, C., Therneau, T.M., Burgoon, L.D., Boverhof, D.R., and Zacharewski, T.R. 2005. Normalization of two-channel microarray experiments: A semiparametric approach. *Bioinformatics*, 21, 1078–1083.

60. Novoradovskaya, N., Whitfield, M.L., Basehore, L.S., Novoradovsky, A., Pesich, R., Usary, J., Karaca, M., et al. 2004. Universal reference RNA as a standard for microarray experiments. *BMC Genomics*, 5, 20.

61. Parmigiani, G., Garrett, E., Anbazhagan, R., and Gabrielson, E. 2004. Molecular classification of lung cancer: a cross-platform comparison of gene expression data sets. *Chest*, 125, 103S.

62. Brown, K.M., Donohue, D.E., D'Alessandro, G., and Ascoli, G.A. 2005. A cross-platform freeware tool for digital reconstruction of neuronal arborizations from image stacks. *Neuroinformatics*, 3, 343–360.

63. Poulsen, C.B., Borup, R., Nielsen, F.C., Borregaard, N., Hansen, M., Gronbaek, K., Moller, M.B., and Ralfkiaer, E. 2005. Microarray-based classification of diffuse large B-cell lymphoma. *European Journal of Haematology*, 74, 453–465.

64. Schlingemann, J., Habtemichael, N., Ittrich, C., Toedt, G., Kramer, H., Hambek, M., Knecht, R., Lichter, P., Stauber, R., and Hahn, M. 2005. Patient-based cross-platform comparison of oligonucleotide microarray expression profiles. *Lab Investment*, 85, 1024–1039.

65. Xu, L., Tan, A.C., Naiman, D.Q., Geman, D., and Winslow, R.L. 2005. Robust prostate cancer marker genes emerge from direct integration of inter-study microarray data. *Bioinformatics*, 21, 3905–3911.

66. Chudin, E., Kruglyak, S., Baker, S.C., Oeser, S., Barker, D., and McDaniel, T.K. 2006. A model of technical variation of microarray signals. *Journal of Computational Biology*, 13, 996–1003.

67. Edelman, E., Porrello, A., Guinney, J., Balakumaran, B., Bild, A., Febbo, P.G., and Mukherjee, S. 2006. Analysis of sample set enrichment scores: Assaying the enrichment of sets of genes for individual samples in genome-wide expression profiles. *Bioinformatics*, 22, e108–e116.

68. Jong, K., Marchiori, E., van der Vaart, A., Chin, S.F., Carvalho, B., Tijssen, M., Eijk, P.P., et al. 2007. Cross-platform array comparative genomic hybridization meta-analysis separates hematopoietic and mesenchymal from epithelial tumors. *Oncogene*, 26, 1499–1506.

69. Pan, F., Kamath, K., Zhang, K., Pulapura, S., Achar, A., Nunez-Iglesias, J., Huang, Y., Yan, X., Han, J., Hu, H., Xu, M., Hu, J., and Zhou, X.J. 2006. Integrative array analyzer: a software package for analysis of cross-platform and cross-species microarray data. *Bioinformatics*, 22(13),1665–1667.

70. Sasik, R., Calvo, E., and Corbeil, J. 2002. Statistical analysis of high-density oligonucleotide arrays: A multiplicative noise model. *Bioinformatics*, 18, 1633–1640.

71. Churchill, G.A. 2002. Fundamentals of experimental design for cDNA microarrays. *Nature Genetics*, 32(Suppl.), 490–495.

72. Bakay, M., Chen, Y.W., Borup, R., Zhao, P., Nagaraju, K., and Hoffman, E.P. 2002. Sources of variability and effect of experimental approach on expression profiling data interpretation. *BMC Bioinformatics*, 3, 4.

73. Sima, C. and Dougherty, E.R. 2006. What should be expected from feature selection in small-sample settings. *Bioinformatics*, 22, 2430–2436.

10 Methods for Assessing Microarray Performance

*Steffen G. Oeser, Shawn C. Baker, Eugene Chudin,
Kenneth Kuhn, and Timothy K. McDaniel*

CONTENTS

10.1 INTRODUCTION

Although carried out with success by hundreds of laboratories, microarray experiments are rather complicated. They involve dozens of steps comprising sample isolation, amplification, labeling, hybridization, washing, signal generation, and scanning [1–3]. While, in experienced hands, results are reproducible, precise, and biologically meaningful, the sheer number of steps means that there are multiple places for variability to creep in and cloud biological meaning. This chapter describes a set of experiments that our group routinely performs to assess the performance of microarrays. As array developers, we use these tests when we have created an array of a new design, change manufacturing processes, or modify experimental protocols. These experiments may be of interest to a broader audience of scientists who perform array experiments. This is because the same approaches can be used to qualify the performance of any particular laboratory, laboratory personnel, reagent lots, and so on. In short, whenever a significant change is made to the array experimental process, it is good to perform some standard set of tests to verify that performance is maintained and to compare performance over different time periods.

We have described the statistical methods underlying these tests in detail in two previous chapters [3,4]. The goal of this review is to explore more broadly their rationale and to describe community-wide efforts to standardize very similar tests, which should allow any research group or core lab to perform similar analyses in their own environment. It should be noted that the tests as described here are generally run to assess the variation of microarrays, not the variation in sample labeling, ribonucleic acid (RNA) isolation, or biology. As such, we tend to perform replicates at the level of hybridization, using aliquots of the same labeled sample, which ensures that variation encountered is due to the array or the array processing and not the upstream steps [3]. The same sorts of tests can be applied to the other levels of the experiment simply by moving the point of replication upstream from the points described here.

10.2   ASSUMPTIONS BEHIND THE TESTS: WHAT MICROARRAYS CAN AND CANNOT MEASURE

The guiding principle of our experiments is that array experiments aim to characterize biological samples by answering two questions about their transcripts:

1. Is the concentration of a given transcript high enough for its signal to be detected above background? In other words, is the transcript expressed?
2. Does the concentration of a given transcript change across two or more samples? In other words, has its expression changed?

A question missing from this very short list is revealing about the limitations of array technology. The missing question is: "What is the absolute concentration of a given transcript?" The reason microarrays cannot answer this effectively is that they consist of *populations* of probes. Each probe has a different sequence, melting temperature, folding property, and so on. Furthermore, the target sequences to which the probes are designed reside in different transcripts, which themselves are of varying lengths, secondary structure, and diffusion rates. The result of this heterogeneity is that the collection of probe/target pairs will span a range of thermodynamic and kinetic properties. In short, the dose–response curve for the binding of each probe to its target will be different from all others, so the conversion between signal and target concentration will also differ. For example, the two probes in Figure 10.1 both show excellent dynamic ranges, linearity, and precision. However, their limits of detection and saturation points differ by ~10-fold. Unless one individually characterizes each of the tens of thousands of probes on an array with defined target, this relationship will be unknown in each case. The ramifications of this are

1. Absolute quantitation is beyond the scope of the technology
2. Performance metrics are not absolute. Limits of detection, dynamic range, and precision will differ by probe. Therefore, an array's performance must be given in terms that summarize populations, such as medians (e.g., "the median limit of detection for all probes is X.")
3. Measurements of "detection" must be stated in terms of probability rather than in absolute terms. We cannot say with certainty if a probe's target is present, but with adequate models, we can estimate the probability that it is detected.

The first limitation is less severe than one might think. Biologists are usually less interested in a transcript's picomolar concentration and more interested in determining if its concentration has changed. This means that precision in measurement is far more important than accuracy, and this has led us to use an unconventional definition of dynamic range, described in more detail later.

10.3   HOW WE CHARACTERIZE MICROARRAYS: THE FOUR BASIC TESTS

We generally characterize our processes by using four basic tests: (i) spiking experiments to determine limits of detection, precision, and dynamic range; (ii) hybridizations to heterologous labeled RNA samples to determine false-positive rates for detection; (iii) replicate hybridizations to

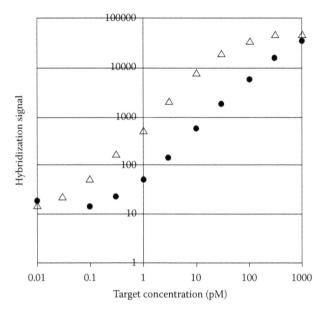

FIGURE 10.1 Example of different dose–response characteristics of two probes to two different spiked targets on the same microarray. The two curves represent the signal responses of probes on the Illumina Mouse-6 microarray for two labeled spiked reporter genes in a series of 12 samples, where the spikes' concentration range from 0.01 to 1000 pM in a constant background of labeled mouse RNA. While both probes show similar background, saturation levels, precision, and dynamic range (~3 logs), the response of the lux gene probe (open triangles) leads that of the e1a probe (black dots) by ~1 log. Because of such differences, microarrays are not used for absolute quantitation of transcript concentration. However, their precision, as indicated by the error bars (standard deviation of measurements from four separate arrays), means they are excellent tools for determining if concentrations have changed between samples.

determine false-positive rates for fold change; and (iv) validations of fold-change detection using orthogonal gene expression technologies.

10.3.1 Spiking Experiments: Limit of Detection, Precision, and Dynamic Range

To characterize the assay's performance we routinely perform dose–response studies. The approach is to spike labeled RNA samples derived from the species under study with known quantities of RNA targets whose sequences are absent from the tested genome. By doping a series of samples with different concentrations of the spiked genes, a dose–response curve can be generated for each gene upon hybridization to the array [3–5]. In our standard test, we employ nine *in vitro*-synthesized polyadenylated spike RNAs [4]. Each target is labeled by our standard assay, purified, measured, and spiked at 12 concentrations (0, 0.01, 0.03, 0.1, 0.3, 1, 3, 10, 30, 100, 300, 1000 pM) across a panel of 12 aliquots of labeled total RNA. Each sample is hybridized to four different microarrays. The intensities and variation across replicates for each gene at each concentration provide the basis of calculating limit of detection, measurement precision, and dynamic range for each probe in question (see Figure 10.1 for two such curves). We then express each performance metric in terms of the median value measured for all probes employed in the test. Under these tests, our typical median limit of detection ($p < 0.01$ based on the signal distribution of >1500 negative control probes) is ≤0.25 pM, and the typical median resolved fold change (defined as the change in concentration that can be distinguished with 95% confidence) is ≤1.3-fold [4].

Probably the most noteworthy feature of how we measure performance is our definition of dynamic range. Unlike typical definitions used in the microarray field, which define the dynamic range semi-quantitatively as the linear portion of a log–log plot of signal versus target concentration beginning at the limit of detection, we define it as the largest range over which twofold increases in target concentration can be resolved with 95% confidence [4]. This has the effect of truncating the lower end of the range, as it is generally possible to distinguish a concentration from background before it is possible to distinguish it from a twofold increase in concentration. It also has the effect of extending the upper end, as the high reproducibility of measurements in that range means that twofold changes are distinguishable even as the best-fit line is saturating. We believe that this is the most relevant way of measuring dynamic range, as it is unambiguously defined based on specific measurements [4] and it gets at the key question of an array experiment: *can one detect with confidence that a transcript's abundance has changed?*

10.3.2 FALSE-POSITIVE RATES

The two types of detection one attempts to measure with microarrays—the presence of a transcript in a sample and the change of its concentration between samples—each have associated false-positive rates. We employ two different experimental designs to estimate these two rates.

10.3.3 FALSE DETECTION OF PRESENCE

We test false detection of presence by two methods: one is to perform a mock hybridization with buffer only and count the number of probes detected with 99% confidence. The second approach is to hybridize with a labeled heterologous RNA sample and count the detected genes. For mammalian arrays we use *Drosophila melanogaster* as the RNA source, which has a similar size transcriptome to rat, human, and mouse, but is sufficiently divergent in sequence that there should be little, if any, true hybridization [4]. The mock buffer hybridizations should reveal probes that are intrinsically fluorescent or that nonspecifically bind the staining reagents. Because the buffer contains no RNA to cross-hybridize to the probes, it likely under-represents false-positive rates. The *Drosophila* hybridizations should additionally identify detection levels caused by nonspecific DNA hybridization. Because there may be some sequence homology between *Drosophila* and the mammalian RNA, this measurement likely over-represents the false-positive rate. For the characterization of our Mouse-6 BeadChip product, the false-positive rates were 0.8% with buffer and 1.4% with *Drosophila*, both within limit given the ideal 1% false-positive value expected at this confidence level.

10.3.4 FALSE DETECTION OF CHANGE

Measuring false detection of change is simple: hybridize the same sample to replicate arrays and count the number of probes that demonstrate a difference in signal greater than a predetermined level of fold-change or statistical significance. In performing this test it is important to filter out probes whose signals reside below levels of system noise, as such signals are by definition irreproducible. For example, if an unexpressed gene gives a signal of one count in one experiment and three counts in another, expression levels will have changed threefold, even though the change is meaningless. In an actual biological experiment, such data would (or should) have been filtered out prior to analysis, so such fluctuations are irrelevant in real life. We generally see very low levels of false detection, with only a handful of expressed genes showing more than twofold change between replicates (Figure 10.2).

10.3.5 CONCORDANCE OF RESULTS BETWEEN ORTHOGONAL TECHNOLOGIES

If two methods based on orthogonal principles produce the same answer, it gives one more confidence in the accuracy of both technologies. In assessing our array results, we generally measure

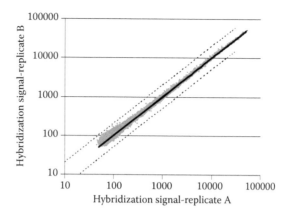

FIGURE 10.2   False-positive rate of differential expression. Replicate hybridizations were performed on 18 arrays on three Human-6 BeadChips using labeled human RNA under standard conditions. The intensity fold change was determined for all genes for all possible 153 pair-wise comparisons of the 18 arrays. The false-positive rate, defined as percentage of genes showing a difference in raw hybridization intensity more than twofold, was 0.00055% ± 0.0015%. A scatter plot of one such comparison is shown above. The thick 45° line marks the best-fit line for all probe signals and the thinner parallel lines mark the thresholds of twofold change in hybridization signal. Note that the data points on this plot are not background subtracted, which avoids the noise issues discussed in the main text.

genes by comparing results to other array platforms [6], or by comparison to quantitative real-time polymerase chain reaction (PCR) [7], such as Applied Biosystems' Taqman® gene expression system. For example, in qualifying our HumanRef-8 BeadChip, we selected 20 genes from a model system comparing human brain versus liver expression. Expression measurements were taken for these 20 genes by TaqMan and compared to those measured on the array. A scatter plot of log-transformed hybridization intensity ratios from BeadChip hybridization and quantitative PCR shows a strong correlation ($r^2 > 0.94$), indicating that this system produces accurate measurements of changes in gene abundance over a wide range of concentrations (Figure 10.3). These high correlation values are consistent with measurements taken in other studies, both on the Illumina® platform and others [6]. In setting up such comparisons, it is important to follow three rules. First, genes must be selected

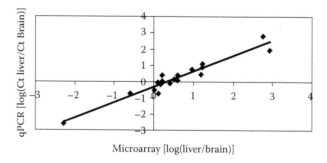

Microarray [log(liver/brain)]

FIGURE 10.3   An independent measure of relative gene expression patterns obtained from an Illumina HumanRef-8 gene expression microarray was assessed using Applied Biosystems' Assays-on-Demand™ quantitative real-time PCR gene expression products. Twenty different genes were chosen from a model system comparing human brain versus liver expression. Genes were selected to span expression ratios from 1:1000 to 1000:1 brain:liver. A scatter plot of log-transformed hybridization intensity ratios from the microarray quantitative PCR shows a strong correlation ($r^2 > 0.94$), cross-validating the accuracy of the array results over a wide range of concentrations.

such that they are present above noise in both tissues in order to avoid ratio skewing due to random fluctuation, such as was noted in the description of the false change detection experiments earlier. Second, the probes to these genes should be bioinformatically cross-mapped to ensure that two different transcripts are not being compared or two different genes with the same names [6]. An extensive list of such cross-mapped probes spanning all commercial human whole-genome array platforms has recently been published [6]. This list should prove invaluable for designing such studies. Third, it is important to select genes spanning a wide range of expected expression ratios [3]. In designing our test, we chose genes showing a brain : liver expression ratio ranging from 1 : 1000 to 1000 : 1 on the basis of previous studies.

10.4   COMMUNITY-WIDE EFFORTS TO STANDARDIZE EVALUATION OF MICROARRAY DATA

The experiments described above are among the most important tests our group has typically performed to characterize our arrays. While the general approaches are likely useful to a broader audience, the specifics of the samples and targets used are probably less so. The difficulty and expense of maintaining the needed reagents require more effort than it is worth for all but the largest array users. Fortunately, there are two community-wide efforts aiming to standardize the reagents and experiments performed to assess array performance. Both of these efforts plan to make available a set of well-characterized reagents available to the larger scientific community.

The first such consortium is the Microarray Quality Control (MAQC) project [6]. The MAQC is an effort led by the United States Food and Drug Administration, but involving more than 100 scientists from the wider government, academic, and industrial communities. Its charter is "to provide quality control tools to the microarray community in order to avoid procedural failures and to develop guidelines for microarray data analysis by providing the public with large reference datasets along with readily accessible reference RNA samples" [8]. The consortium has generated a series of graded RNA mixtures that have been characterized against all major array platforms as well as orthogonal RNA profiling technologies such as TaqMan quantitative real-time PCR [6]. This dataset is publicly available and aliquots of the RNA samples used in the study are available commercially. This means that the simple studies described in the consortium's publications can be replicated at any site as a means of self-qualification. (How well do my results compare with the Consortium's results on my platform? How do my results compare to qRT-PCR results that the consortium generated on several hundred genes?)

The External RNA Control Consortium (ERCC) is an ad hoc group involving academic, government, and industry representatives, including the participation of the Unites States National Institute of Standards and Technology [9]. The group states its aim as being "to develop a set of external RNA control transcripts that can be used to assess technical performance in gene expression assays. The external RNA controls will be added after RNA isolation, but prior to cDNA synthesis. They are being designed to evaluate whether the results for a given experiment are consistent with defined performance criteria" [10]. As with the MAQC, the reagents created by the ERCC will be prepared in large batches and sold commercially to the scientific community [9]. The controls will be useable much in the same way the spike RNAs are used by our group to determine limit of detection, dynamic range, and precision, with the key difference being that, as planned, more than 100 of these targets will be available for much more extensive characterizations.

10.5   CONCLUSION

With the data and reagents provided by the MAQC and ERCC groups, the sorts of experiments described in this review will be easy to perform by any group that seeks to monitor its own performance. The added benefit of having standardized, well-characterized reagents is that each group will be able to compare their own performance against benchmarks established across many platforms. We predict that these efforts will increase the general level of quality produced from microarray experiments and accelerate the acceptance of microarrays as standard research and diagnostic tools.

REFERENCES

1. Schena M, Shalon D, Davis RW, and Brown PO. Quantitative monitoring of gene expression patterns with a complementary DNA microarray. *Science* 1995; 270: 467–470.
2. Lockhart DJ, Dong H, Byrne MC, Foliette MT, Gallo MV, Chee MS, Mittmann M, et al. Expression monitoring by hybridization to high-density oligonucleotide arrays. *Nat Biotechnol* 1996; 14: 1675–1680.
3. Kuhn K, Baker SC, Chudin E, Lieu M-H, Oeser S, Bennett H, Rigault P, Barker D, McDaniel TK, and Chee MS. A novel, high-performance random array platform for quantitative gene expression profiling. *Genome Res* 2004; 14: 2347–2356.
4. Chudin E, Kruglyak S, Baker SC, Oeser S, Barker D, and McDaniel TK. A model of technical variation of microarray signals. *J Comp Biol* 2006; 13: 996–1003.
5. Choe SE, Boutros M, Michelson AM, Church GM, and Halfon M. Preferred analysis methods for Affymetrix GeneChips revealed by a wholly defined control dataset. *Genome Biol* 2005; 6: R16.
6. Shi L, Reid LH, Jones WD, Shippy R, Warrington JA, Baker SC, Collins PJ, et al. The MicroArray Quality Control (MAQC) project shows inter- and intraplatform reproducibility of gene expression measurements. *Nat Biotechnol* 2006; 24: 1151–1161.
7. Heid CA, Stevens J, Livak KJ, and Williams PM. Real time quantitative PCR. *Genome Res* 1996; 6: 986–994.
8. Available at: http://www.fda.gov/nctr/science/centers/toxicoinformatics/maqc/.
9. Baker SC, Bauer SR, Beyer RP, Brenton JD, Bromley B, Burrill J, Causton H, et al. The External RNA Controls Consortium: A progress report. *Nat Methods* 2005; 2: 731–734.
10. Available at: http://www.cstl.nist.gov/biotech/Cell&TissueMeasurements/GeneExpression/ERCC.htm.

11 Comparison of Different Normalization Methods for Applied Biosystems Expression Array System

Catalin Barbacioru, Yulei Wang, Roger Canales, Yongming Sun, David Keys, Frances Chan, Kathryn Hunkapiller, Karen Poulter, and Raymond R. Samaha

CONTENTS

11.1 INTRODUCTION

Deoxyribonucleic acid (DNA) microarray technology provides a powerful tool for characterizing gene expression on a genome scale. Although the technology has been widely used in discovery-based medical and basic biological research, its direct application in clinical practice and regulatory decision-making has been questioned [1,2]. A few key issues, including the reproducibility, reliability, compatibility, and standardization of microarray analysis and results, must be critically addressed before any routine usage of microarrays in clinical laboratory and regulated areas can occur. Considerable effort has been dedicated to investigate these important issues, most of which focused on the compatibility across different laboratories and analytical methods, as well as the correlation between different microarray platforms. In this study, we investigate some of these issues for Applied Biosystems Human Genome Survey Microarrays.

The Applied Biosystems Human Genome Survey Microarray contains 31,700 60-mer oligonucleotide probes representing 29,098 individual human genes, and uses chemiluminescence (CL) to identify and measure gene expression levels in cells and tissues. In addition to the unique 60-mer probe, an internal control probe (a 24-mer oligonucleotide) is co-spotted with the 60-mer probe on the microarray and labeled with a complementary oligo, containing the fluorescent LIZ® dye (FL) during the hybridization of the microarray. The FL signal, which has a close spatial correlation with the CL signal, locates all features on the microarray, even in the absence of gene expression products. Multiple images are taken to bring the microarray into focus and to measure the FL and CL signals. The microarray is imaged in both short (5 s) and long (25 s) read times to extend the linear dynamic range of the chemiluminescent signals (>1000-fold).

AB microarray data are generated through an automated algorithm, processing the input image set to an output feature table that includes extracted feature signals, signal uncertainties, quality metrics, and confidence values. This process divides into (i) image processing and (ii) quantification steps.

(i) Image processing step comprises three major stages: gridding, image correction, and primary image analysis. Gridding locates and identifies features in the image with regard to the known microarray design. Accurate gridding is ensured by using the FL image which has been designed in such a way that all features "light up" with a high signal-to-noise (S/N) ratio, and also with easily automatically locatable landmark fiducials. A number of corrections are applied to improve both the raw chemiluminescent and florescent images prior to image analysis: image calibration, spectral cross-talk correction, long–short CL images merger, and pixel alignments for images in the two channels. Fully calibrated and registered CL and FL images along with grid positions are fed to the image analysis algorithm. Tasks that are accomplished here include background correction in both channels, feature integration and feature normalization, flagging of problematic features due to detector saturation and other issues, rejection of outlying feature pixels associated with artifacts on the array, estimation and propagation of uncertainties in background, and signal at both pixel and integrated signal level into the extracted normalized signal standard deviation estimate.

(ii) During quantification feature signals are extracted from the image and chemiluminescent signals are then normalized to fluorescent feature intensities at the pixel level, eliminating variations due to spotting inaccuracies, together with confidence, quality metrics, and some intermediate quantification results, reported in the final table.

In this study, we analyzed the gene expression profiles of two human tissues: brain and universal human reference (UHR) sample using Applied Biosystems Human Genome Survey Microarrays. Five technical replicates in three different sites were performed on the same total ribonucleic acid (RNA) samples according to manufacturer's standard protocols. Five different methods, quantile [3,4], median [5], scale [6,7], variance stabilization normalization (VSN) [8], and cyclic Loess [6] were used to normalize AB microarray data within each site. One thousand genes spanning a wide dynamic range in gene expression levels were selected for real-time polymerase chain reaction (PCR) validation. Using the TaqMan® assays data set as the reference set, the performance of the five normalization methods was evaluated focusing on the following criteria: (1) sensitivity and reproducibility in detection of expression; (2) fold-change correlation with real-time PCR data; (3) sensitivity and specificity in detection of differential expression; and (4) reproducibility of differentially expressed gene lists.

11.2 RESULTS

11.2.1 TARGET SELECTION FOR REAL-TIME PCR VALIDATION

To conduct a comprehensive and unbiased survey of the Applied Biosystems Expression Array System performance, we selected the gene targets for real-time PCR validation based on the following strategies: (1) ensure a large enough number of validation targets to provide representative overviews of the microarray performance; (2) select genes with expression levels spanning a wide range of expression levels; and (3) fold changes (Figure 11.1). One thousand TaqMan Gene Expression Assays are used in this study, covering 997 genes (three genes had more than one assay) (MAQC project [9]). Over 90% of these genes were selected from a subset of 9442 RefSeq common to the various microarray platforms (Affymetrix, Agilent, GE Healthcare, and Illumina). This selection was designed so that the genes would cover the entire intensity and fold-change ranges and include any bias due to RefSeq itself. A subset of (~100) genes were included based on tissue-specificity (UHR versus brain).

11.2.2 SIGNAL DETECTION

The dynamic range for AB microarray platform spans 3–4 orders of magnitude [10], whereas TaqMan-based real-time PCR can achieve 7–8 orders of magnitude dynamic range [11,12]. The larger dynamic range imparts TaqMan assays with superior detection sensitivity (limit of detection ~1–5 copies per reaction [11]); we therefore used the TaqMan assays data set as the reference to evaluate the performance of AB microarray platform in terms of detection sensitivity and accuracy for the list of genes selected for real-time PCR validation. First, genes that were detectable (positives: above detection threshold) and not detectable (negatives: below detection threshold) were determined for each sample

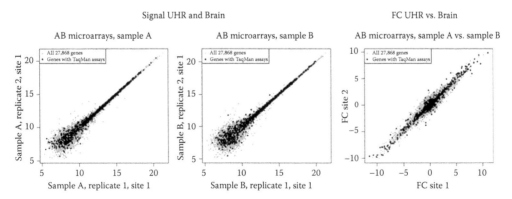

FIGURE 11.1 Gene selection: 1000 gene targets were selected for TaqMan assay validation in order to span a wide dynamic range in expression level and fold changes. Scatter plots between two technical replicates for UHR (sample A) and brain (sample B) samples were shown for the 29,098 genes represented on AB micro-arrays. The 1000 gene targets are represented in black, and show a wide dynamic range of expression levels and fold change.

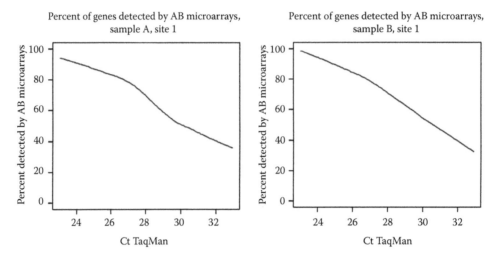

FIGURE 11.2 Detection concordance: only genes detected as present in more than half of the replicates of individual samples by TaqMan assays are used in these plots. A sliding window containing 100 consecutive genes was constructed and moved one gene at a time to cover the whole range of Ct values. Within each sliding window, the percent of genes detected as present in at least half of the replicates of individual samples by AB microarray platform was computed and plotted as a function of mean Ct value of the 100 genes in the given window.

according to the manufacturer's recommendations (see Section 11.4 for detailed descriptions). Figure 11.2 shows the relationship between percent genes detected by AB microarrays out of the ones detected by TaqMan assays as a function of cycle threshold (Ct) measurement (the number of template transcript molecules is inversely related to Ct—the more template transcript molecules at the beginning, the lower the Ct). Gene expression levels were ordered according to TaqMan assay measurements (average Ct within each sample). A sliding window containing 100 consecutive genes was constructed and one gene at a time was moved to cover the whole range of Ct values. Within each sliding window, the percent of genes detected as present in at least half of the replicates of individual samples by AB microarray platform was computed out of those detected by TaqMan assays, and plotted as a function of mean Ct value of the 100 genes in the given window. The overall sensitivity (true-positive rate, TPR) and specificity (1 − false-positive rate, FPR) are presented in Table 11.1, and 76.6% and 81.3%, respectively, are present in UHR sample measured in test site 1.

TABLE 11.1

Detection Concordance between AB Microarrays and TaqMan Assays in UHR and Brain

	TP	TN	FP	FN	Sensitivity	Specificity
			UHR, Site 1			
TaqMan	789	59	0	0	100	100
AB microarrays	605	48	11	184	76.68%	81.36%
			Brain, site 1			
TaqMan	744	104	0	0	100	100
AB microarrays	581	79	25	163	78.09%	75.96%

Notes: For each platform, sample, and site, a gene is declared detected (present) if it is detected according to the platform specifications, in more than half of the replicates. Concordance between detection calls are presented in these tables for UHR and brain samples in site 1 (sensitivity = TPR, specificity = 1 − FPR).

11.2.3 SIGNAL REPRODUCIBILITY

For each normalization method, parallel gene expression data were collected from each of the five technical replicates for the two total RNA samples (human brain and UHR) from the three different testing sites. Genes detected by TaqMan assays were used to evaluate the impact of the five normalization methods on signals levels of the microarrays (Figure 11.3). There are almost no differences in the signal levels when each of these normalization methods is used with the exception of VSN which results in a small increase in signal level for low expressing genes. We use coefficient of variation (CVs) of $\log_2$(signal) to evaluate the effect of the five normalization methods on signal reproducibility, both within and between sites (Figures 11.4 and 11.5). Reproducibility of technical replicates for the five normalization approaches for site 1 is illustrated in Figure 11.4 for both brain and UHR samples. Panel A, where all 29,069 genes are represented, shows the CV across the five technical replicates, as a function of the expression level when the data are normalized using the quantile approach. Panel B shows the CVs only for genes with TaqMan assays as a function of TaqMan Ct. Lines represent the Lowess smoothing fitting curves [13] of all data points from each normalization method. As expected, CVs show a strong dependence on expression level, decreasing from 10% for low expressers or absent genes, to 1% for high expressers. A small improvement in reproducibility of low expression level genes is observed in VSN normalization, in both representations, and for both samples. Signal reproducibility between the three testing sites is presented in Figure 11.5. Within-sites CVs (dotted lines) and between-sites CVs (solid lines) of all 29,098 genes show similar trends for all five normalization methods. The VSN normalization, which shows some improvement in within-sites variability of low expression level genes, performs similarly to the other normalization methods when between-sites variability is considered.

11.2.4 FOLD-CHANGE CONCORDANCE WITH TAQMAN ASSAYS

To evaluate the concordance of fold changes between microarray and real-time PCR data, we performed regression analysis of fold differences between UHR sample (A) and brain sample (B).

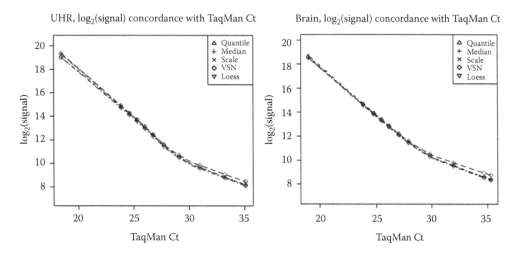

FIGURE 11.3 Signal concordance: genes detected (present) by TaqMan assays are used to represent the relationship between expression levels measured by AB microarrays and TaqMan assays. The average $\log_2$(signal) of the five replicates from site 1 for all five normalization methods are plotted as functions of gene expression level measured by TaqMan assays. Lines represent Lowess smoothing fitting curves to the set of data points corresponding to one normalization method.

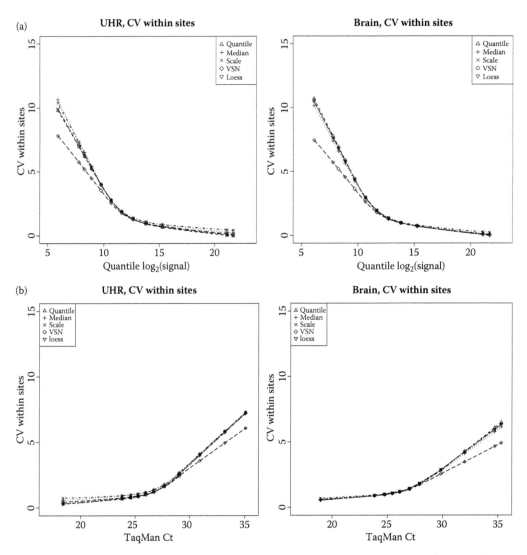

FIGURE 11.4 Reproducibility within sites: CVs are used to evaluate the impact of the five normalization methods on data reproducibility. (a) presents the CVs, of log$_2$(signal), within site 1 for all 29,098 genes as a function of expression level measured by quantile normalization; (b) presents the CVs within site 1 for genes with TaqMan assays as a function of TaqMan Ct values. Lines represent Lowess smoothing fitting curves of all data points from each normalization method.

Fold-change metrics tend to cancel out systematic platform biases in absolute signal values, and in addition, it is the most biologically meaningful metric. Fold change (log$_2$) was computed as the difference in mean expression level of the five technical replicates measured within each site for each sample. Genes were filtered based on real-time PCR detection thresholds (detectable in at least three out of four technical replicates in both samples), and only genes detected in both samples (848) were used. Fold changes between brain and UHR samples were estimated and plots between (log$_2$) fold changes determined from AB microarray data and TaqMan assays ($\Delta\Delta$CT) are presented in Figure 11.6. The linear regression fitting curve for all data points was plotted for each scatter plot (Panel A). The R^2, slope, and intercept for TaqMan assays versus microarray data are presented in Table 11.2. The five normalization methods used for AB microarrays showed similar linear correlation characteristics with real-time PCR measurements with R^2 values ranging between

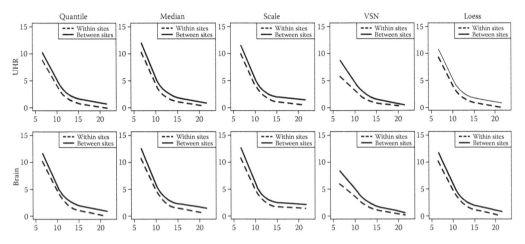

FIGURE 11.5 Variability between sites: CVs are used to evaluate the impact of the five normalization methods on data reproducibility. One-way (site) ANOVA is used to estimate variability within/between sites. CVs within sites (dotted lines) and between sites (solid lines) are plotted against quantile normalized data.

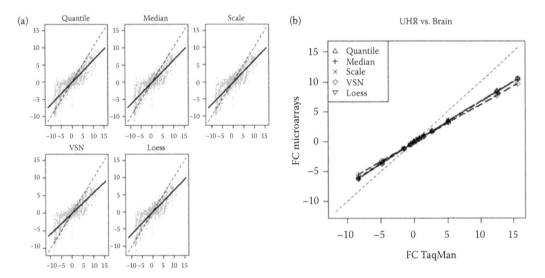

FIGURE 11.6 Fold-change concordance: fold changes between UHR and brain, determined by each normalization method applied to AB microarray data (*y*-axis), were plotted against those determined by TaqMan assays (*x*-axis). Genes were filtered based on real-time PCR detection thresholds (detectable in at least three out of four technical replicates in both samples). (a) Linear regression lines (solid lines) are presented in each plot. (b) Lines represent Lowess smoothing fitting curves to the 2550 data points (data from all three sites) of each normalization method.

TABLE 11.2
Fold-Change Concordance: Linear Regression Parameters

	Quantile	Median	Scale	VSN	Lowess
Intercept	−0.095	−0.111	−0.154	−0.127	−0.168
Slope	0.639	0.641	0.646	0.589	0.640
R^2	0.744	0.745	0.744	0.734	0.744

0.73 and 0.75 and slopes ranging from 0.65 to 0.59, suggesting the existence of some ratio compression in the microarray data. VSN shows slightly higher compression than the other normalization methods, most likely caused by generating higher signal levels for low expressers (as seen in Figure 11.3). A similar fold-change comparison between the five normalization methods and TaqMan assays was also performed using Lowess smoothing (Figure 11.5, Panel B), which does not assume a linear relationship of fold-change values between platforms. The estimated range of fold changes (on $\log_2$ scale), for TaqMan assays, is from −8 to 15, whereas for AB microarrays it is from −4 to 10. To better understand the cause of fold-change compression an additional analysis was performed on genes with similar expression levels in the two samples. Genes were binned into low/medium/high according to TaqMan assays Ct measurements (the Ct cut-offs are set to 23:29:35). Only genes having expression level in the same bin in both samples A and B are included. Boxplots of the fold change for each normalization method and TaqMan assays are presented in Figure 11.7.

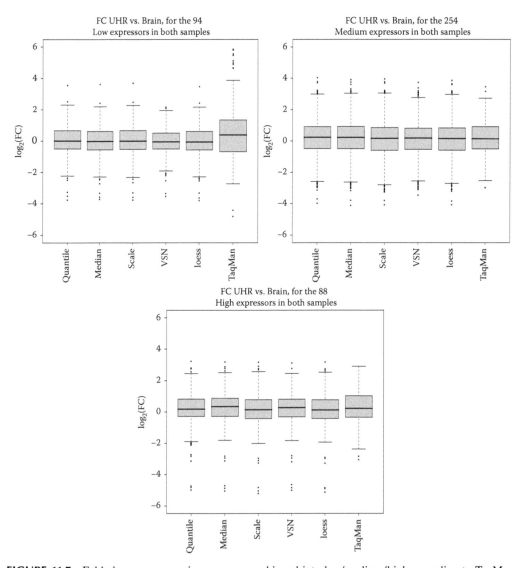

FIGURE 11.7 Fold-change compression: genes were binned into low/medium/high according to TaqMan assays Ct measurements (the cut-offs are set to 24:29:35). Only genes having expression level in the same bin in both samples A and B are included. Boxplots of fold changes for each normalization method and TaqMan assays are presented.

The range of fold changes measured by the microarrays is significantly lower for low expression level genes and a little lower for high expression level genes when compared to the range of fold changes measured by TaqMan assays. For medium expression level genes, the two platforms have a higher agreement on the magnitude of fold differences between the two samples.

11.2.5 DIFFERENTIALLY EXPRESSED GENES

We evaluated the performance of the different normalization methods in detecting differential expression between the two samples using multiple statistical approaches. Traditionally, analysis of accuracy is carried out by analyzing the TPR and false discovery rate (FDR) [14]. In this case, the actual rates are unknown. For this reason, we used TaqMan as the reference platform. Only genes detected by TaqMan assays in both samples were used for this. Using TaqMan assay calls as the reference, we constructed contingency tables against microarray data in which the concordance was determined, and both the p-value significance of the t-test controlling FDR at 5% level [14] and fold-change directionality (up- or down-regulation) were taken into consideration. Specifically, true positives (TP) are genes differentially expressed (significant p-value for the t-test) in both TaqMan and microarray platforms with similar direction of the fold change; true negatives (TN) are genes not differentially expressed in both platforms; false positives (FP) consist of two sets of genes: (i) genes not differentially expressed in TaqMan and differentially expressed in microarrays, or (ii) genes differentially expressed in both platforms with opposite fold change direction; false negatives (FN) genes are differentially expressed for TaqMan and not for microarrays. Genes were first ranked according to their average Ct value in UHR and brain samples. For each bin of 50 consecutive genes (according to the ranking), we compared the results from each normalization method with the ones from TaqMan assays. TPR defined as $TPR = TP/(TP + FN)$ represents the percentage of genes detected differentially expressed in microarray data out of the ones detected by TaqMan assays. FDR was defined as $FDR = FP/(TP + FP)$ and represents the percentage of differentially expressed genes detected only by the microarrays out of all genes differentially expressed in microarrays. As shown in Figure 11.8, at the highest expression level, all normalization methods display reasonably

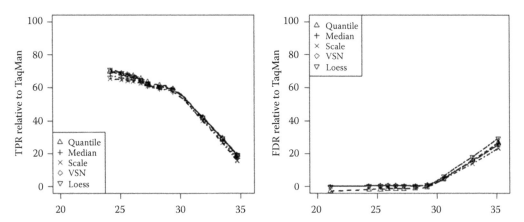

FIGURE 11.8 Significantly differentially expressed genes concordance: genes detected in both samples by TaqMan assays were first ranked according to their average Ct value in UHR and brain. We use t-test to detect significantly differentially expressed genes, controlling FDR at 5% level. For each bin of 50 consecutive genes (according to the ranking), we compare the results from each normalization method with the ones from TaqMan assays. We keep track of up/down-regulation in each platform. TPR represents the percentage of genes detected differentially expressed in microarray data out of the ones detected by TaqMan assays. FDR was defined as $FP/(TP + FP)$, where FP is false positive in microarray data and represents the percentage of differentially expressed genes detected only by microarray out of all genes differentially expressed in microarray.

good sensitivities: 70–75% TPR; the performance drops as the expression level decreases, and at the lowest expression level, the TPR is 20% (Figure 11.8, Panel A). A reversed change can be seen for the FDR plots (Figure 11.8, Panel B): a relative constant level of false findings (FDR 1–2%) is observed for genes with high and medium expression levels (Ct <30), after which FDR increases up to 25–30% for genes with low expression levels. The overall accuracy of each of the five normalization methods is presented in Table 11.3. TPR and FDR indicate that the performance of the AB microarray system is not dependent on the normalization method used. The observed FDR of 7.1% is slightly higher to the one expected from the FDR control we used to select differentially expressed genes (5%), the 2% overestimate (approximately 28 genes) being possibly explained by genes incorrectly called differentially expressed by TaqMan assays.

Furthermore, we investigated four additional methods for the identification of differentially expressed genes for microarrays: simple t-test ($p < 0.05$), t-test combined with fold-change ($p < 0.05$ and FC > 1.5), t-test with FDR and FC control (FDR = 5% and FC < 1.5), and significant analysis of microarrays (SAM) ($q < 0.05$) to determine their impact on the detection of differentially expressed genes. Only data from site 1 were used for this case. Figure 11.9 shows TPR and FDR plots comparing genes differentially expressed for each statistical method applied to each normalization approach with TaqMan data used as a reference as described previously (t-test controlling FDR at 5% level). Table 11.4 presents the overall concordance for quantile normalized data. The performance of the five methods does not change with the normalization method used. It is important, however, when interpreting these results to take into consideration the specificities of this experiment. The two samples compared, human brain and UHR, being extremely divergent tissue types, display big differences in gene expression levels, and so the vast majority of genes used (~90%) show significant changes (TaqMan assays). From this perspective, it is expected that FDR control, for genes found differentially expressed by microarrays, will have little impact on FP rates. This explains why the results obtained from the simple t-test, and the t-test with FDR control, are similar. In a previous comparative study [15] where samples with smaller differences were used, we have seen that these two methods show bigger differences in specificity. However, restrictions on the magnitude of gene fold change reduce both the number of TPs and FPs. Finally, SAM method produces even more specific results (as expected) penalizing some of the low expressers that TaqMan assays find differentially expressed.

11.2.6   Reproducibility of Differentially Expressed Gene Lists

A fundamental step in most microarray experiments is determining the lists of differentially expressed genes that distinguish biological conditions. Reproducibility of differentially expressed

TABLE 11.3
Significantly Differentially Expressed Genes Concordance

	TP	TP Rate	TN	FP	FDR	FN
TaqMan	2283	0	261	0	0	0
Quantile	1415	61.98	140	101	7.14	888
Median	1374	60.18	142	99	7.79	929
Scale	1354	59.31	144	97	7.76	949
VSN	1393	61.02	142	99	7.69	910
Loess	1402	61.41	139	102	7.84	901

Notes: Only genes detected in both samples by TaqMan assays (761) are used for this comparison; t-test is used to detect significantly differentially expressed genes, controlling FDR at 5% level. We compare the results from each normalization method, for all sites, with the ones from TaqMan assays keeping track of up/down-regulation in each platform.

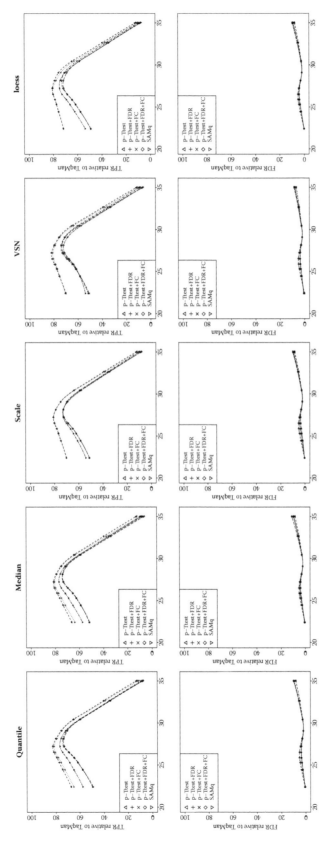

FIGURE 11.9 Differential expression t-test, t-test + FDR, t-test + FC, t-test + FDR + FC cut, and SAM applied to quantile normalization: we use different methods to detect significantly differentially expressed genes for different normalization methods: (1) t-test ($p < 0.05$), (2) t-test controlling FDR at 5% level, (3) t-test ($p < 0.05$) and FC < 1.5, (4) t-test controlling FDR at 5% level and FC < 1.5, or (5) SAM $q < 0.05$. We compare the results for data generated by site 1, from each normalization method, with the ones from TaqMan assays for which differential expression is detected using t-test and controlling FDR at 5% level. We keep track of up/down-regulation in each platform.

TABLE 11.4
Differential Expression *t*-test, *t*-test + FDR, *t*-test + FC, *t*-test + FDR + FC Cut, and SAM Applied to Quantile Normalized Data

Method	TP	TP Rate	TN	FP	FDR	FN
TaqMan	763	0	85	0	0	0
p-test	558	73.13	55	30	5.10	205
p-test + FDR	549	71.95	56	29	5.02	214
p-test + FC	483	63.30	64	21	4.17	280
p-test + FDR + FC	479	62.78	64	21	4.20	284
SAM *q*	508	66.58	63	22	4.15	255

Notes: We use different methods to detect significantly differentially expressed genes for data quantile normalized: (1) *t*-test ($p < 0.05$), (2) *t*-test controlling FDR at 5% level, (3) *t*-test ($p < 0.05$) and FC < 1.5, (4) *t*-test controlling FDR at 5% level and FC < 1.5, or (5) SAM $q < 0.05$. We compare the results for data generated by site 1, from each normalization method, with the ones from TaqMan assays for which differential expression is detected using *t*-test and controlling FDR at 5% level. We keep track of up/down-regulation in each platform.

genes across highly similar experiments becomes one of the important aspects in assessing reliability of microarray results (MAQC study). We use percentage of overlapping genes (POG) between differentially expressed genes lists as the measure of reproducibility [9]. For each testing site and each normalization method, we declare genes differentially expressed again using *t*-test and controlling FDR at 5% level. In this way, for each site we generate a list of differentially expressed genes. Figure 11.10 shows the overlap between these lists of genes for each normalization method. Table 11.5 summarizes both percentages and counts of gene overlapping between either pairs of sites or all three sites. One can see that the POG obtained from different normalization methods are similar, ranging from 69.87% (for scale normalized data) to 74.01% (for data Loess normalized) when all three sites are compared. Site 2 shows some differences compared to the other two sites, whereas the comparison between sites 1 and 3 shows consistently 83% POG. Almost similar results were observed when the other four statistical methods were used for generating gene lists for microarray data (data not shown). It is also important to note that no gene showed discordant results between brain and UHR, that is, a significant fold change in opposite direction, when different normalization methods were used.

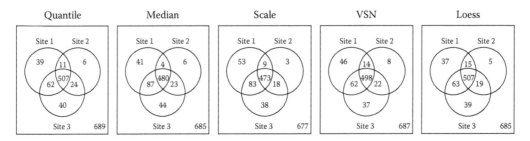

FIGURE 11.10 Reproducibility of differentially expressed gene lists: we use *t*-test controlling FDR at 5% level within each site to detect significantly differentially expressed genes for normalized data. We compare the results from each normalization method, across the three sites.

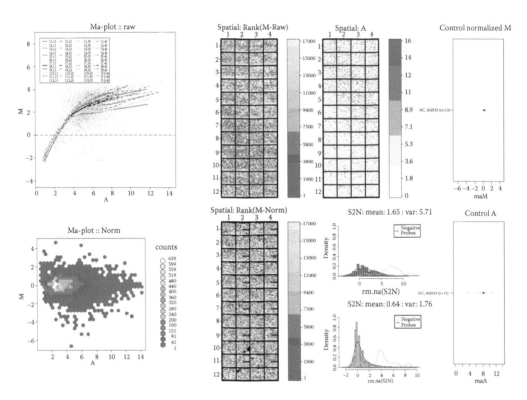

FIGURE 2.1 Diagnostic plots for two-channel array quality assessments.

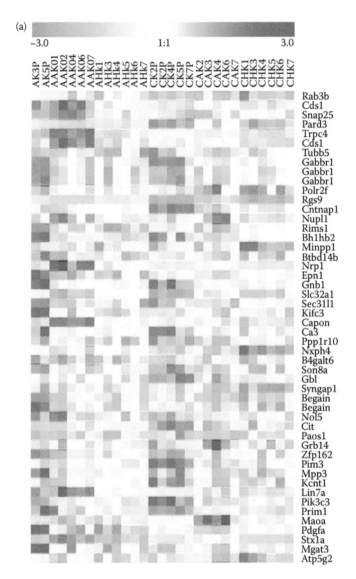

FIGURE 2.3 Hierarchical clustering analysis applied to gene expression data. Hierarchical clustering method (a) has been applied to an unordered dataset (b) the hierarchical clustering clearly reveals underlying patterns that can help identify coexpressed genes.

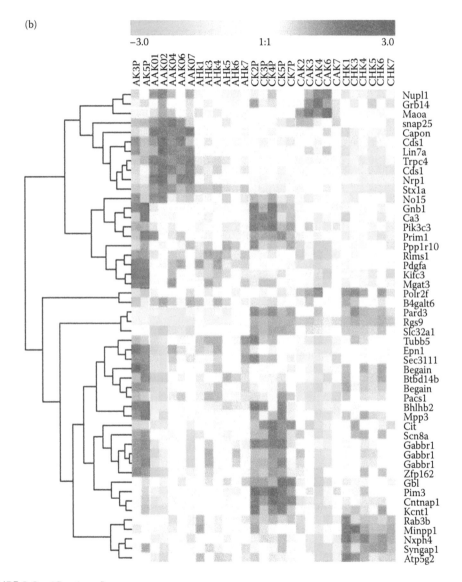

FIGURE 2.3 (*Continued*)

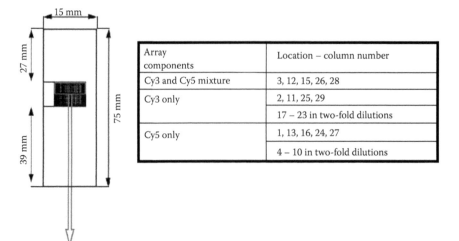

Array components	Location – column number
Cy3 and Cy5 mixture	3, 12, 15, 26, 28
Cy3 only	2, 11, 25, 29
	17 – 23 in two-fold dilutions
Cy5 only	1, 13, 16, 24, 27
	4 – 10 in two-fold dilutions

Colours number   1 2 3 4 5 6 7 8 9 10 11 12 13 14 15 16 17 18 19 20 21 22 23 24 25 26 27 28 29

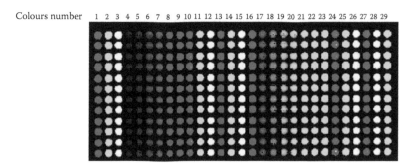

FIGURE 5B.1   FMB Scanner Validation Slide layout.

I. Increase of excitation

(a)

λ_{ex}

Optical interference coating

Organic coating

BF33 glass

(b)

2E

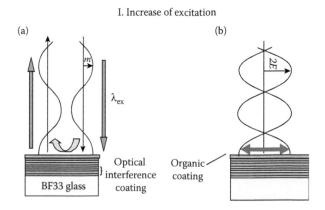

II. Maximizing detection towards reflector

λ_{ex}

Fluorescence emission

Spots with labeled target

λ_{ex}

BF33 glass

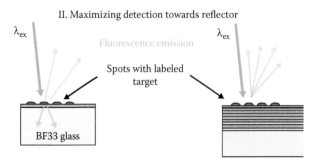

FIGURE 6.2   Schematic illustration of the principle of the optical interference coating. (a) Reflecting wave overlaps incoming wave; (b) constructive interference results in amplification of the excitation energy.

Slide E   HiSens E

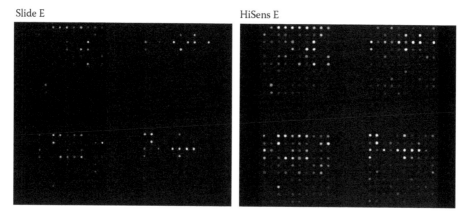

FIGURE 6.4 The same amount of miRNA (bladder versus lung samples) target was hybridized to oligo probes printed on HiSens E and Slide E. More signals (low expressor) and higher signals are visible on HiSens E.

(a) Spot diameter on slide E depending on detergent concentration in LEB

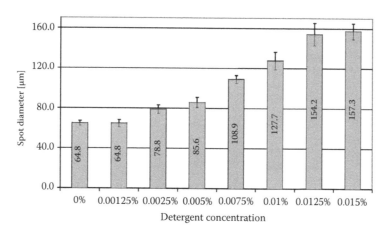

(b) Evaporation rate of LEB versus standard print buffer

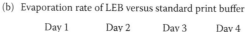

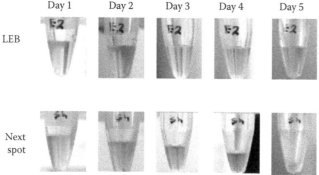

FIGURE 6.5 (a) Effect of detergent concentration on spot size in low evaporation buffer (LEB); (b) evaporation features of LEB versus standard spot buffer.

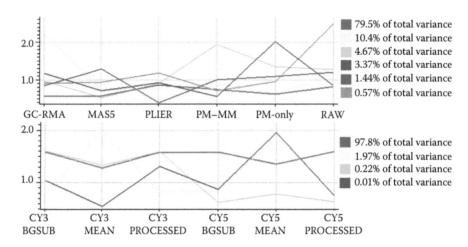

FIGURE 9.2 Principal components analysis of 2215 probes for Affymetrix and 18,513 probes for Agilent. X-axis is the mean for each tissue and normalization, Y-axis is the underweighted dot products for each condition's expression profile. Top panel (Affymetrix) shows the six most probably components of variance. The fourth component shown in blue accounts for ~3% of the total variance and indicates that GC-RMA and dChip PM are the furthest away from the other normalizations, as borne out by analysis of variance (ANOVA) overlap and clustering. Bottom panel (Agilent) shows that the second and fourth components in yellow and blue, respectively, account for the differences between raw (MEAN) and processed signal for both channels, and only amount to ~2% of the total variance. The cyan line is the variance component caused by the difference between channels, and is remarkably low.

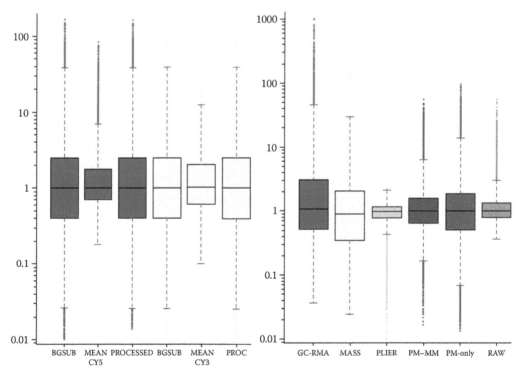

FIGURE 9.3 Boxplot showing the range of values for all samples per normalization. Left panel (Agilent) shows the compression and right-shift skew of near-raw data. Normalization creates a Gaussian distribution that is significantly different from raw data. Left panel (Affymetrix) shows the compression of PLIER (left-skewed) and RAW (right-skewed) data. Only dChip methods have a near-Gaussian distribution.

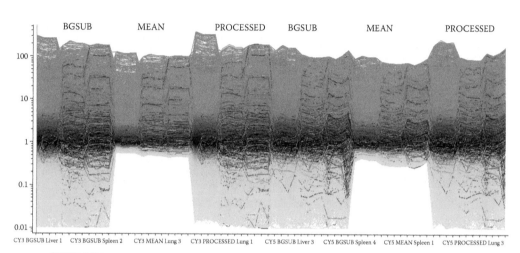

CY3 BGSUB Liver 1   CY3 BGSUB Spleen 2   CY3 MEAN Lung 3   CY3 PROCESSED Lung 1   CY5 BGSUB Liver 3   CY5 BGSUB Spleen 4   CY5 MEAN Spleen 1   CY5 PROCESSED Lung 3

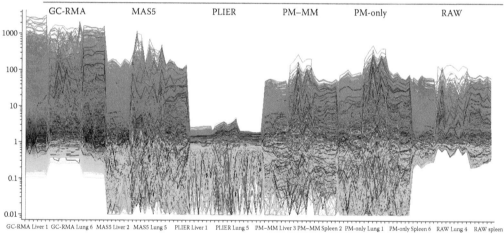

GC-RMA Liver 1 GC-RMA Lung 6 MAS5 Liver 2 MAS5 Lung 5 PLIER Liver 1 PLIER Lung 5 PM–MM Liver 3 PM–MM Spleen 2 PM-only Lung 1 PM-only Spleen 6 RAW Lung 4 RAW spleen

FIGURE 9.4 Visualization of signal dynamic range per tissue and normalization. *X*-axis is sample, right panel is $\log_{10}$ signal. Top panel (Agilent) shows the compression of low-signal strength probes that are expanded by the normalization methods. Bottom panel (Affymetrix) again emphasizes the left compression of PLIER, the general compression of RAW signals, and the right skew of GC-RMA.

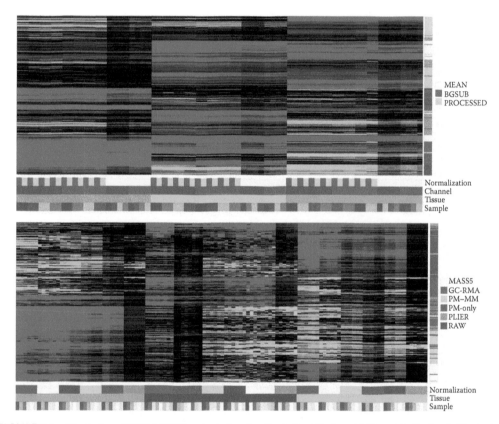

MEAN
■ BGSUB
■ PROCESSED

Normalization
Channel
Tissue
Sample

MASS5
■ GC-RMA
PM–MM
PM-only
PLIER
RAW

Normalization
Tissue
Sample

FIGURE 9.5 Clustering of 1000 ANOVA-selected probes for Agilent (top) and Affymetrix (bottom). Clustering is not tissue-weighted because the Model 1 ANOVA selected only those probes that differed from the total sums of squares per probe. As seen in the top panel the difference between the tissues dominates, and PROCESSED and BSUB normalizations are virtually indistinguishable. Channel has a slight effect in the intensity scale, but little in the clustering results. Bottom panel shows the different alignments to which each normalization approach conforms. Occasionally MAS5 and dChip PM–MM cluster, showing the differential effect of background-subtraction methods versus variance stabilization methods. Note also that groupings are not consistent per tissue.

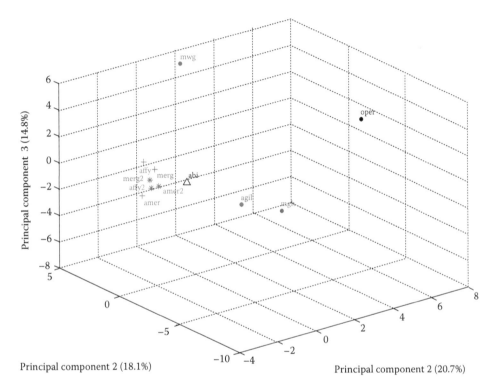

FIGURE 12.3 Cross-platform PCA plot. The plot illustrates PCA performed on $\log_2$ ratios corresponding to 130 RS identifiers common to eight of the 10 platforms. As academic cDNA and Compugen had few RS identifiers in common with the other platforms, we chose to exclude these from this analysis in order to increase the number of genes applicable to PCA without missing values. For Affymetrix (affy2), Amersham (amer2), and Mergen (merg2) expression profiles obtained from a second laboratory were included in the analysis. The other expression profiles are labeled with abbreviations of the platform names used elsewhere. Each expression profile is plotted according to the first, second, and third principal components. For each axis, the number in parenthesis gives the amount of variation (in percent of total) accounted for by the corresponding principal component.

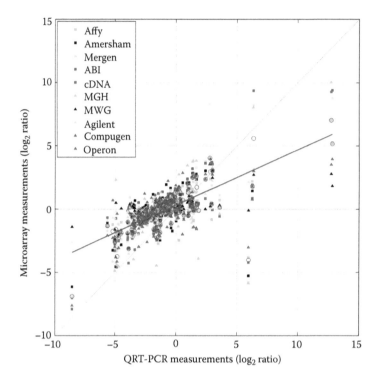

FIGURE 12.5 Scatter plot of QRT-PCR versus all microarrays. Log$_2$ ratios from the microarray platforms are plotted (y-axis) versus the corresponding log$_2$ ratios from QRT-PCR (x-axis). From each platform, all log$_2$ ratios based on probes that could be mapped to any of the 153 RS identifiers used in QRT-PCR were used. The regression line between the median microarray log$_2$ ratios (over all platforms including a given gene) and the QRT-PCR log$_2$ ratios is shown in blue. The slope of the line is 0.437, indicating a smaller dynamic range for the microarrays as compared to QRT-PCR. The Pearson correlation coefficient between the median measurements and QRT-PCR was 0.76 (p-value 1.36 E-30), indicating relatively good albeit highly significant correlation.

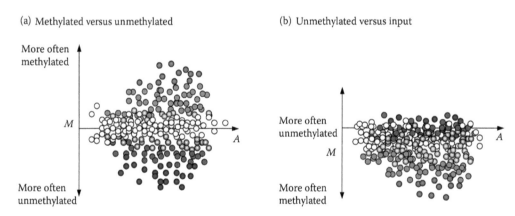

FIGURE 13.4 Illustrations of MA-plots for DNA methylation array data for experiments where (a) the methylated and unmethylated fractions from a single sample are cohybridized and (b) the unmethylated fraction from a single sample is cohybridized with the input (unenriched) fraction. M represents the log-ratio of the Cy5 and Cy3 channel intensities and A represents the log geometric mean of the Cy5 and Cy3 channel intensities. Dots represent M and A values for probes on a microarray. Methylation status is based on an average over the genomic DNA from all cells in the sample. Blue dots correspond to sequences that are mostly methylated in the sample, whereas red dots correspond to sequences that are mostly unmethylated in the sample. Unfilled circles correspond to sequences whose methylation status is ambiguous. Methylated and unmethylated sequences are more easily distinguished in (a) than (b). Theoretically, there should be no positive M values in (b).

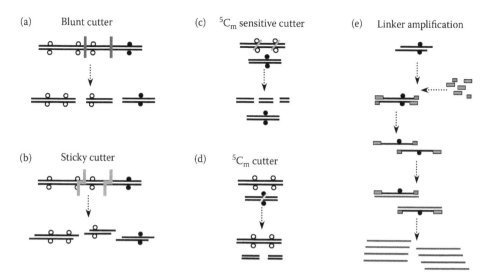

FIGURE 13.5 Restriction endonuclease enzymes cut double-stranded DNA at given recognition sites. These enzymes have different properties and activities, including (a) blunt cutting enzymes that leave no overhanging ends; (b) sticky cutters that leave overhanging ends; (c) methylation-sensitive enzymes (that only cut if their recognition site is unmethylated); and (d) methylation-dependent enzyme cutters. Linkers can be ligated to sticky or blunt ends and used to prime PCR amplification (e).

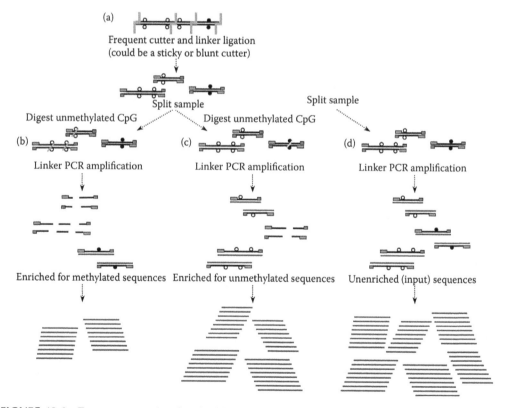

FIGURE 13.6 Enzyme approaches for obtaining methylated (b), unmethylated (c), and input (d) fractions from a single sample. Genomic DNA is cut with a frequent sticky cutter (a), linkers are ligated and the sample is split. Then separate fractions can be subjected to (b) unmethylated sites are digested such that intact methylated sequences are amplified; (c) methylated sites are digested such that intact unmethylated sequences are amplified; and (d) the input sample (with ligated linkers) is amplified. Linker PCR amplification involves denaturation step to make single-stranded DNA that is used as a template for amplification. Any of the fractions from (b), (c) or (d) can then be cohybridized to an array.

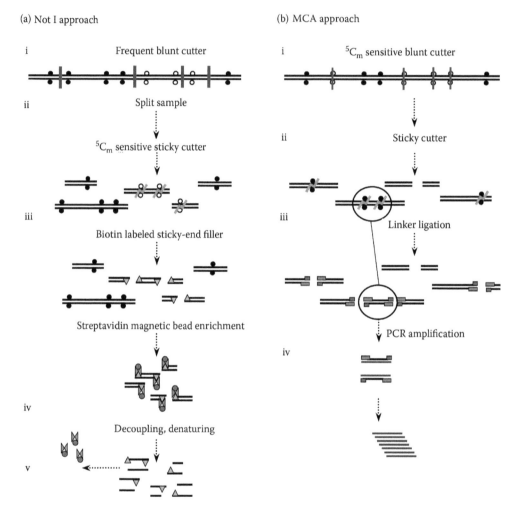

(a) Not I approach

i Frequent blunt cutter

ii Split sample

^5C_m sensitive sticky cutter

iii

Biotin labeled sticky-end filler

Streptavidin magnetic bead enrichment

iv

Decoupling, denaturing

v

(b) MCA approach

i ^5C_m sensitive blunt cutter

ii Sticky cutter

iii Linker ligation

PCR amplification

iv

FIGURE 13.7 Enzyme approaches (a) that do not require amplification and (b) that require consecutive methylated recognition site for enrichment. In (a), genomic DNA is digested with a blunt cutter (i) and the sample is split (ii). One half is further digested with a methylation-sensitive sticky cutter (iii), the sticky ends are filled with biotin-labeled nucleotides and strong binding streptavidin-coated magnetic beads are added. Sequences bound to the beads are extracted (using a magnet) (iv) and a decoupling reaction is used to release the bound sequences (v). In (b), a methylation-sensitive blunt cutter digests the genomic DNA (i) and the sample is split (ii). One half is further digested with an enzyme [with the same recognition sequence as in (i)] whose activity is not inhibited by methylation (ii). After linker ligation (iii), only sequences with both 3′ and 5′ linkers are amplified (iv).

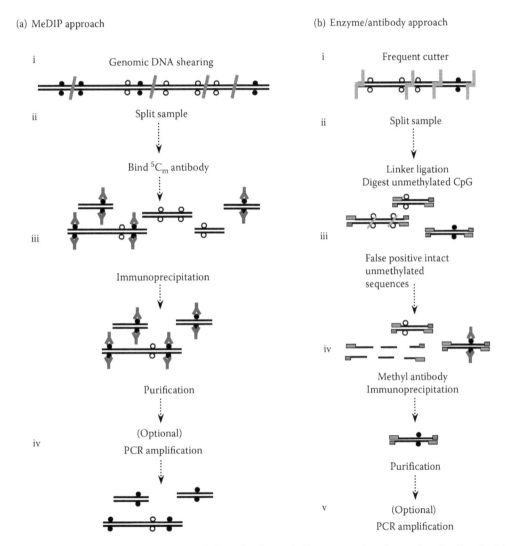

(a) MeDIP approach

i Genomic DNA shearing

ii Split sample

Bind ^5C_m antibody

iii

Immunoprecipitation

Purification

(Optional)
iv PCR amplification

(b) Enzyme/antibody approach

i Frequent cutter

ii Split sample

Linker ligation
Digest unmethylated CpG

iii

False positive intact
unmethylated
sequences

iv

Methyl antibody
Immunoprecipitation

Purification

v (Optional)
PCR amplification

FIGURE 13.8 Approaches using (a) MeDIP antibodies and (b) enzyme digestion and antibodies. In (a), genomic DNA is sheared (i) and the sample is split (ii). Antibody proteins that bind to methylated CpGs are added to one half (iii) and sequences with bound antibodies are precipitated and this enriched sample is purified. PCR amplification may be used (iv) depending on the amount of sample required for hybridization. In (b), genomic DNA is digested with a sticky cutter (i), linkers are ligated and the sample is split (ii). One half is further digested with a methylation-sensitive enzyme leaving methylated sequences intact. Any unmethylated sequences that remain intact (escaping digestion) are subsequently removed using the MeDIP approach (iv). Finally, the sample is purified and (v) amplification may be used.

(a) Genomic DNA shearing

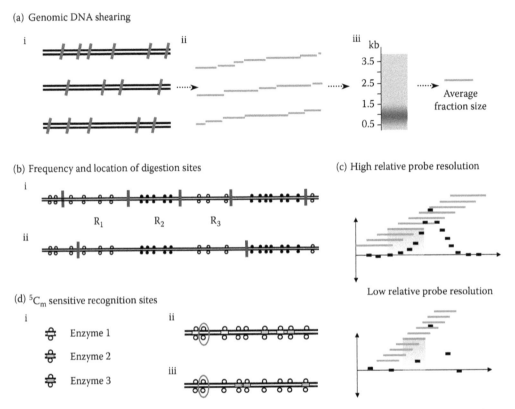

(b) Frequency and location of digestion sites

(c) High relative probe resolution

(d) ^5C_m sensitive recognition sites

Enzyme 1

Enzyme 2

Enzyme 3

FIGURE 13.9   (a) Genomic DNA shearing; depicted are three genomes (i) which (after shearing) results in random overlapping fragments (cyan segments) (ii) of varying size depending on the amount of shearing applied. The overall size of the fragments should be confirmed by running the sample on an electrophoresis gel (iii). Bearing in mind the overlapping fragments resulting from genomic shearing, data obtained from consecutive probes along the genome (c) will be correlated. The extent of correlation between neighboring probes depends on the size of the average target fragments (cyan segments) relative to the probe resolution (spacing). In (c) where there is a high relative probe resolution, a single fragment may potentially bind to five neighboring probes, but where there is a low relative probe resolution, a given fragment may only bind to one probe. (b) If enzyme digestion is used to fragment the genome, infrequent cutting (ii) results in regions of altered methylation status (R1–R3) with ambiguous methylation. In contrast, where restriction sites are just as frequent as regions with altered methylation status (i), target fragments will be informative for methylation. (d) Often a combination of enzymes (i) is required for effective and informative digestion of DNA. For a given fragment, enzyme 1 has three recognition sites, (ii), enzyme 2 has two recognition sites (ii), and enzyme 3 has three recognition sites (iii). The combination of enzymes 1 and 2 results in five of the nine CpGs being included in a recognition site (ii).

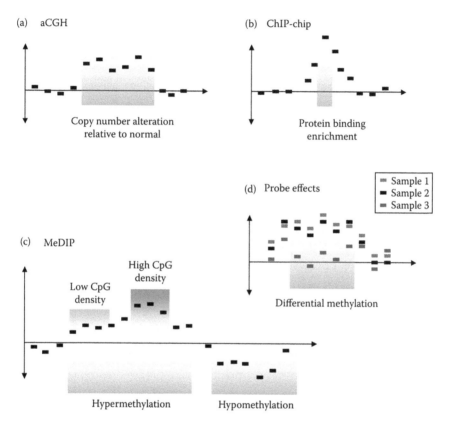

FIGURE 13.10 Log-ratio data for probes along the genome illustrating typical changes that are observed in (a) array CGH experiments, (b) ChIP-chip experiments, and (c) MeDIP methylation array experiments. Changes in log-ratio data are expected to appear like (a) gained or lost segments, (b) positive symmetric peaks, and (c) variably peaked regions (positive or negative, after normalization). In (c), the variable peak heights of a methylated region depend on the density of methylated CpGs. In addition to CpG density effects in MeDIP data, there will be (d) probe effects. Real changes in methylation (sample 3, blue data points) can be distinguished from changes due to probe effects when there are multiple samples. Similar probe effects are observed in samples 1, 2, and 3. However, sample 3 is differentially methylated across a five-probe region compared to the other samples.

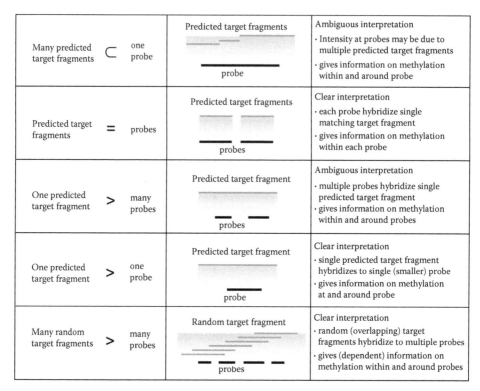

		Predicted target fragments	Ambiguous interpretation
Many predicted target fragments ⊂ one probe			· Intensity at probes may be due to multiple predicted target fragments · gives information on methylation within and around probe
		probe	
Predicted target fragments = probes		Predicted target fragments	Clear interpretation · each probe hybridize single matching target fragment · gives information on methylation within each probe
		probes	
One predicted target fragment > many probes		Predicted target fragment	Ambiguous interpretation · multiple probes hybridize single predicted target fragment · gives information on methylation within and around probes
		probes	
One predicted target fragment > one probe		Predicted target fragment	Clear interpretation · single predicted target fragment hybridizes to single (smaller) probe · gives information on methylation at and around probe
		probe	
Many random target fragments > many probes		Random target fragment	Clear interpretation · random (overlapping) target fragments hybridize to multiple probes · gives (dependent) information on methylation within and around probes
		probes	

FIGURE 13.11 Relationship between target fragments and probes in the interpretation of methylation measurements across probes. Fractionation of the genome by enzyme digestion results in predicted target fragments. Fractionation by shearing yields random (i.e., overlapping) target fragments. Target fragments that hybridize to a given probe determine the amount of methylation measured at that probe.

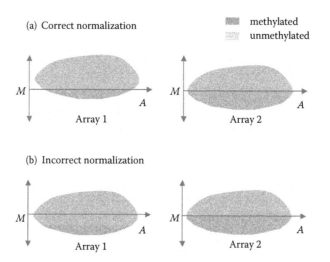

FIGURE 13.12 Diagram of MA-plots for two arrays after (a) correct normalization and (b) incorrect normalization. Shaded regions in the plots represent the areas of the plot where data would be observed. Blue areas represent data from truly methylated sequences. Orange areas represent data from truly unmethylated sequences. The sample hybridized to array 1 has considerably more methylated sequences than the sample hybridized to array 2. After correct normalization (a), methylated and unmethylated regions coincide with the $M = 0$ line for both arrays; positive M-values correspond to true methylation and negative M-values correspond to true unmethylation. After incorrect normalization, array 2 (which has been globally centered) has some of the methylated region below the $M = 0$ line, falsely represented as unmethylated.

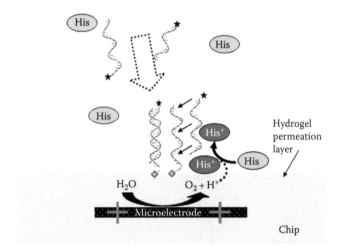

FIGURE 14.5 Illustration demonstrating the interactions between the electrochemistry at the electrode and DNA hybridization. Water oxidation at the positive electrode generates H^+ ions which protonate histidine. The protonated histidine acts to shield the negative charge on DNA thereby allowing the two single strands of DNA to approach and hybridize.

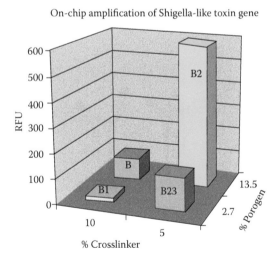

FIGURE 14.8 Comparison of assay results on four different hydrogel permeation layers. The hydrogels were formulated to optimize the performance of an on-chip SDA reaction. In this experiment, the effects of the acrylamide crosslinker and the porogen were examined. Formulation B2 gave the best performance with this type of assay.

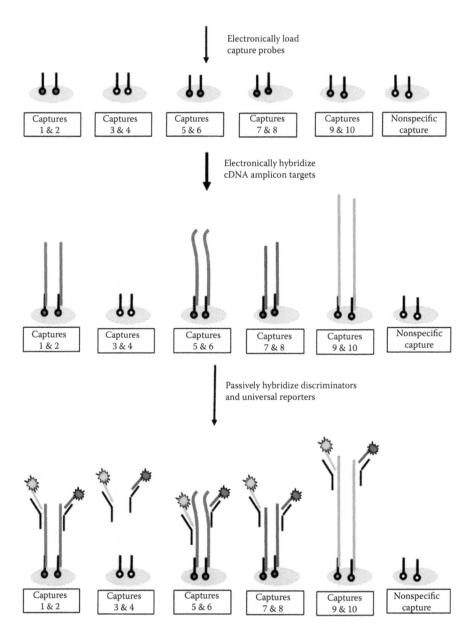

FIGURE 14.14 Assay procedure for CAP/sepsis assay. Capture probes are loaded two at a time on five array sites. The sixth array site is addressed with a nonspecific capture and used as a background control. After capture addressing, all of the cDNA amplicon targets are simultaneously hybridized to all six array sites. Next, discriminator probes and universal reporter probes are passively hybridized to the capture-amplicon complex. Finally, the red/green fluorescence signals are measured and the data analyzed.

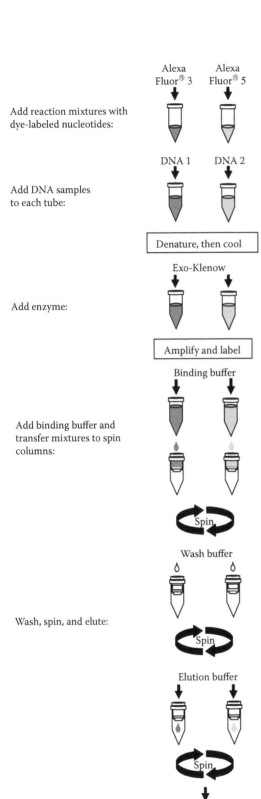

Add reaction mixtures with dye-labeled nucleotides:

Alexa Fluor® 3   Alexa Fluor® 5

Add DNA samples to each tube:

DNA 1   DNA 2

Denature, then cool

Exo-Klenow

Add enzyme:

Amplify and label

Binding buffer

Add binding buffer and transfer mixtures to spin columns:

Spin

Wash buffer

Wash, spin, and elute:

Spin

Elution buffer

Spin

Proceed to hybridization

FIGURE 15.2 BioPrime Total Array CGH Genomic Labeling Protocol.

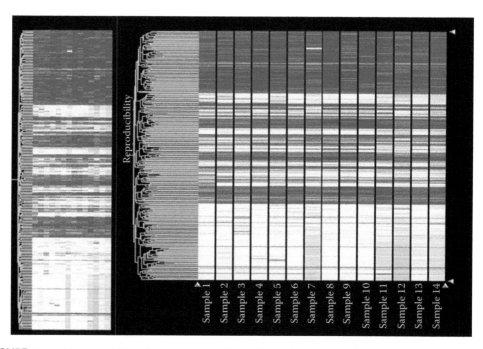

FIGURE 15.9 Reproducibility of microarray platform. The heat map and table to the right demonstrate the precision of the NCode miRNA analysis platform comparing 18 replicate homotypic arrays of 10 µg of pooled human RNA. Data were background corrected and normalized. The %cv calculated from mean $\log_2$ expression values from one array to another.

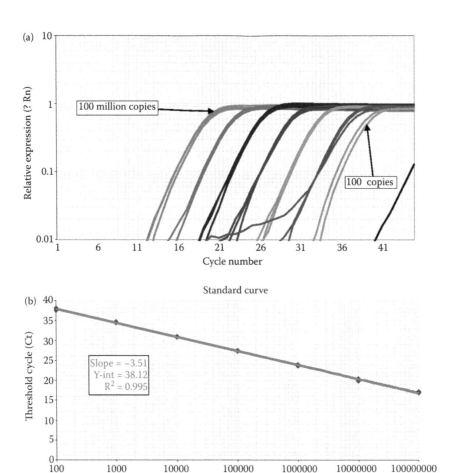

FIGURE 15.12 qRT-PCR analysis of miRNAs. The cDNA synthesis module of the NCode SYBR Green miRNA qRT-PCR Kit was used to polyadenylate a synthetic miRNA oligo and generate cDNA from the tailed miRNA. The cDNA was then amplified in duplicate qPCR reactions on an ABI PRISM® 7700 using SYBR Green I dye, a universal qPCR primer (provided in the kit), and a forward primer specific for the oligo sequence. The amount of starting material ranged from 1×10^8 to 1×10^2 copies. The assay demonstrated a wide dynamic range of seven orders of magnitude and sensitivity down to 100 copies of template.

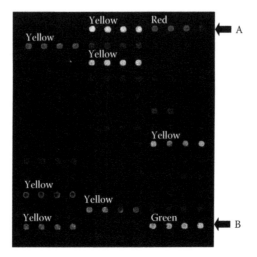

FIGURE 16.1 Cy3- and Cy5-labeled samples were used for CDH and then visualized together as in cDNA microarray analysis. DNA from sample A with a mutation in position A was labeled with Cy5, and DNA from sample B with a mutation in position B was labeled with Cy3. When the microarray was analyzed with a ScanArray 5000 (PerkinElmer, Boston, Massachusetts, USA), the signal for mutant A appeared red, that of B appeared green, and the wild-type signals shared by the two samples appeared yellow. This cDNA-like image provides a rapid and simple method for detecting mutations.

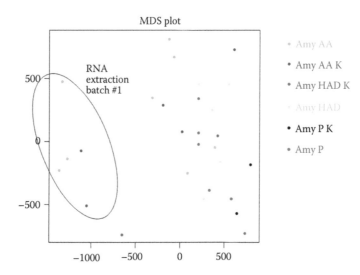

FIGURE 18.2 The multidimensional scaling plot (MDS) displays 30 samples from 6 different experimental groups (color coded). Distances within the plot represent approximately the sum of absolute differences over all genes between any two samples. Overlaying the plot with the experimental information reveals that RNA extraction is a major source of variance.

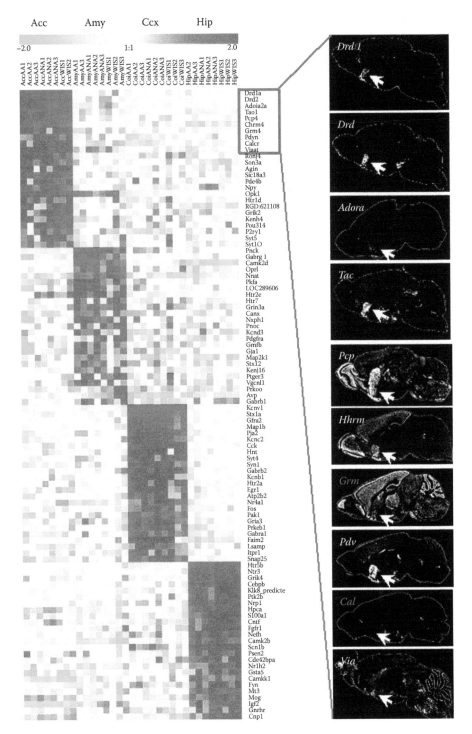

FIGURE 18.3 Clustering of 33 microarray experiments from rat forebrain. Left panel: Four brain regions (Acc: nucleus accumbens, Amy: amygdala, Ccx: cingulate cortex, Hip: hippocampus) from three rat lines (AA, ANA, Wistar) were analyzed for common region-specific gene expression patterns. Only the top 25 genes per cluster are shown. Data are reanalyzed from a previously published experiment [125]. Right panel: Verification of accumbens-specific gene expression using the Allen Brain Atlas (http://mouse.brain-map.org; [135]). For 8 of the 10 selected genes, the atlas shows increased *in situ* hybridization signals in the ventral striatum including the nucleus accumbens. Two genes (red labels) could not be verified.

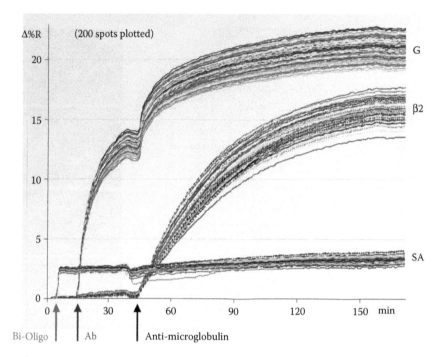

FIGURE 19.7 Kinetic binding assay as a function of percent change in reflectivity in a high-density SPR microarray assay. The change in reflectivity is calculated based on a comparison to time zero. In a flow cell attached to the microarray, a biotinylated oligonucleotide was independently added until it reached a binding equilibrium as indicated by the flat line. Afterwards IgG was introduced, until reflectivity was stable, then a microglobulin-specific IgG was added which caused an increase of reflectivity change for both Protein G and beta-2-microglobulin.

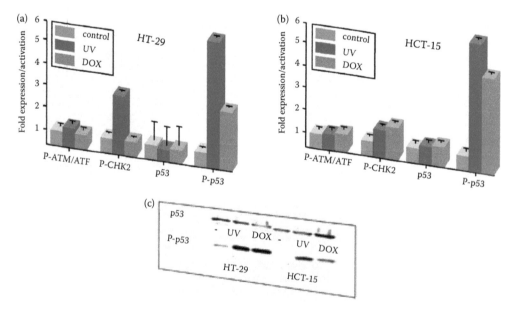

FIGURE 19.8 Biomarker detection using reverse phase arrays. Protein levels measured across three conditions (control—blue, UV—red, DOX—orange). (a) Expression profile in HT-29 cells; (b) expression profile in HCT-15 cells; and (c) Western blot data used to validate the specificity of p53 and P-p53 antibodies.

TABLE 11.5
Reproducibility of Differentially Expressed Gene Lists

Method	Sites 1 and 3	Sites 2 and 3	Sites 1 and 2	Sites 1, 2, and 3
Quantile	83.3	81.6	79.8	73.58
Median	83.5	78.1	75.5	70.07
Scale	82.5	78.6	75.4	69.87
VSN	82.4	81.7	78.8	72.49
Loess	83.8	81.2	80.8	74.01
Method	**Site 1**	**Site 2**	**Site 3**	**Common**
Quantile	619	548	633	507
Median	612	513	634	480
Scale	618	503	612	473
Loess	622	546	628	507
VSN	620	542	619	498

Notes: We use *t*-test controlling FDR at 5% level within each site to detect significantly differentially expressed genes for normalized data. We compare the results for each normalization method, from the three sites.

11.3 DISCUSSION

One unanswered question in the microarray field has always been the effect of various normalization as well as statistical methods on the end results of a profiling experiment and more explicitly whether using different normalization or statistical approaches results in different gene lists of less concordance between different microarray platforms. In this study, we have assessed five different methods used for normalization of Applied Biosystems Expression Array System data for their performance. Our results show a high level of concordance between these normalization methods. This is true, regardless of whether signals, variation, or fold change measurements were interrogated. In addition, these five normalization methods showed similar performance of signal reproducibility between the three testing sites used for this study. Furthermore, we used TaqMan assays as a reference to generate TPR and FDR plots for the various normalization methods across the assay range (Figure 11.8). TPR was directly correlated to gene expression levels, whereas FDR was inversely correlated. This is not completely surprising as the two platforms have different dynamic ranges and sensitivity levels, with the detection levels of the microarrays being lower than those of TaqMan assays. These differences more than likely explain the lower TP rates and higher FP rates for the genes at the low expression levels. These effects were also observed for several other microarray platforms in a separate study [16]. One conclusion of this study is that, at least for the Applied Biosystems Gene Expression Arrays, the current normalization approaches have little impact on the signal, detection levels as well as TP and FP rates in detection of differentially expressed genes. In addition, we also explored the contribution of several statistical approaches commonly used in the field on the TP and FP rates. As expected in this case, the consequences of varying stringencies in differential expression detection are that although a relaxed stringency can lead to better detection and differential expression concordance, there will be a higher percentage of FPs. At the opposite end of the spectrum, FDR control, and SAM methods, which are more restrictive in detection of differential expression, produce gene lists with fewer FPs. SAM method, as expected, shows a reduced number of FPs for low expression level genes, on the expense of missing some differentially expressed genes, which is also an effect of the lower sensitivity observed for microarrays. The expected percentage of FPs in these lists is close to the one observed when comparing results to TaqMan assays. We believe that the full strength of these methods is obscured by the fact that the vast majority of the genes chosen for TaqMan validation (90%) show significant fold changes between samples. More importantly however, applying the different normalization approaches to the various statistical methods tried, had no significant impact on identifying differentially expressed genes.

Finally, when comparing the overlap in gene lists generated by each of these statistical methods, a concordance of 69.7–74.01% was observed between all three sites, and 82.4–83.8% between sites 1 and 3, indicating little effect of the analysis approach used on the final gene list obtained. This result is, however, sensitive to the cut-offs used in determining the gene lists and can affect the degree of overlap observed [9]. We were pleasantly surprised, however, at the little effect observed by the various normalizations on the statistical approaches analyzed, which indicates a certain robustness of the analysis methods currently in use in the field, particularly when used in conjunction with the Applied Biosystems Gene Expression System.

11.4   METHODS

11.4.1   RNA Samples

11.4.1.1   Sample Definition
Sample A was UHR RNA (Stratagene) and sample B was human brain total RNA (Ambion).

11.4.1.2   Selection of Genes for Validation by TaqMan Assays
A list of 1297 RefSeqs was selected by the MAQC consortium. Over 90% of these genes were selected from a subset of 9442 RefSeq common to the four platforms (Affymetrix, Agilent, GE Healthcare, and Illumina) used in the MAQC Pilot-I Study (RNA Sample Pilot), based on annotation information provided by manufacturers in August 2005. This selection ensured that the genes would cover the entire intensity and fold-change ranges and include any bias due to RefSeq itself. One thousand TaqMan gene expression assays were used in the study that matches with the MAQC gene list. These 1000 assays covered 997 genes (three genes had more than one assay).

11.4.2   Applied Biosystems Expression Array Analysis

The Applied Biosystems Human Genome Survey Microarray (P/N 4337467) contains 31,700 60-mer oligonucleotide probes representing 29,098 individual human genes. Digoxigenin-UTP-labeled complementary ribonucleic acid (cRNA) was generated and amplified from 1 μg of total RNA from each sample using Applied Biosystems Chemiluminescent RT-IVT Labeling Kit v 1.0 (P/N 4340472) according to the manufacturer's protocol (P/N 4339629). Array hybridization was performed for 16 h at 55°C. CL detection, image acquisition, and analysis were performed using Applied Biosystems Chemiluminescence Detection Kit (P/N 4342142) and Applied Biosystems 1700 Chemiluminescent Microarray Analyzer (P/N 4338036) following the manufacturer's protocol (P/N 4339629). Images were auto-gridded and the chemiluminescent signals were quantified, background subtracted, and finally, spot and spatially normalized using the Applied Biosystems 1700 Chemiluminescent Microarray Analyzer software v 1.1 (P/N 4336391). Five technical replicates were performed on each sample, at three different testing sites, for a total of 30 microarrays.

11.4.3   TaqMan Gene Expression Assay-Based Real-Time PCR

11.4.3.1   TaqMan Assays
Each TaqMan Gene Expression Assay consists of two sequence-specific PCR primers and a TaqMan assay—FAM™ dye-labeled mirror groove binder (MGB) probe. Each TaqMan assay was run in four replicates for each RNA sample. A 10-ng total cDNA (as total input RNA) in a 10-μl final volume was used for each replicate assay. Assays were run with 2 × Universal PCR Master Mix without uracil-N-glycosylase (UNG) on Applied Biosystems 7900 Fast Real-Time PCR System using universal cycling

conditions (10 min at 95°C; 15 s at 95°C, and 1 min at 60°C, 40 cycles). The assays and samples were analyzed across a total of 44–384 well plates. Robotic methods (Biomek FX) were used for plate setup and each sample and assay replicate was tracked on a per-well and per-plate basis.

11.4.4   DATA ANALYSIS

Statistical analyses were performed using the open source and open development software project R together with the Bioconductor packages ab1700, limma, multtest, and affy (http://www.bioconductor.org).

11.4.5   NORMALIZATION METHODS

When running experiments that involve multiple high-density long-oligonucleotide arrays, it is important to remove sources of variation between arrays of nonbiological origin. Normalization is a process for reducing this variation. We present five methods of performing normalization at the probe intensity level.

11.4.5.1   Scale Normalization

Scale normalization was proposed by Yang et al. [6], and is further explained by Smyth and Speed [7]. The idea is to scale the log-ratios to have the same median-absolute-deviation (MAD) across arrays.

11.4.5.2   Global Median

The idea is to scale the log-ratios to have the same median across arrays [5].

11.4.5.3   Quantile Normalization

Quantile normalization was proposed by Bolstad et al. [3] for Affymetrix-style single-channel arrays and by Yang and Thorne [4] for two-color cDNA arrays. This method ensures that the intensities have the same empirical distribution across arrays.

11.4.5.4   VSN

Based on a function (arsinh) that calibrates for sample-to-sample variations through shifting and scaling, and transforms the intensities to a scale where the variance is approximately independent of the mean intensity [8].

11.4.5.5   Cyclic Loess

This approach is based on the idea of the M versus A plot, where M is the difference in log expression values and A the average of the log expression values, presented in Dudoit et al. [17]. However, rather than being applied to two-color channels on the same array, as is done in the cDNA case, it is applied to probe intensities from two arrays at a time.

11.4.6   SIGNAL DETECTION ANALYSIS

Detection thresholds are defined according to each platform manufacturer's recommendation. For TaqMan Gene Expression Assays, detection threshold is set as $Ct < 35$ and standard deviation (of the four technical replicates) <0.5; for Applied Biosystems Expression Arrays, detection threshold is set as $S/N > 3$ and quality flag <5000. Detection in each sample was defined as detectable in three out of four technical replicates for TaqMan assays and three out of five technical replicates within each site for microarrays. Using TaqMan Gene Expression Assays calls as the reference, contingency tables were constructed against microarrays, in which TPRs (genes detectable by both TaqMan assay and microarrays as a percentage of all genes detectable by TaqMan assays) were plotted against TaqMan Ct values (Figure 11.1).

11.4.7 VARIABILITY WITHIN AND BETWEEN SITES FOR DIFFERENT NORMALIZATION METHODS FOR APPLIED BIOSYSTEMS MICROARRAY SYSTEM

CV is used to measure variability within each site. In Figure 11.4, we present the dependency between CV of site 1, with TaqMan Ct measurements for each normalization method and each sample [18]. These curves represent the Lowess approximation of the CV between the five technical replicates of all genes against the Ct measurement.

To quantify the variability between the sites these normalization methods produce, we perform one-factor (site) analysis of variance (ANOVA) on all 29,098 genes. In this way, we estimate the percent variability from the total variability (of each gene) that can be explained from site variability (Figure 11.5). For each gene, CVs are plotted against median expression level measured by quantile normalized data, and Lowess fitting curves are used to approximate the all points generated from one normalization method.

11.4.8 TPRs AND FDRs IN DETECTION OF DIFFERENTIALLY EXPRESSED GENES FOR DIFFERENT NORMALIZATION METHODS

To have a comprehensive understanding of the performance of these five normalization methods, detection of differentially expressed genes between UHR and brain samples is a key issue. Only genes detected in both samples A and B by TaqMan assays were used in this comparison. Significantly differentially expressed genes between samples were defined as $p < 0.05$ based on a student's t-test controlling FDR at 5% level (BH). Using calls from TaqMan Gene Expression Assays as the reference, contingency tables were constructed against the different normalization methods in which we are taking into considerations both p-value significance and fold-change direction (up- or down-regulation). Based on this matrix, the TPR, FPR, FDR, and accuracy were calculated for each normalization method. Results are presented in Table 11.3. A more detailed representation of TPRs and FDRs, as functions of Ct measurements are presented in Figure 11.8. Genes were first ranked according to their average value in the tissue comparison. For each bin of 50 consecutive genes (according to the ranking), we compare the results from each normalization method with the ones from TaqMan assays. We keep track of up- or down-regulation in each platform. The average value of these 50 genes in the two samples is plotted against TPR or FDR of the concordance between the two platforms in detecting differentially expressed genes.

REFERENCES

1. Hackett JL and Lesko LJ. Microarray data—the US FDA, industry and academia. *Nat Biotechnol* 2003; 21: 742–743.
2. Petricoin EF, 3rd, Hackett JL, Lesko LJ, Puri RK, Gutman SI, Chumakov K, Woodcock J, Feigal DW, Jr., Zoon KC, and Sistare FD. Medical applications of microarray technologies: A regulatory science perspective. *Nat Genet* 2002; 32(Suppl.): 474–479.
3. Bolstad BM, Irizarry RA, Astrand M, and Speed TP. A comparison of normalization methods for high density oligonucleotide array data based on variance and bias. *Bioinformatics* 2003; 19: 185–193.
4. Yang YH and Thorne NP. Normalization for two-color cDNA microarray data. In: DR Goldstein (ed.), *Science and Statistics: A Festschrift for Terry Speed*, Institute of mathematical statistics, Lecture Notes—Monograph Series, 2003, vol. 40, pp. 403–418.
5. Hartemink A, Gifford D, Jaakkola T, and Young R. Maximum likelihood estimation of optimal scaling factors for expression array normalization. In: M Bittner, Y Chen, A Dorsel, and E Dougherty (eds), *Microarrays: Optical Technologies and Informatics*. Proc. SPIE, 2001; vol. 4266, pp. 132–140.
6. Yang YH, Dudoit S, Luu P, Lin DM, Peng V, Ngai J, and Speed TP. Normalization for cDNA microarray data: A robust composite method addressing single and multiple slide systematic variation. *Nucleic Acids Res* 2002; 30(4): e15.
7. Smyth GK and Speed T. Normalization of cDNA microarray data. *Methods* 2003; 31(4): 265–273.

8. Huber W, Heydebreck A, Sueltmann H, Poustka A, and Vingron M. Variance stabilization applied to microarray data calibration and to the quantification of differential expression. *Bioinformatics* 2002; 18(Suppl. 1): S96–S104.

9. MAQC consortium, Shi L, Reid LH, Jones WD, Shippy R, Warrington JA, Baker SC, et al. The MicroArray Quality Control (MAQC) project shows inter- and intraplatform reproducibility of gene expression measurements. *Nat Biotechnol* 2006; 24(9): 1151–1161.

10. Applied Biosystems Expression Array System Available at: http://www3.appliedbiosystems.com/cms/groups/mcb_marketing/documents/generaldocuments/cms_040420.pdf.

11. Heid CA, Stevens J, Livak KJ, and Williams PM. Real time quantitative PCR. *Genome Res* 1996; 6: 986–994.

12. Gibson UE, Heid CA, and Williams PM. A novel method for real time quantitative RT-PCR. *Genome Res* 1996; 6: 995–1001.

13. Cleveland WS and Devlin SJ. Locally-weighted regression: An approach to regression analysis by local fitting. *J Am Stat Assoc* 1988; 83(403): 596–610.

14. Benjamini Y and Hochberg Y. Controlling the false discovery rate—a practical and powerful approach to multiple testing. *J Roy Statist Soc Ser B Met* 1995; 57(1): 289–300.

15. Wang Y, Barbacioru C, Hyland F, Xiao W, Hunkapiller KL, Blake J, Chan F, Gonzalez C, Zhang L, and Samaha RR. Large scale real-time PCR validation on gene expression measurements from two commercial long-oligonucleotide microarrays. *BMC Genomics* 2006; 7: 59.

16. Canales RD, Luo Y, Willey JC, Austermiller B, Barbacioru CC, Boysen C, Hunkapiller K, et al. Evaluation of DNA microarray results with alternative quantitative technology platforms. *Nat Biotechnol* 2006; 24(9): 1115–1122.

17. Dudoit S, Yang YH, Callow MJ, and Speed TP. Statistical methods for identifying genes with differential expression in replicated cDNA microarray experiments. *Stat Sin* 2002; 12(1): 111–139.

18. Wong ML and Medrano JF. Real-time PCR for mRNA quantitation. *Biotechniques* 2005; 39: 75–85.

12 A Systematic Comparison of Gene Expression Measurements across Different Hybridization-Based Technologies

Winston Patrick Kuo, Fang Liu, Tor-Kristian Jensen,
Shari L. Benson, Connie L. Cepko, Eivind Hovig, and
Lucila Ohno-Machado

CONTENTS

12.1   INTRODUCTION

High-throughput ribonucleic acid (RNA)-oriented technologies have become powerful tools in both basic and applied areas of biomedical research and have matured significantly over the past decade. The application of these tools is expected to become a means to explore, classify, and predict the biological processes underlying human diseases, justifying claims for "personalized medicine" and "targeted drug development." Many variants of these technologies exist and new implementations continue to be developed in the hope of providing more precise relative and absolute transcript abundance measurements. Generally, they can be grouped into two categories: "hybridization" and "sequencing"-based approaches. Hybridization approaches include all forms of oligonucleotide-based microarrays [1] and complementary deoxyribonucleic acid (cDNA) microarrays [2], whereas sequencing approaches include serial analysis of gene expression (SAGE) [3], massively parallel signature sequencing (MPSS) [4], and next-generation sequencing technologies [5,6]. In the past few years, data generated by the above technologies have been overwhelming. Efficient access to these data is expected to allow for comparison and integration of data obtained in related biological systems that will provide clinicians and researchers with an opportunity to address complex questions in an effective way.

The current expectation is that this type of technology will extend its current role as an experimental tool for basic science research and be increasingly applied in clinical practice. Recently, there have been several large-scale efforts to create standardized protocols for different aspects of a microarray experiment (from probe annotation to data analysis) for commonly used commercial platforms and even for platforms developed "in-house." The Minimum Information About a Microarray Experiment (MIAME) standard has required microarray users to report their experimental design, description of the sample, hybridization protocol, data imaging conditions, and data analysis using a predetermined level of detail [7]. The External RNA Controls Consortium (ERCC) has focused on developing a universal RNA standard to aid in the normalization of intensities from different microarray platforms. Another recent initiative, the MicroArray Quality Control (MAQC) project [8], has created a community-wide collaborative effort initiated and led by Food and Drug Administration (FDA) scientists involving 137 participants from 51 organizations to assess the performance of different microarray platforms. All these initiatives aimed at improving the quality of microarray data by creating standards.

Currently, many gene expression data sets have been stored in well-known public repositories, such as Gene Expression Omnibus (GEO) [9] and ArrayExpress [10], which have become major portals for deposition and retrieval of data for researchers. The usefulness of such repositories rests in sufficient reliability on the experiments and their annotation so that meaningful results can be extracted. In the early stage of microarray development, it was not clear whether discrepancies within platforms were inherent to the technology. Given the diversity of platforms and the corresponding plethora of microarray data, an important issue is whether the platforms measure gene expression differently, and, if so, how data from different platforms can be compared or combined. Several attempts have been made to address these issues [11–47], though the reported results have been mixed and had continued to be debated, with data to support both those who claim that it is possible to obtain comparable results across platforms and those who claim that it is not. Owing to the diversity of technical and analytical sources that can affect the results of an experiment, the scientific community has been trying to establish guidelines for standardizing certain experimental procedures that are involved in generating and analyzing microarray data. Although an increasing body of information is being created, at least one of the following factors may have biased the results of previous studies: (1) nonidentical samples were processed in different laboratories; (2) the biological samples were not chosen well enough to be sufficiently distinct; (3) samples were processed using different protocols; (4) there were no replicates within each type of platform; (5) data preprocessing steps were not standardized; (6) few types of platforms were directly compared; (7) measurements were matched according to annotations such as UniGene (UG) clusters, rather than probe sequences; (8) "agreement" was not unambiguously quantified; or (9) biological validation was not performed or was limited to a very small number of genes. While some of the above conditions may be reflective of the anticipated use of these platforms in practice, they make it impossible to measure the magnitude of disagreement that can be attributed to differences within or between platforms per se. In a recent special issue in Nature Methods, several comparison studies involving microarrays have justified "guarded optimism" for the reproducibility of measurements across platforms and expressed the need for further large-scale comparison studies [12,15,18,48].

In establishing this comparative framework, we have tried to find a balance between what is optimal for a single platform and what is optimal for a cross-platform comparison. In this study, data sets from 10 different microarray platforms were compared and the results were validated with quantitative real-time polymerase chain reaction (QRT-PCR) [49]. The different platforms included one- and two-dye platforms, encompassing both cDNA and oligonucleotide microarrays that were either fabricated "in-house" or were commercially available. To minimize mistakes that can occur during mapping of genes across platforms solely using gene annotations, we have implemented the example of recent studies [21,50,51] that have demonstrated that mapping at the probe sequence level improves cross-platform consistencies and obtained the probe sequences for all 10 platforms. The sequence information was used to obtain mappings both at the "gene" and "exon" levels. This was the first study to map probes across the different microarray platforms at the "exon" level.

Each laboratory received aliquots from two different RNA samples, mouse retina (MR), and mouse cortex (MC), that were prepared in the Cepko Laboratory, Department of Genetics at Harvard Medical School. Pooling tissue from many animals prior to extracting the RNA minimized the biological variations within tissue RNA preparations. Large pools of each sample were collected, which allowed for the inclusion of emerging technologies into the study as they developed. Microarray hybridizations were conducted in replicates of five to enhance statistical reliability [52]. Additionally, as part of this framework, the variability of the same platform in cross-laboratory experiments was evaluated for three platforms [15,43,53,54]. Although it would have been ideal to have all hybridizations conducted at one facility, this was not always practical given resource constraints; therefore, some experiments were conducted at multiple sites. This permitted us to evaluate the level of the variation of different platforms among laboratories and concurrently, the variation of similar platforms used at different sites. Using this approach, if the variation exists between platforms, one can develop a general assessment of whether it was attributable to a particular platform. Furthermore, this replicates the real-world scenario in which data from different laboratories are sent to a large repository for secondary data analyses.

A number of factors identified in previous studies as being important to achieve reproducible and high-quality results, both within and between platforms, have provided us with background knowledge that guided the study design. One of the most important factors was the approach in which measurements were matched across platforms. To date, there has been no other reported large-scale study that evaluated microarrays utilizing "actual" probe sequences. Furthermore, in studies that have been done, the protocols for comparison were either not sufficiently well described or not sufficiently general. In this study, data analysis combines well-described commonly used and publicly available analytical approaches in a framework that can be utilized every time the reliability of a new platform needs to be assessed.

12.2 METHODS

12.2.1 SAMPLE COLLECTION AND ISOLATION

RNA samples used for all platforms were aliquoted from two pools of samples: C57/B6 adult MR and Swiss-Webster postnatal day 1 (P1) MC. MR and MC were chosen due to their availability and biological interest. MR samples were obtained from a pool of C57/B6 mice ($n = 350$) and the MC was obtained from P1 Swiss-Webster mice ($n = 19$), which were both purchased from Charles River Laboratories (Charles River Laboratories, Inc., Wilmington, Massachusetts). The animal experiments were approved by the Institutional Animal Care Facility at Harvard University. The MC was used as a reference sample for "two-dye" platforms.

12.2.2 MICROARRAY EXPERIMENTS

Sample preparation and hybridization steps were conducted following the protocols provided for each platform. Eight of the 10 microarray platforms evaluated are currently commercially available: Affymetrix, Agilent, Applied Biosystems (ABI), GE Amersham (now Applied Microarrays), Compugen (now Sigma-Genosys), Mergen (stopped production in 2005), MWG Biotech (now Ocimum Biosolutions), and Operon. The remaining two platforms are from academic laboratories: (1) cDNA arrays provided from the Cepko Laboratory, herein referred to as "academic cDNA" arrays; and (2) long oligonucleotide arrays from the Microarray Core Facility at Massachusetts General Hospital (MGH), herein referred to as "MGH long oligo" arrays. Oligonucleotides from both Compugen and Operon were purchased by the Division of Biology at California Institute of Technology and were printed together onto the same slide. A total of eight research laboratories were involved in this collaboration.

To evaluate cross-laboratory consistency, a subset of the platforms was conducted independently at a second laboratory using identical samples. The results from Affymetrix, Amersham, and Mergen platforms were completed for the study. Each laboratory provided the raw data sets and scanned images for analysis.

Six of the 10 microarray platforms (Agilent, academic cDNA, Compugen, MGH long oligo, MWG, and Operon) are considered to be "two-dye" platforms as they require the hybridization of two samples, whereas the others (ABI, Affymetrix, Amersham, and Mergen) are "one-dye" platforms. Since data based on a single array are often considered insufficient to obtain conclusive results [52], five replicates of each sample were used to assess the degree of variation in the expression data within each platform. The number five was chosen as a reasonable compromise between the wish to reduce the effect of array-to-array variability and resource limitations. A total of 91 hybridizations were completed in this study. The experimental design is shown in Figure 12.1.

12.2.3 LABELING AND HYBRIDIZATION METHODS

All labeling and hybridization methods were completed as specified by each manufacturer's labeling and hybridization protocol. Image processing of the scanned images were conducted using the manufacturer's recommended scanners and settings. Detailed description of the protocols used for each platform is provided below.

12.2.3.1 Affymetrix

12.2.3.1.1 Probe Labeling, Hybridization, and Scanning

Sample labeling, hybridization, and staining were carried out according to the Eukaryotic Target Preparation protocol in the Affymetrix Technical Manual (701021 rev-1) for GeneChip® Expression Analysis (Affymetrix, Santa Clara, California). In summary, 5 μg of purified total RNA were used in a 20-μL first-strand reaction with 200 U SuperScript II (Invitrogen Life Technologies, Carlsbad, California)

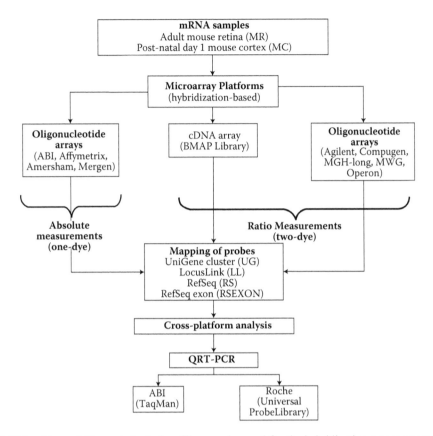

FIGURE 12.1 Cross-platform study design. The samples used for the hybridization process were MR and MC samples. The different microarray platforms were grouped based on their dye approach that is "one-dye" versus "two-dye". Comparisons of the platforms were conducted using four probe mapping options: UniGene (UG) clusters, LocusLink (LL) identifiers, RefSeq (RS) identifiers, and RefSeq exon (RSEXON) position, followed by biological verification using two QRT-PCR technologies.

and 0.5 µg (dT)-T7 primer [5′-GGCCAGTGAATTGTAATACGACTCACTATAGGGAGGCGG(T)$_{24}$] in 1X first-strand buffer (Invitrogen Life Technologies, Carlsbad, California) with a 42°C incubation for 1 h. Second-strand synthesis was carried out by the addition of 40 U *Escherichia coli* DNA polymerase, 2 U *E. coli* RNase H, and 10 U *E. coli* DNA ligase in 1X second-strand buffer (Invitrogen Life Technologies, Carlsbad, California) followed by incubation at 16°C for 2 h. The second-strand synthesis reaction was purified using the GeneChip Sample Cleanup Module according to the manufacturer's protocol (Affymetrix, Santa Clara, California). The purified cDNA was amplified using BioArray High Yield RNA Transcription Labeling Kit (Enzo Life Sciences, Inc., Parmingdale, New York) according to the manufacturer's protocol to produce 70–120 µg of biotin-labeled complementary RNA (cRNA).

Mouse Genome U74Av2 GeneChip probe arrays were prehybridized in a GeneChip Hybridization Oven 640 (Affymetrix, Santa Clara, California) according to the manufacturer's protocol. 15 µg of labeled cRNA were fragmented in 30 µL 1X fragmentation buffer containing 100 mM KOAc and 30 mM MgOAc at 95°C for 35min. The fragmented labeled cRNA was resuspended in 300 µL 1X hybridization buffer containing 100 mM 2-(N-morpholino) ethanesulfonic acid (MES), 1 M [Na⁺], 20 mM ethylenediaminetetraacetic acid (EDTA), 0.01% Tween 20, 0.5 mg/mL aceylated bovine serum albumin (BSA), 0.1 mg/mL herring sperm DNA, control oligonucleotide B2, and control transcripts bioB 1.5 pM, bioC 5 pM, bioD 25 pM, and cre 100 pM. 200 µL of the hybridization cocktail (containing 10 µg of labeled cRNA) was hybridized to GeneChip probe arrays according to the manufacturer's protocol (Affymetrix, Santa Clara, California).

The hybridized GeneChip probe arrays were washed and stained using streptavidin–phycoerythrin (Molecular Probes, Eugene, Oregon) and amplified with biotinylated anti-streptavidin (Vector Laboratories, Burlingame, California) (Sigma, Saint Louis, Missouri) GeneChip Fluidics Station 400 (Affymetrix, Santa Clara, California) using an antibody amplification protocol.

The GeneChip probe arrays were scanned using GeneArray Scanner (Hewlett Packard, Corvallis, Oregon).

12.2.3.2 Agilent

12.2.3.2.1   cDNA Synthesis and In Vitro Transcription

To generate the Cy3 and Cy5 fluorescently labeled cRNA for Agilent Oligonucleotide Microarrays (Agilent Technologies, Palo Alto, California), the Agilent Fluorescent Linear Amplification (Agilent Technologies, Palo Alto, California) from total RNA protocol was used with minor modifications implemented by the Center for Expression Arrays (http://ra.microslu.washington.edu/). One of the modifications was the use of a thermocycler with a heated lid for all incubations and icing steps, to obtain optimal cRNA yield. The other modification was the use of the Ambion MEGAClear Kit (Ambion Inc., Austin, Texas) and the protocol for purification of labeled cRNA, as opposed to the lithium chloride precipitation recommended by Agilent (Agilent Technologies, Palo Alto, California). The reagents and enzymes used for cDNA synthesis and *in vitro* transcription (IVT) came from the Linear Amplification Kit (G2554A) by Agilent Technologies. The Cy3-Cytidine 5′-triphosphate (CTP) and Cy5-CTP were ordered through PerkinElmer (PerkinElmer, Boston, Massachusetts).

In brief, in a 0.2-µL PCR tube, 5 µg of total RNA was added in a volume of 4.5 µL or less. Then, 5 µL of T7 Promoter Primer from the amplification kit was added, and the total volume in the tube was brought up to 9.5 µL by adding nuclease-free H_2O. The primers and templates were denatured by heating the reactions at 65°C for 10 min in a thermocycler. The samples were then cooled to 4°C for 5 min. After cooling, 10.5 µL of cDNA mix was added to each reaction and incubated at 40°C for 4 h, then heated to 65°C for 15 min. The cDNA mix consists of 5X first-strand buffer, 0.1M dithiothreitol (DTT), 10 mM deoxynucleotide triphosphate (dNTP) mix, random hexamers, Moloney Murine Leukemia Virus Reverse Transcriptase (MMLV-RT), RNaseOUT, and 0.3% Triton X-100.

After heating, the samples were again cooled to 4°C for 5 min, and then spun briefly in a microcentrifuge for a few seconds to bring the contents of the tube down to the bottom. Next, 4 µL of 6 mM Cy3-CTP and 4 mM Cy5-CTP were added to the appropriate reactions followed by 56 µL of IVT mix to each reaction. The reactions were incubated for 1 h at 40°C. After the addition of Cy3 and Cy5 CTPs, the samples were protected from light as much as possible to avoid degradation of dyes.

12.2.3.2.2   Purification of cRNA

After completion of the IVT, the samples were purified using Ambion MEGAclear Kit (Ambion Inc., Austin, Texas, Catalog #1908). Most of the reagents needed are in the kit, accept for the 100% ethanol required for binding cRNA to filter cartridge and for preparation of wash solution. After the IVT, the reactions were brought to a total volume of 100 µL with elution solution and mixed gently by pipetting. Then 350 µL of binding solution concentrate was added to each reaction and mixed gently. Finally, 250 µL of 100% ethanol was added to each sample, mixed well, and 700 µL of each sample was applied to the filter cartridge placed on top of a collection/elution tube. The samples were then spun at 12,000 rpm for 1 min, the flow-through was discarded, and the collection/elution tubes were used again for the washing steps. Before the wash solution concentrate could be used, 20 mL of 100% ethanol were added and mixed. Approximately, 500 µL of wash solution concentrate was added to each sample tube and centrifuged for 1 min at 12,000 rpm. The flow-through was discarded; the collection/elution tubes were reused and the wash step was repeated once more as above. To elute the cRNA, the filter cartridges were moved into new collection/elution tubes. We used RNA elution option 2 from the protocol, where 110 µL of elution solution per reaction was heated to 95°C in a heat block for 7–10 min, and then 50 µL of the solution was applied to each sample filter cartridge and centrifuged at room temperature for 1 min at 14,000 rpm. This step was repeated with another 50 µL of preheated elution solution, such that, each tube had about 100 µL of eluted cRNA, either pink (Cy3 labeled) or blue (Cy5 labeled).

12.2.3.2.3   Quantitating Fluorescent-labeled cRNA

We used the Shimadzu UV-1601 spectrophotometer to quantify the labeled cRNA products. The spectrophotometer was baselined with 100 µL of TE buffer pH 8.0 and each sample was diluted 1:20 with TE buffer pH 8.0 to obtain readings at A260 and A280 for calculating the cRNA yield.

12.2.3.2.4   Hybridization, Washing, and Scanning

Hybridization of the Agilent oligonucleotide arrays for this study was carried out according to the Agilent Oligonucleotide Microarray Hybridization Protocol (Agilent Technologies, Palo Alto, California). For 2X cRNA target preparation, 0.75 µg of Cy3 and 0.75 µg of Cy5-labeled cRNA were combined with 50 µL of 10X control targets and nuclease-free H_2O was added to bring the total volume of the target to 250 µL. Prior to hybridizations, the 2X target solutions were fragmented by adding 10 µL of 25X fragmentation buffer and heating the reactions in the dark in a heat block at 60°C for 30 min. After the incubation, 250 µL of 2X hybridization buffer were added to the reactions to stop the fragmentation process. The reagents for hybridization are included in the Agilent *In Situ* Hybridization Kit (Agilent Technologies, Palo Alto, California, Part No. 5184-3568). The arrays were hybridized with 500 µL of hybridization solution in hybridization chambers in the hybridization oven on a rotating rack at 60°C for 17 h.

The arrays were washed the following morning in Wash 1 and Wash 2 solutions according to the Agilent protocol.

12.2.3.3   Applied Biosystems

Experiments were conducted at the microarray facility of Applied Biosystems (Applied Biosystems, Foster City, California). Digoxigenin-UTP-labeled cRNA was generated and linearly amplified from 1 µg of total RNA using Applied Biosystems Chemiluminescent RT-IVT Labeling Kit v2.0 and the manufacturer's protocol. cRNA from five independent labeling reactions of each sample were pooled and split across five microarrays for hybridization. Array hybridization, chemiluminescence detection, image acquisition, and analysis were performed using Applied Biosystems Chemiluminescence Detection Kit (Applied Biosystems, Foster City, California) and Applied Biosystems 1700 Chemiluminescent Microarray Analyzer (Applied Biosystems, Foster City, California) following the manufacturer's protocol. Briefly, each microarray was first prehybridized at 55°C for 1 h in hybridization buffer with blocking reagent. 20 µg of labeled cRNA targets was first fragmented by incubating with fragmentation buffer at 60°C for 30 min, mixed with internal control target (ICT, 24-mer oligonucleotide labeled with LIZ fluorescent dye) and hybridized to each prehybridized microarray in a 1.5-mL volume at 55°C for 16 h. After hybridization, the arrays were washed with hybridization wash buffer and chemiluminescence rinse buffer. Enhanced chemiluminescent signals were generated by first incubating arrays with anti-digoxigenin-alkaline phosphatase, enhanced with chemiluminescence enhancing solution and finally adding chemiluminescence substrate. Images were collected for each microarray using the 1700 analyzer. Images were auto-gridded and the chemiluminescent signals were quantified, corrected for background and spot, and spatially normalized.

12.2.3.4   cDNA

cDNA microarrays were constructed at the BioDiscovery Center at Biogen Idec (Biogen Idec, Cambridge, Massachusetts). RNA samples were reverse transcribed and cDNAs were amplified for 16–20 cycles using the SMART system (Clontech, Mountain View, California). To generate probes for array hybridization, 10 µg cDNA was labeled by incorporation of either Cy5- or Cy3-dCTP (Amersham Pharmacia) during oligo-dT-primed or random hexamer-primed primer extension in the presence of Klenow DNA polymerase (Roche). Arrays were prehybridized with poly-adenylic acid (Sigma) and mouse Cot-1 DNA (Life Technologies) before overnight hybridization with paired Cy3- and Cy5-labeled probes in the presence of poly-adenylic acid and mouse Cot-1 DNA. Pairs of labeled probes were hybridized at 42°C overnight. Arrays were washed at room temperature in 0.2 × SSC, 0.1%SDS for 2 min, followed by two washes in 0.2 × sodium chloride sodium citric acid (SSC) for 2 min each.

12.2.3.5   Compugen and Operon

The 16K mouse oligonucleotide arrays were inkjet-printed by Agilent Technologies (Palo Alto, California. This oligonucleotide array consists of 13,536 probes of 70-mers (Operon Technologies Inc., Alameda, California) and 2304 probes of 65-mers (Sigma-Genosys, The Woodlands, Texas). The Agilent Fluorescent Linear Amplification Kit (Agilent Technologies, Palo Alto, California) was used for the preparation of fluorescently labeled target samples. Briefly, both first and second-strand cDNAs were synthesized by incubating 3 µg of total RNA with T7 promoter primer (5'-GG CCAGTGAATTGTAATACGACTCACTATAGGGAGGCGG-(dT)$_{24}$-3') using SuperScript II (Invitrogen Life Technologies, Carlsbad, California). Fluorescently labeled cRNA was amplified by adding 4 µL of Cy3-dCTP (6 mM) or 4 µL of Cy5-dCTP (4 mM) (PerkinElmer, Inc., Boston, Massachusetts) to the 20 µL of cDNA template and mixing with 56 µL of reaction mixture containing, in the following order: 20.1 µL of nuclease-free water, 20 µL of 4X transcription buffer, 6 µL of 0.1M DTT, 8 µL of nucleotide *triphosphate* (NTP) mix, 0.5 µL of RNaseOUT, 0.6 µL of inorganic pyrophosphatase, and 0.8 µL of T7 RNA polymerase. After incubating at 40°C for 3 h, cRNAs were cleaned up using RNeasy Mini Kit (Qiagen Inc., Valencia, California). 5 µg of each labeled cRNA target was used for hybridization on the oligonucleotide-arrayed slide after being fragmented by fragmentation buffer (Agilent Technologies, Palo Alto, California). After overnight hybridization at 65°C, the oligonucleotide arrays were washed and scanned by the Agilent Scanner G2505A (Agilent Technologies, Palo Alto, California).

12.2.3.6   GE Amersham CodeLink

The RNA was processed using a modified Amersham CodeLink (GE Healthcare Amersham Biosciences, Chandler, Arizona) protocol. Briefly, 2 µg of high-quality total RNA was reverse transcribed (Invitrogen Life Technologies, Carlsbad, California, Invitrogen reagents) producing first-strand cDNA, and second-strand cDNA was then synthesized (Invitrogen Life Technologies, Carlsbad, California, Invitrogen reagents). The double-stranded cDNA was cleaned using the QIAquick Purification Kit (Qiagen Inc., Valencia, California), all of which were used as templates for IVT utilizing Ambion's T7 MEGA script reagents and Biotin-11-UTP (PerkinElmer/NEN). Resulting biotin-labeled cRNA was recovered and purified with RNeasy (Qiagen Inc., Valencia, California) column and once again quantified. cRNA yield was quantified by measuring the ultra-violet (UV) absorbance at 260 nm, and fragmented in fragmentation buffer at 94°C for 35 min. This protocol usually yields 50–100 µg of cRNA, enough to hybridize to several arrays. 10 µg was applied to each CodeLink Bioarray, and hybridized overnight at 37°C, then fluorescently tagged according to the manufacturer's protocol. The arrays were imaged using the PerkinElmer ScanArray scanner (PerkinElmer, Boston, Massachusetts) and spot intensity was quantitated using PerkinElmer's ScanArray software (PerkinElmer, Boston, Massachusetts). Expression data files were generated from array images using the CodeLink Image analysis software.

12.2.3.7   Mergen

The ExpressChip MO3 DNA Microarray System (Mergen Ltd., San Leandro, California) was used for this study. The assay was carried out according to the manufacturer's instructions. In brief, DNase-treated total RNA (20 µg) was reverse-transcribed using an oligo[(dT)$_{24}$ T7 promoter]$_{65}$ primer (consisting of the nucleotide binding sequence for the T7 RNA polymerase followed by 24 thymidine nucleotides) followed by second-strand synthesis. The double-stranded cDNA was cleaned using the QIAquick Purification Kit (Qiagen Inc., Valencia, California), all of which were used as templates for IVT utilizing Ambion's T7 MEGA script reagents and Biotin-11-UTP

(PerkinElmer/NEN). These probes were hybridized to the arrays overnight at 30°C with continuous agitation. The arrays were then washed, and hybridized probes were detected using a streptavidin/ Cy3 fluorescent dye-conjugated antibody. Chips were imaged using a ScanArray scanner (PerkinElmer, Boston, Massachusetts) and spot intensity was quantitated using ScanArray software (PerkinElmer, Boston, Massachusetts).

12.2.3.8 MGH (Academic Oligonucleotide Array)

Protocol used for the study can be found at the MGH microarray website: https://dnacore. mgh.harvard.edu/microarray/protocols.shtml. Material sections can be found at the beginning of each protocol.

12.2.3.9 MWG Biotech

Briefly, RNA was prepared using the RNeasy Midi Kit according to the manufacturer's instructions (Qiagen Inc., Valencia, California). First-strand complementary DNA synthesis from 5 μg total RNA was done at 42°C for 1 h, followed by second-strand cDNA synthesis at 16°C for 2 h. RNase I digestion was followed by proteinase K treatment. The resulting *double*-stranded cDNA was purified using the RNeasy Mini Kit according to the manufacturer's instructions (Qiagen Inc., Valencia, California). Fluorescence-labeled RNA was generated by carrying out an IVT reaction (40 mM Tris-HCl, pH 7.5, 7 mM MgCl2, 10 mM NaCl, 2 mM spermidine, 5 mM ditihothreitol, 7.5 mM each of adenosine-5′-*triphosphate* (ATP), cytidine-5′-*triphosphate* (CTP), and guanosine-5′-*triphosphate* (GTP), 5 mM UTP, 20 units RNase inhibitor, and 1000 units T7 polymerase) at 37°C for 4h. This reaction was done in the presence of fluorescence-labeled nucleotide (Amersham Pharmacia Biotech) to generate Cy3- or Cy5-labeled RNA. The labeled RNA was subsequently purified (Qiagen RNeasy Mini Kit) and chemically fragmented at 94°C for 15 min in fragmentation buffer (20 mM Tris-acetate, pH 8.1, 50 mM potassium acetate, and 15 mM magnesium acetate). The fragmented, Cy3- or Cy5-labeled cRNA (20 μg) was lyophilized, resolubilized in hybridization buffer (50% formamide, 50 mM sodium phosphate, pH 8.0, 6X SSC, 5X Denhardt's Solution, 0.5% SDS), and hybridized to MWG Biotech oligonucleotide mouse 11K microarrays (MWG Biotech AG, Ebersberg, Germany) at 42°C for 24h. T7 RNA polymerase was purchased from Ambion (Ambion Inc., Austin, Texas): primers, RNase inhibitors, and all other enzymes were purchased from Roche Molecular Biochemicals (Roche Applied Sciences, Indianapolis, Idaho).

12.3 DATA ANALYSIS: DATA ACQUISITION

Raw intensities from all platforms generated using the recommended scanners as specified in the vendor's protocol were used for the study. Each Affymetrix GeneChip microarray (Affymetrix, Santa Clara, California) was scanned using Affymetrix's GeneArray Scanner (Affymetrix, Santa Clara, California). The raw data were collected after analysis by using Affymetrix's Microarray Analysis Suite 5.0 (MAS 5.0) software (Affymetrix, Santa Clara, California). Amersham CodeLink and Mergen Mouse (MO3) arrays were both scanned using ScanArray Express Software and ScanArray Express HT scanner (Packard BioScience, Meriden, Connecticut). The scanned image files from Amersham were analyzed using CodeLink image and data analysis software. The scanned image files from Mergen arrays were analyzed using Mergen's ExpressWare® (Mergen Ltd., San Leandro, California) image and data analysis software. Applied Biosystem's arrays were completed at Applied Biosystems microarray facility (Applied Biosystems, Foster City, California) and raw data files were provided for the analyses. Academic cDNA and MGH long oligo arrays, along with the MWG Biotech arrays were scanned using the ScanArray 500 confocal laser scanner (GSI Lumonics, Marina Del Ray, California), followed by signal digitalization with Axon Instrument, GenePix 4.0 quantification software (Axon Instruments, Union City, California). Agilent Mouse Development, Compugen, and Operon microarrays were scanned using the Agilent Scanner G2505A (Agilent Technologies, Palo Alto, California) at the default scanner settings. Agilent's Feature Extraction software (Version A.6.1.1) was used to quantify the scanned images.

12.3.1 Preprocessing (Normalization, Transformation, and Filtering)

Preprocessing methods included normalization, transformation, and filtering steps. Specific normalization methods were chosen based on past microarray studies that have indicated their potential advantages over other methods in one- and two-dye platforms [55–57]. In the case of microarray data from one-dye platforms, normalization was performed using quantile normalization, where 10 arrays (five for retina samples and five for cortex) were considered as one group [56]. Data from two-dye platforms were normalized using locally weighted scatterplot smoothing (LOWESS) normalization [55,57]. Since probes from Compugen and Operon platforms were printed onto the same slide, LOWESS normalization was performed on the whole chip before they were separated and analyzed in the study. We also examined and confirmed that when this normalization was performed for each platform independently, the results were similar (data not shown).

Data transformation included both linear and percentile scaling of the raw intensities, as well as $\log_2$ ratios between the two samples. The scaling transformations were needed to allow comparison of raw intensities quantified by different software packages. Linear scaling mapped the intensities of each slide/channel into a scale of 1–100, linearly and analogously. This method was used in measuring intra-platform coefficient of variations (CV) of the intensities. Percentile transformation projected the data to a hundred discrete levels (i.e., 1–100) according to percentiles of the intensity values. Beyond making the measurements among various platforms comparable, percentile scaling may be useful to correct the artifacts introduced by different intensity distribution characteristics among various platforms, as well as to purposefully neglect some minor fluctuations in expression levels. Percentile transformation was mainly used in the inter-platform comparisons.

$\log_2$ ratios were computed to allow the comparison of one- and two-dye platforms. When we evaluated intra-platform variations, five $\log_2$ ratios were obtained from five technical replicates of each two-dye platform. For one-dye platforms, $\log_2$ ratios were obtained from five randomly paired arrays across samples without replacement. The averaged $\log_2$ ratios of technical replicates for each platform were used to assess inter-platform variation.

Stringent filtering for spot quality has been reported to improve consistency across different platforms [27,58]. The filtering criteria chosen in the study were either recommended by the vendors or have been broadly adopted by the research community. Filtering was conducted at the spot (image) level, taking into account both quality flags and signal-to-noise ratio (SNR) thresholds. Ideally, all the platforms should have been scanned and quantitated using the same scanner, with similar scanner settings. Owing to the diversity of the technical approaches of the various platforms, different scanners were used, and this limited our ability to apply the same filtering criteria to all the platforms. In the case of the Affymetrix and Amersham platforms, probe set and spot quality flags were referenced, respectively. These meant only "present" and "good" calls were adopted, for Affymetrix and Amersham, respectively. The SNR threshold of three was used for ABI, in addition to removal of flagged spots as recommended by the vendor. An SNR threshold was set to two for Agilent, Compugen, Mergen, and Operon platforms. For academic cDNA, MGH long oligo, and MWG arrays, the images were scanned using GenePix software 3.0 (Molecular Devices, Sunnyvale, California). The software automatically generated flags at default settings for poor and missing spots, which were removed.

12.3.2 Gene Matching across Different Platforms

Gene mapping was conducted using two approaches: annotation-based and sequence-based. For the annotation-based approach, MatchMiner [59] was used to map UG clusters (UniGene Build 136) and LocusLink (LL) identifiers by using GenBank accession numbers that were provided by each platform.

For the sequence-based approach, the February 2003 version of the mouse reference sequences (UCSC version mm3) was downloaded from the UCSC Genome Site (http://hgdownload.cse.ucsc.edu/downloads.html mouse) and used for mapping the probe sequences. The probe sequences from each microarray platform were mapped to the mouse genome using the BLAT stand-alone program (www.genome.ucsc.edu/cgi-bin/hgBlat) [60]. The sequence alignment results were also parsed so that only probe-to-exon matched pairs were extracted. "Probe-to-exon" meant only aligned sequences positioned completely within an exon were considered as a match. In the instances where multiple within-exon matches for a probe sequence occurred, the best match in terms of the length of "hit" was selected. If no match was found, then that probe was excluded. In this way, the probes from different platforms were matched both at the gene level by RefSeq (RS) identifiers and at the exon level by RefSeq exon (RSEXON). The probe sequences used for mapping ABI and Affymetrix had lengths of 180 and 255 base pairs, respectively. Affymetrix uses 11 probe pairs to measure the expression level for each gene and the 255 base pairs correspond to the length of the sequences spanned by the 11 probe pairs (complete probe). The context sequences for Affymetrix were obtained from their NetAffx analysis center (http://www.affymetrix.com/analysis/index.affx). For ABI, the probe sequence for each gene on the array lies within the 180 base pairs used in the mapping.

12.3.3 EVALUATION OF INTRA- AND INTER-PLATFORM DATA CONSISTENCY

We chose to measure data consistency by calculating CVs, correlation coefficients, and standard deviations (SDs) of the difference between measurements. Principal component analysis (PCA) was performed to allow the display of axes corresponding to the largest variance in multiple platforms. Additionally, the degree of deviations of each platform from others was quantified by defining outliers across various platforms' measurements for each gene.

The CV is defined as the variation among multiple measurements in proportion to their mean. We used CV to measure the reproducibility among multiple replicate experiments within each platform. Besides the conventional use of CV on channel-specific intensities, we also defined a segmental function for the CV of $\log_2$ ratios. When the mean of $\log_2$ ratios was between -1 and $+1$, the CV equals to the SD, otherwise, the conventional definition of CV was applied. This was to avoid including small denominators to distort the CVs considerably when a large proportion of probes having a mean of $\log_2$ ratio close to zero are expected in microarray experiments.

Pearson and Spearman correlation coefficients were calculated for both intra- and inter-platform comparisons. Intra-platform correlations consisted of computing the correlations for both linearly transformed intensities within each sample and their $\log_2$ ratios. For inter-platform comparisons, the correlations were calculated based on the averaged $\log_2$ ratios. As the expression data were not normally distributed (data not shown), we conducted two permutation tests on inter-platform correlations, aiming to estimate the significance of the correlation coefficients for the cross-platform probe matching. In both tests, for any pair of platforms, averaged log-ratios and paired measurements were randomly selected. The first test involved measurements chosen from the whole data set for the given platforms. In the second test, the measurements were chosen from a subset of data included in the list of matched probes between the platforms. In both cases, 10,000 randomized sets of matched measurements were created for each pair of platforms. Thus, empirical distributions of correlation coefficients were calculated and empirical confidence intervals of correlation coefficients were obtained to assess statistical significance.

SDs of the differences between matched measurements were computed as another measure of data consistency. In the case of intra-platform agreement, technical replicates were referred to as "matched measurements," whereas each pair of platforms was considered in the case of inter-platform agreement.

PCA was performed on data from eight of the platforms, excluding the academic cDNA and Compugen platforms. In addition, PCA included second lab data from three one-dye platforms. PCA was conducted after standardization so that each gene had a zero mean and unit SD.

To examine which platforms were more prone to have measurements that were markedly different from the others, we computed the frequency of outliers for each platform. For a given gene that has been measured in at least five platforms, if a platform's measurement lies outside of the range of the mean expression ratios ±1 SD, it was identified as an outlier.

The analyses were conducted using the R software environment (www.R-project.org), BioConductor packages [61], and MATLAB (The Math Works, Natick, Massachusetts).

12.3.4 SELECTION OF GENES FOR VALIDATION

Molecular confirmation of microarray results is important when checking for consistencies of expression measurements across different platforms. Since the gene coverage varied across the platforms, we decided to select genes that were common across a minimum of six platforms based on RSEXONs. The six platforms had to include four common microarray platforms: ABI, Affymetrix, Agilent, and Amersham, and any additional two others. Based on this criterion, 399 genes were identified where two groups of genes were created based on their intensity. The groups were derived from the percentile-transformed data, where three categories of expression measurements were created: high (67–100 percentiles), medium (34–66 percentiles), and low (1–33 percentiles). The first group included genes that had combinations of (1) high–high, (2) high–medium, and (3) medium–medium expression measurements for both samples. In the second group, the expression measurements for both samples included genes that had combinations of (1) high–low, (2) medium–low, and (3) low–low expressions. A total of 158 genes were validated by quantitative real time-polymerase chain reaction (QRT-PCR) from these groups.

12.4 VALIDATION OF RESULTS

Biological validations for this study were conducted using QRT-PCR. Samples identical to the ones used for the microarray experiments were used for the biological validation step. The validation methods were conducted using Exiqon ProbeLibrary (http://exiqon.com/SEEEMS/11.asp), now Roche Universal ProbeLibrary (www.roche-applied-science.com/sis/rtpcr/upl/), on two different Roche LightCyclers® (Roche Applied Sciences, Indianapolis Indiana), and TaqMan® Gene Expression Assays on ABI PRISM 7900 HT Sequence Detection System (Applied Biosystems, Foster City, California). Seventy-four and 91 genes were verified using Universal ProbeLibrary and ABI TaqMan Gene Expression Assays, respectively. Protocols for both approaches are discussed below.

A total of 165 genes were validated by QRT-PCR. Pearson correlation coefficients were computed for the $\log_2$ ratios for the set of genes validated by QRT-PCR and the corresponding platform. Expression ratios measured by QRT-PCR were calculated as follows:

$$\log_2 \mathrm{ratio}\!\left(\frac{\mathrm{MR}}{\mathrm{MC}}\right) = -\!\left(\overline{\mathrm{Ct}}_{\mathrm{MR}} - \overline{\mathrm{Ct}}_{\mathrm{MC}}\right),$$

where $\overline{\mathrm{Ct}}_{\mathrm{MR}}$ and $\overline{\mathrm{Ct}}_{\mathrm{MC}}$ correspond to the mean cycle thresholds for MR and MC, respectively.

The primer sequences for Roche's Universal ProbeLibrary and TaqMan assay identifiers are provided in Tables 12.1a and 12.1b.

12.4.1 ROCHE UNIVERSAL PROBELIBRARY

Mouse Universal ProbeLibrary probes and target-specific PCR primers were selected using the ProbeFinder assay design software (https://www.roche-applied-science.com/sis/rtpcr/upl/adc.jsp). We attempted to design the primers to lie within the exon corresponding to the matched probe sequences, but this was not possible for all the genes using this software. All assays were prepared using standard conditions in a master mix solution without any effort toward assay optimization. For each data point, there were three replicates. cDNA was synthesized from 15 µg of total RNA for each sample using

Roche reverse transcriptase (Roche Applied Sciences, Indianapolis Indiana). The primers were used to confirm relative changes in mRNA levels by QRT-PCR, using both Roche 2.0 LightCycler and 480 LightCycler (Roche Applied Sciences, Indianapolis Indiana). Reactions for the Roche 2.0 LightCycler were performed in 20 μL reaction volumes for 33 genes using 1 μL of cDNA under the following conditions: 95°C for 10 min, 45 cycles at 95°C for 10 s, and 60°C for 30 s. For the Roche 480 LightCycler, the reactions were performed in 20 μL reaction volumes for 34 genes under the following conditions: 95°C for 5 min, 50 cycles at 95°C for 10 s, 60°C for 15 s, and 72°C for 1 s.

TABLE 12.1a
List of Forward and Reverse Primer Sequences of Genes that were Designed using Roche's ProbeFinder Software Program and used for QRT-PCR

RefSeq	Forward	Reverse
NM_011170	caatttaggagagccaagcag	gccgacatcagtccacatag
NM_009455	ctctccaccagcgctaagag	gcggagggtctaaagtgatg
NM_031494	tgagtcatccctgtgtgtcatt	cttggggaagggcaagac
NM_009871	gcggagggtttcctcttc	ccgccttcctcactgtagc
NM_011579	cagatcaaggtcaccactgc	aagaatgcatcaaagctggag
NM_007487	cagccaatgtccctccatt	tttctcttcctgggtgatcg
NM_008086	tctatgatgtttgttctgttgtcct	agaccctgataggggcagag
NM_008686	ggagagactggggagagctt	cactgcttctgggatgctg
NM_018798	caccaccacgaccacaag	gctgaagatgctagccacgta
NM_009717	tccttcgaggaaagagcatt	tcctcctcttctttctcggttt
NM_010267	tggatcaggtggaaactgaa	cagagccaaggctggttg
NM_010894	gcagaaggcaaggtgtcc	tttggtcatgtttccacttcc
NM_011462	gcatcatgcctgattccaa	ggcatattccacttgcttgc
NM_011607	gggctatagaacaccgatgc	catttaagtttccaatttcaggttc
NM_015814	gcctgaaggagctttggac	ggcttgcacatgtacaccag
NM_016797	gagttcgttgctcgagtgc	ttcttttgagctgtcttcagga
NM_016801	tgccatctttgcctctgg	atctcactgagggcctgct
NM_007553	gtaccgcaggcactcagg	cctccacggcttctagttga
NM_007559	cctgtatgaactccaccaacc	ggggatgatatctggcttca
NM_007591	acagattccaagcctgagga	gcttcttagcatcagggtcag
NM_008010	cacttcagtgtgcgtgtaacag	gggcgagtccaataaggag
NM_008089	gaatcctctgcatcaacaagc	gggcaagggttctgaggt
NM_008122	acaggagttctggtgaacagg	ctagcaggcgagtcaggaag
NM_008348	gctcccattcctcgtcac	aagggcttggcagttctgt
NM_008452	ctaaaggcgcatctgcgta	tagtggcgggtaagctcgt
NM_008748	gacagtgtggccatcctca	aggcttgccctcacagag
NM_008756	tccgtgaggcctttgaa	ggtgcataatgattgggtttg
NM_008952	gtaccacctggccaaacact	gaatggggaagaaagaactgc
NM_009309	cagcccacctactggctcta	gagcctggggtgatggta
NM_009360	aaggatgattcggctcagg	aagctgaatatatgcctgcttttc
NM_009469	cacccccatcaagaaatcc	ccagaccctgagcttggata
NM_009471	ccttggccactggaaactac	aaattcacagtgctcctcagc
NM_009559	tggctagaagcagtctggaat	ggatggctgggaagactgt
NM_009608	gaagagctatgaacttcctgacg	gaatgccagcagattccatac
NM_009697	cctcaaagtgggcatgagac	tgggtaggctgggtaggag

continued

TABLE 12.1a (continued)

RefSeq	Forward	Reverse
NM_009804	gcgaccagatgaagcagtg	gtggtcaggacatcaggtctc
NM_009811	tggttggaaaacccaagatattta	gggaaccacaggtacgtca
NM_009917	gagacatccgttccccctac	gtcggaactgacccttgaaa
NM_009922	tcaatgagtcaactcagaactgg	cccatacttggtaatggctttg
NM_009971	cctggagcaagtgaggaaga	cagcttgtaggtggcacaca
NM_010052	cgggaaattctgcgaaatag	tgtgcaggagcattcgtact
NM_010055	ggggatcctataggcagtacg	cctctttcaccgacactgg
NM_010139	tacgagaaggtcgaggatgc	tcagatgcctcagacttgaaga
NM_010193	ggtgcgcttgctgttagaa	tggcaccgtcaatgacatac
NM_010228	ggcccgggatatttataagaac	ccatccattttaggggaagtc
NM_010262	tttgtttctattttggctttttga	ccgacatggctcagatagg
NM_010453	caagctgcacattagtcacga	gcggttgaagtggaattctt
NM_010698	tgcctgaagacctgtctgttt	gccttgtgggttctgctg
NM_010756	tctttgaggggccactagg	ggaagggaaatgaaaggaaga
NM_010786	acgagctctcagatgaggatg	tctccttcaaaagagtctgtatcg
NM_010807	ggcagccagagctctaagg	tcacgtggccattctcct
NM_011030	tcaaagacttagcaaaaccaagg	gctaattctgtaatgtaccgtctcc
NM_011341	gactcagagccacggagaag	ccggtcagtgttcacgtcta
NM_011403	tgtgaccgagatgcaggac	gggacgtagtctgtggctgt
NM_011494	gaacgaggtgctaagcatga	tcattccacagtgtacctttcttg
NM_011589	atatggctccctcgacagtg	cttcctcagggctccacag
NM_011877	gtcaggagggactcatcctct	ggaagaagcaaatggctctg
NM_013598	tcaacattaggtcccgagaaa	actgctactgctgtcattcctaag
NM_013634	cccagctgacaacaaatgg	tggggaaatttcaagaagga
NM_013727	tctgcagctaagaacagaggtg	tgagggtgggtttaaattcct
NM_013791	ctacgcgctgcataagtgg	catttggtttatccactaaaatgttc
NM_016769	tccgtatgagcttcgtcaaa	ggtgctggtcactgtctgtc
NM_016963	aagaccgggacgattatgtg	tggcttctgtttggggataa
NM_016969	ctgtcgtccttgtttgtgctac	cacacctggaaaaggctgat
NM_019764	atggtggcgaaactgctt	cagcagtgccatctctcg
NM_019821	gcccatcaaggcagacata	ggctgggtcggtgtcata
NM_023061	gcgaggcagaaagtaaccag	cagcagctggcctgtctt
NM_023595	tggcttggctgtaaagcact	ccccaacgtttcctctgtaa
NM_024242	aaagtcaccgtaaaattgagacaa	tgctctgtcggccttatctt
NM_025889	agaaaaaccgccgggtagt	aagatggcactggagacacc
NM_033561	tcaggaaaggtggacctgat	tcccatccacctctagattctc
NM_033565	cagaagaagactgaaggcaagg	ggagctatttggagccttctc
NM_080575	ccaccaagatcgccaagta	atctggttttggggagacg
NM_133998	tcagccgaatgaccacagt	ctgaaaccgggtcctttctt
NM_138606	ccttcgagagagaccaggag	gattagggcacagcaatcg
NM_145564	caataacctcaaggcgttcc	gtcgatgtaaacggcaccttt
NM_145737	ccttcgagagagaccaggag	gattagggcacagcaatcg
NM_172397	atgaagccaaaggcagca	cccgcaagctaaaggactta
NM_173442	acagattcaggcttcctgtga	ccttgaatgccaatgatgttt
NM_008141	catgtccacactaggcattga	gccaggttgttgagctgtct
NM_016701	gttgcagcccactgaaaagt	gaccctgcttctcctgctc
NM_031249	gagtgggaaatcctggaggt	catgcctgctccttgcat

TABLE 12.1b
Applied Biosystem's TaqMan Assay Identifiers that were used for QRT-PCR in the Study

Assay ID	Assay ID
Mm00436337_m1	Mm00726782_s1
Mm00436757_m1	Mm00777369_m1
Mm00438851_m1	Mm00779233_s1
Mm00440701_m1	Mm00784521_s1
Mm00441825_m1	Mm00786385_s1
Mm00442663_m1	Mm00786736_s1
Mm00443474_m1	Mm00815479_s1
Mm00443538_m1	Mm00834297_m1
Mm00443561_g1	Mm00834570_s1
Mm00443805_m1	Mm00850305_s1
Mm00445180_m1	Mm01158982_g1
Mm00447576_m1	Mm01163545_g1
Mm00450664_m1	Mm01191618_g1
Mm00452302_g1	Mm01193537_g1
Mm00465631_s1	Mm01194707_m1
Mm00469135_m1	Mm01196307_m1
Mm00474685_g1	Mm01210482_m1
Mm00476162_m1	Mm01211708_m1
Mm00486497_g1	Mm01218728_g1
Mm00486943_m1	Mm01219608_m1
Mm00488044_m1	Mm01225849_s1
Mm00491391_s1	Mm01232271_g1
Mm00491444_m1	Mm01232724_m1
Mm00492606_m1	Mm01237743_s1
Mm00493897_s1	Mm01242767_g1
Mm00494322_g1	Mm01247138_m1
Mm00495681_m1	Mm01273696_m1
Mm00498596_g1	Mm01274547_m1
Mm00499128_g1	Mm01285682_m1
Mm00501348_m1	Mm01292910_g1
Mm00501659_m1	Mm01297292_m1
Mm00502265_s1	Mm01298543_m1
Mm00502291_m1	Mm01299984_g1
Mm00502351_g1	Mm01305963_g1
Mm00502938_g1	Mm01309843_g1
Mm00504790_g1	Mm01318925_m1
Mm00507743_m1	Mm01329634_s1
Mm00512034_m1	Mm01329822_m1
Mm00512275_m1	Mm01329922_m1
Mm00514572_m1	Mm01341688_gH
Mm00517192_m1	Mm01620599_g1
Mm00517414_m1	Mm01621873_s1
Mm00599712_m1	Mm01622471_s1
Mm00627900_m1	Mm01700423_m1
Mm00656724_m1	Mm01946604_s1
Mm00658175_m1	

12.4.2 ABI TAQMAN GENE EXPRESSION ASSAYS

For the TaqMan results, the samples were sent to Applied Biosystems (Applied Biosystems, Foster City, California), where the expression of mRNA for 91 genes was measured in each of the two samples by RT-PCR using TaqMan Gene Expression Assays on ABI PRISM 7900 HT Sequence Detection System (Applied Biosystems, Foster City, California). The primers for each gene were designed to lie within the same exon as that of the matched probe sequences across the platforms. Approximately 2 µg of total RNA of each sample was used to generate cDNA using the ABI High Capacity cDNA Archiving Kit (Applied Biosystems, Foster City, California) and the QRT-PCR reactions were carried out following the manufacturer's protocol. Four technical replicates were performed for each gene for each sample in a 384-well format plate. On each plate, three endogenous control genes (*RPS18*, *ActinB*, and *GAPDH*) and one no-template control (NTC) were also performed in quadruplicates.

12.5 RESULTS

12.5.1 MAGNITUDE OF THE RELATIVE MEASUREMENTS

The dynamic range of a platform provides information about how a particular platform can reliably identify the magnitude of gene expression changes between two samples. The distribution (dynamic range) of the mean $\log_2$ ratios for each platform is illustrated in Figure 12.2. The gene expression ratios were computed for the MR and MC. ABI and MGH arrays had the widest ranges, whereas Agilent and MWG arrays displayed the highest ratio compression. Among the two-dye platforms,

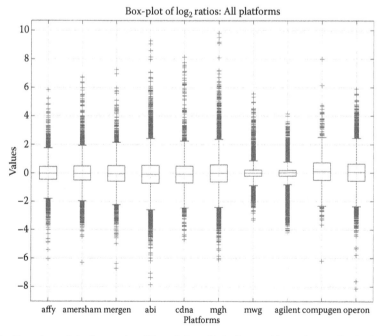

FIGURE 12.2 Box-plots of the $\log_2$ ratios (filtered) for all platforms. Along the *x*-axis, the box-plots of the $\log_2$ ratios after filtering are shown for the 10 platforms in the order of Affymetrix, Amersham, Mergen, ABI, cDNA, MGH, MWG, Agilent, Compugen, and Operon.

MGH had the widest range, followed by Operon, whereas Agilent and MWG had the smallest ranges. For the one-dye platforms, ABI had the widest range, followed by Mergen, Amersham, and Affymetrix. Overall, the dynamic ranges were comparable for most of the platforms, except for Agilent and MWG, which had less than half the dynamic range of the others.

12.5.2 Consistency within Platforms

An important assessment before conducting a cross-platform evaluation is to measure the level of reproducibility across replicate experiments within platforms. The mean Pearson and Spearman correlation coefficients and their SDs were computed for the intensities across the technical replicates for each sample within each platform (data not shown). The Pearson correlations for the intensities were high ($r > 0.91$) across the technical replicates for both samples for all non-cDNA arrays. The correlations of intensities for both samples in the one-dye platforms were higher than those of the two-dye platforms. For example, for the MC, the one-dye platforms had correlations >0.98, whereas for the two-dye platforms, the correlations ranged from 0.9124 to 0.9917.

For the relative measurements, the two-dye platforms had similar or higher correlations, except the cDNA platform, which had the lowest correlations among the replicates. Within the one-dye platforms, ABI and Mergen platforms had the highest and lowest correlations for $\log_2$ ratios, respectively. The effect of spot quality filtering increased the intra-platform correlations for all platforms, particularly for $\log_2$ ratios, except for the cDNA platform.

Using CVs, we also examined the variability of replicated measurements within each platform (data not shown). The CVs in channel-separate intensities were calculated after the transformation that projects signals of different platforms linearly onto the same range of 1–100. For both MR and MC, most platforms have the mean CV of all genes across five replicates below 0.20, except for cDNA and Operon, which showed a higher variability in MC with CVs of 0.20 and 0.21, respectively. We noticed that the one-dye platforms in general have better CVs for intensity measurements, as expected. Among two-dye platforms, Agilent's CV is comparable to the CVs of one-dye platforms.

A modified definition of CV was used to analyze the variation in the microarray expression ratios; that is, the SD and ordinary CV were used to indicate the variations in low ($\log_2$ ratio$^2 \leq 1$) and high ($\log_2$ ratio$^2 > 1$) relative expression, respectively (see Section 12.2). In terms of the degree of variations, the two subclasses of relative measurements revealed similar patterns among the multiple platforms. The lowest CVs of $\log_2$ ratios for both subclasses occurred in Agilent (0.12 and 0.08 for low and high ratios, respectively), and the highest in cDNA (0.76 and 0.56 for low and high ratios, respectively).

Based on the CVs, Affymetrix, Amersham, ABI, and Agilent had generally good and consistent intra-platform reproducibility, both for absolute and relative measurements.

12.5.3 Consistency across Platforms

Comparison across platforms was performed by mapping the gene expression measurements for UG, LL, RS, and RSEXON identifiers. If there was more than one probe matched to an identifier, the values were averaged. In most instances, one probe represented only one gene. The number of overlapping probes decreased by more than 40% and 50% when the probes were mapped by RS and RSEXON, respectively, compared to UG and LL mappings.

The assessment of reproducibility across platforms was performed by computing the Pearson and Spearman (data not shown) correlation coefficients of the $\log_2$ ratios for unfiltered data. The range of correlations between the one-dye platforms was much higher than correlations between the two-dye platforms. For example, when platforms were matched by UG, 0.71–0.79 was the range of Pearson correlations for one-dye platforms and 0.49–0.67 was the range for two-dye platforms (cDNA and Compugen measurements were not included because they were outliers). Poor pair-wise correlations were seen with any pair-wise comparisons with the cDNA and Compugen arrays, with Compugen correlations slightly higher than the cDNA correlations. Reasonable Pearson correlations were seen in pair-wise comparisons between the one-dye platforms and the following two-dye platforms: Agilent, MGH, MWG, and Operon (0.54–0.68). In general, the correlation between platforms was reasonably high and similar for UG and LL identifiers, but improved using RS and RSEXON matching options. For example, the Pearson correlations for the pair-wise comparison of Affymetrix and Amersham were 0.76, 0.76, 0.81, and 0.85 for UG, LL, RS, and RSEXON, respectively.

Spot quality filtering had a profound effect for all the inter-platform correlations, except cDNA arrays (Table 12.2). For each matching option, in general there was an increase in correlation coefficients in the filtered data compared to the unfiltered data. In the case of the cDNA arrays, the correlations remained essentially the same, whereas with Compugen, the pair-wise correlations improved, though not as much as the other two-dye platforms. The correlations remained higher for the comparisons within one-dye platforms than within two-dye platforms. However, Agilent, MGH, and Operon platforms had comparable correlations to the one-dye platforms. Additionally, there was a sequential improvement in the correlations as the probe matching option was more specific, that is, mapping probes across platforms at the exon level (RSEXON) produced higher correlations than mapping at the UG, LL, or RS levels. The computed randomization tests demonstrated that the correlations for the RSEXON matched data were highly significant for all pairs across platforms (data not shown).

PCA was used to illustrate the overall similarity of expression profiles. Figure 12.3 shows the PCA plot based on the first three principal components calculated from 130 genes common to eight of the platforms. We see that one-dye platforms are clustered together, whereas two-dye platforms are more spread apart.

12.5.4 Consistency across Laboratories

Another important aspect of microarray experiments is the consistency of gene expression measurements that are generated from the same platform, but conducted at multiple sites. Data from three platforms, Affymetrix, Amersham, and Mergen, were analyzed and compared in a cross-laboratory evaluation. As expected, the intra-platform Pearson and Spearman correlations for the intensities were high for both samples ($r > 0.95$). Among the three platforms, Amersham had the highest correlation for $\log_2$ ratios, followed by Affymetrix and Mergen. When examining the RSEXON matched probes among the three platforms, the Pearson correlation coefficients between different labs were 0.89, 0.93, and 0.79, for Affymetrix, Amersham, and Mergen, respectively. The highest selected pair-wise combination of cross-platform Pearson correlations were 0.81 (Affymetrix versus Amersham), 0.76 (Affymetrix versus Mergen), and 0.78 (Amersham versus Mergen). Similar results were obtained when using Spearman correlations. Overall, these preliminary results suggest that cross-laboratory variations are significantly smaller than the cross-platform variations.

12.5.5 Mapping of Probe Sequences to Exon Positions

The probe sequences obtained from each platform were used to help minimize mistakes in matching genes across platforms. Mapping by sequences ensured that the annotation of the probes provided by the vendors was correct. We found that Amersham, Mergen, and Compugen had the highest percentage (>72%) of "exon-matched" probes. The remainder of the platforms had <63% of their probes mapping to an exon. As expected, the longer cDNA probes had the lowest percentage of exon-matched probes: only 1024 out of 11,832 unique probes were mapped to an exon.

To visualize the location of the probe sequences within the exon and their relationship to other platforms, a customized MATLAB (The Math Works, Natick, Massachusetts) script was created to

Table 12.2
Pair-Wise Cross-Platform Pearson Correlation Coefficients using Four Alternative
Probe Matching Strategies

	Affymetrix	Amersham	Mergen	ABI	cDNA	MGH	MWG	Agilent	Compugen	Operon
Panel A: UniGene Clusters										
Affymetrix	1	0.76	0.71	0.79	0.31	0.68	0.60	0.67	0.39	0.64
Amersham	0.78	1	0.73	0.79	0.28	0.64	0.59	0.67	0.38	0.65
Mergen	0.74	0.75	1	0.73	0.32	0.58	0.54	0.65	0.35	0.59
ABI	0.84	0.80	0.79	1	0.29	0.66	0.62	0.70	0.38	0.63
cDNA	0.35	0.31	0.37	0.29	1	0.30	0.35	0.25	0.05	0.26
MGH	0.67	0.66	0.59	0.68	0.30	1	0.53	0.63	0.34	0.51
MWG	0.59	0.60	0.61	0.64	0.36	0.55	1	0.67	0.34	0.49
Agilent	0.72	0.69	0.73	0.73	0.25	0.64	0.66	1	0.35	0.58
Compugen	0.36	0.37	0.37	0.39	0.07	0.40	0.33	0.33	1	0.35
Operon	0.66	0.65	0.61	0.66	0.30	0.55	0.50	0.61	0.33	1
Panel B: LocusLink Identifiers										
Affymetrix	1	0.76	0.70	0.80	0.31	0.69	0.61	0.68	0.39	0.63
Amersham	0.84	1	0.73	0.80	0.28	0.65	0.59	0.68	0.37	0.64
Mergen	0.86	0.83	1	0.73	0.31	0.58	0.55	0.65	0.35	0.59
ABI	0.87	0.86	0.85	1	0.29	0.69	0.63	0.71	0.41	0.65
cDNA	0.30	0.29	0.30	0.29	1	0.30	0.31	0.24	0.15	0.24
MGH	0.81	0.76	0.79	0.79	0.32	1	0.54	0.64	0.34	0.54
MWG	0.70	0.67	0.68	0.68	0.33	0.62	1	0.67	0.35	0.49
Agilent	0.79	0.75	0.73	0.76	0.24	0.69	0.68	1	0.33	0.58
Compugen	0.66	0.61	0.64	0.67	0.31	0.59	0.57	0.53	1	0.36
Operon	0.80	0.79	0.81	0.80	0.28	0.72	0.63	0.70	0.70	1
Panel C: RS										
Affymetrix	1	0.81	0.75	0.84	0.34	0.65	0.61	0.73	0.38	0.67
Amersham	0.87	1	0.78	0.83	0.34	0.66	0.59	0.68	0.37	0.67
Mergen	0.91	0.85	1	0.81	0.40	0.62	0.62	0.74	0.38	0.61
ABI	0.91	0.87	0.89	1	0.30	0.67	0.63	0.74	0.39	0.67
cDNA	0.30	0.33	0.34	0.25	1	0.40	0.19	0.26	0.08	0.34
MGH	0.81	0.77	0.78	0.79	0.34	1	0.57	0.68	0.44	0.52
MWG	0.68	0.67	0.71	0.68	0.32	0.62	1	0.65	0.30	0.49
Agilent	0.81	0.76	0.79	0.78	0.23	0.68	0.66	1	0.40	0.63
Compugen	0.57	0.62	0.64	0.58	0.18	0.63	0.56	0.53	1	0.37
Operon	0.83	0.80	0.82	0.83	0.31	0.72	0.62	0.72	0.63	1
Panel D: RSEXON Position										
Affymetrix	1	0.84	0.87	0.85	0.29	0.80	0.70	0.79	0.67	0.81
Amersham	0.89	1	0.84	0.84	0.30	0.75	0.68	0.75	0.61	0.80
Mergen	0.91	0.87	1	0.86	0.31	0.78	0.68	0.74	0.65	0.81
ABI	0.92	0.88	0.90	1	0.29	0.77	0.68	0.74	0.65	0.80
cDNA	0.28	0.38	0.36	0.28	1	0.31	0.36	0.26	0.22	0.30
MGH	0.82	0.79	0.77	0.79	0.38	1	0.61	0.68	0.60	0.69
MWG	0.71	0.66	0.71	0.68	0.14	0.63	1	0.68	0.56	0.64
Agilent	0.82	0.77	0.80	0.79	0.24	0.72	0.65	1	0.54	0.69
Compugen	0.65	0.62	0.67	0.55	0.21	0.68	0.53	0.61	1	0.69
Operon	0.83	0.82	0.82	0.84	0.34	0.74	0.63	0.73	0.64	1

Notes: For the four probe matching strategies: (a) UG clusters, (b) LL identifiers (LL), (c) RS identifiers, and (d) RSEXON position, pair-wise between platform Pearson correlations of mean normalized $\log_2$ ratios is shown in the upper triangle matrix for unfiltered data, and the lower triangle for filtered data.

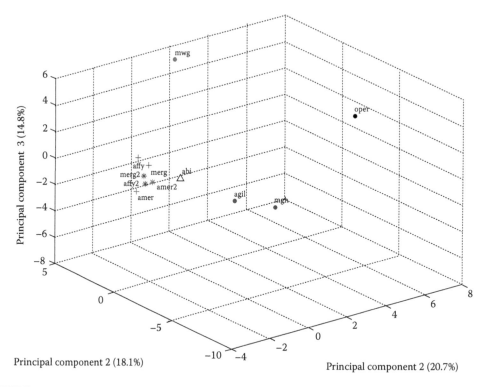

FIGURE 12.3   (**See color insert following page 138.**) Cross-platform PCA plot. The plot illustrates PCA performed on log_2 ratios corresponding to 130 RS identifiers common to eight of the 10 platforms. As academic cDNA and Compugen had few RS identifiers in common with the other platforms, we chose to exclude these from this analysis in order to increase the number of genes applicable to PCA without missing values. For Affymetrix (affy2), Amersham (amer2), and Mergen (merg2) expression profiles obtained from a second laboratory were included in the analysis. The other expression profiles are labeled with abbreviations of the platform names used elsewhere. Each expression profile is plotted according to the first, second, and third principal components. For each axis, the number in parenthesis gives the amount of variation (in percent of total) accounted for by the corresponding principal component.

generate plots of exon positions for probe sequences across the different platforms. Figure 12.4 illustrates an example in which the probe sequences from all 10 platforms lie within the same exon. In total, there were only four RSEXONs that were common across all 10 platforms: NM_008086:1 (Growth arrest specific 1), NM_008686:1 (Nuclear factor, erythroid derived 2), NM_018798:1 (Ubiquilin 2), and NM_018871:1 (3-monooxygenase/tryptophan 5-monooxygenase activation protein). All or most of the sequences available for oligonucleotide design resided in one exon for these genes. In this diagram, when probe sequences were mapped within the same exon, the intensities were found to be very similar across the platforms, even though the probe sequences were not overlapping. Overall, mapping of probe sequences (RS and RSEXON) improved inter-platform correlations compared to mappings based on UG and LL for all the platforms. Annotation-based identifiers had a higher number of matched probes compared to sequence-based mappings.

    To verify these findings, evaluation of the variation among the different probe matching options was carried out. Since there were very few overlaps across 10 platforms, the relative measurements that were extracted for each matching option had to be observed in at least six platforms, that is, the four widely used platforms: ABI, Affymetrix, Agilent, and Amersham, and any additional two platforms. Based on this criterion, the distributions of the SDs for matched transcripts based on each set of identifiers were computed. The results demonstrate that there was more variation between the

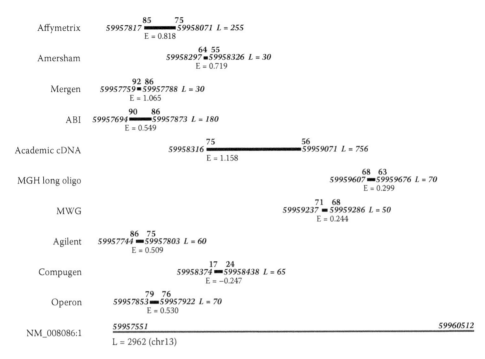

FIGURE 12.4 Cross-platform agreement of probes matched within 1 exon. For the *Gas1* gene (RefSeq NM_008086, LocusLink 14451, UniGene Mm.22701), all 10 platforms had probes that could be mapped completely within the boundaries of the first exon. The diagram shows the location of the probes from the different platforms. The complete exon is indicated at the bottom, with the 3′ end on the left-hand side. The start and end positions on chromosome 13 are given just above the left-hand and the right-hand ends, respectively, of the bar representing the exon. The probes are indicated with black bars, flanked by the start and end coordinates, as given by the sequence alignments of the probes to the genome. The length of the alignment between the probe and the exon (L) is shown to the right of the probe bars. The relative gene expression (E) shown below each probe bar is the log$_2$ ratio of MR versus MC. The percentile-transformed intensities from each platform are shown above the respective probe bars. For each platform, the number in red on the left-hand side is the intensity from MR and the number in green on the right-hand side is the intensity from MC.

measurements for annotation-based identifiers when compared to the sequence-based identifiers. Overall, RSEXON matched data had the lowest variation across platforms, followed by RS, LL, and UG (data not shown).

12.5.6 QUANTITATIVE BIOLOGICAL VALIDATIONS

As an independent validation strategy, RNA levels for a total of 160 unique genes were obtained from QRT-PCR with two methods. Log$_2$ ratios for 91 genes were obtained using ABI TaqMan assays and for 74 genes using Roche Universal ProbeLibrary. As a replacement for a true gold-standard, we considered log$_2$ ratios from QRT-PCR as nominal values and used the slope of the regression line of the log$_2$ ratios from each microarray platform against QRT-PCR results as an accuracy measure to evaluate the platforms. By this statistic, ABI would be ranked the highest, followed by Affymetrix and Operon, whereas cDNA, MWG, and Agilent were the poorest. These findings were confirmed by correlation coefficients on QRT-PCR data paired with data from the microarray platforms (data not shown). We observed slightly lower correlations for ProbeLibrary results than for TaqMan results when investigating the two subsets of QRT-PCR separately.

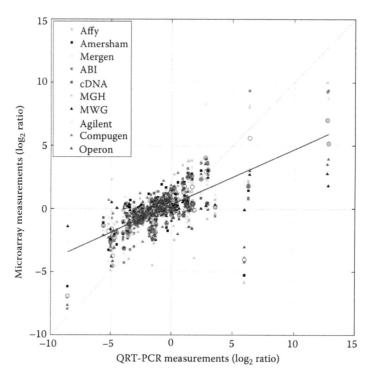

FIGURE 12.5 **(See color insert following page 138.)** Scatter plot of QRT-PCR versus all microarrays. Log$_2$ ratios from the microarray platforms are plotted (*y*-axis) versus the corresponding log$_2$ ratios from QRT-PCR (*x*-axis). From each platform, all log$_2$ ratios based on probes that could be mapped to any of the 153 RS identifiers used in QRT-PCR were used. The regression line between the median microarray log$_2$ ratios (over all platforms including a given gene) and the QRT-PCR log$_2$ ratios is shown in blue. The slope of the line is 0.437, indicating a smaller dynamic range for the microarrays as compared to QRT-PCR. The Pearson correlation coefficient between the median measurements and QRT-PCR was 0.76 (*p*-value 1.36 E-30), indicating relatively good albeit highly significant correlation.

Overall, the measurements from most microarray platforms accord well with QRT-PCR. However, the dynamic range for QRT-PCR was noticeably larger than for the microarrays (Figure 12.5). The median of all the microarray measurements had a Pearson correlation with QRT-PCR of 0.76, indicating reasonably good agreement. In terms of log$_2$ ratio difference, one-dye platforms had lower distances compared to two-dye platforms. Within the two-dye platforms, Agilent and Operon had the shortest average distances.

Most of the platforms had results consistent with QRT-PCR for high and medium expressed genes. For highly expressed genes, one-dye platforms had markedly better agreement with QRT-PCR than two-dye platforms. In the case of low expressed genes, the agreement with QRT-PCR was much poorer for all platforms. For a subset of seven retina-related genes (confirmed to be highly expressed in retina from multiple studies using classical techniques), high expression in MR versus MC was found in almost all platforms with probes for these genes and also confirmed by QRT-PCR.

12.6   DISCUSSION AND FUTURE WORK

In this study we compared gene expression data from 10 different microarray platforms using identical samples. Although there was good agreement between measurements from some of the platforms, others exhibited marked differences. Motivated by the large number of both commercial and noncommercial microarray platforms in particular, and technologies for gene expression measurements in general, it has been an emphasized goal during this effort to develop a sound and consistent framework for cross-platform comparisons. Our goals were to (1) provide unbiased results with clear metrics for evaluations of performance, preferably using established analytical techniques; (2) conduct the experiments for different platforms as systematically and as similarly as possible, yet allow flexibility to accommodate for single platform restrictions; and (3) be general enough to allow inclusion of novel academic and commercial platforms as they develop. The latter is of importance, as this is an ongoing comparative effort and we hope to be able to include future platforms as they become available.

To make the conditions for the different platforms as similar as possible, we used pooled RNA as the sample sources. Ideally, one would like to have two unlimited sources of RNA, both with a diverse set of transcripts having a wide dynamic range of expression. For the general usefulness of the comparison, the RNA should be selected from a commonly used organism. At the beginning of this study, such sources were not available. We therefore chose two samples that had some of the above features. We extracted RNA from tissues of the cortex and retina from the well-studied *Mus musculus*, selecting inbred mice to eliminate genetic variability. The MC was chosen because brain tissues are generally considered to have broad expression profiles. The retinal tissue was selected because there is generally little variability in this tissue, and it also has some well-known tissue-specific transcripts, such as cone- and rod-related genes that are expressed at a high level. Both tissue samples can be considered as replenishable sources of RNA with little variability, as demonstrated by laser-based capillary electrophoresis of labeled samples (data not shown). Recently, mouse universal reference RNA sources have become commercially available, such as those from Ambion (http://www.ambion.com/catalog/CatNum.php?6050) and Stratagene (http://www.stratagene.com/manuals/740000.pdf) and are being included in our ongoing work.

We tried to include all platforms claiming to perform whole-genome scale profiling of mRNA for the mouse that could provide us the probe sequence information. Despite including one-dye and two-dye, cDNA-based and oligonucleotide-based arrays, and commercial as well as academic platforms, this set of hybridization-based platforms should be considered homogenous relative to digital-based platforms such as SAGE and MPSS. Each platform, however, may have a distinct set of laboratory or quality control features that affect the ease of inclusion for comparison purposes, including external spikes, alien probes, and positive and negative controls. The differences with respect to such features among the included platforms pose constraints on the choices of both intra- and inter-platform comparative approaches and metrics. Generally, it is preferable that both preprocessing and evaluation of performance are based on internal standards, for example, spikes and repeated elements on the arrays. However, for cross-platform comparisons, the layout of the array is in the hands of the provider, and for this study we chose to use internal controls where available. This may introduce biases in the comparisons that are not easily compensated for, but it also reflects the current usage of this platform in laboratory environments.

The variations in technical and instrumental choices among platforms make direct comparison based on raw (intensity) signals impossible. In particular, various image analysis software packages may have proprietary algorithms to digitize and segment hybridization signals with special purpose criteria and scales. To compensate for the different ranges of signal intensities of the platforms, we applied both a linear scaling transformation and a percentiles transformation. The effect of both methods was to bring the signal ranges to a uniform scale, which was found useful in comparing intra-platform variations. In spot quality filtering procedures, we chose to prioritize the usage of quality flags generated by commercial software, in addition to considering the recommendations platform vendors. It has previously been shown that stringent spot quality filtering can improve data consistency [27,58], and this result was confirmed in our study.

As emphasized in a recent review article [62], the many factors affecting sensitivity and accuracy of DNA microarrays must be carefully controlled to ensure reliable results. During our analysis, we have identified possible confounding factors biasing the results for three of the platforms: academic cDNA, Compugen, and Agilent. The academic cDNA platform from our lab had the lowest ranking both in intra- and inter-platform evaluations. These results may not be indicative of spotted cDNA arrays in general. One factor could be the retina-specific cDNA clones that have been selected. Another factor that could be related to false probe identity that is about 50% of the probes on the arrays were sequence-verified and 15–20% were found to be incorrect (data not shown). A more general issue with cDNA arrays is the expected lower specificity of the probes, as they are possibly more prone to cross-hybridization of highly homologous transcripts [63]. Lack of optimized printing or hybridization protocols for this platform may have also been relevant, as low signal intensities were consistently produced. We also found that after the PCR and purification steps, the actual amount of DNA varied from spot to spot, which could have caused, when the data were filtered, the deletion of some good spots. The arrays using the Compugen spotted oligonucleotides also had low performance scores in general. A reason for this could be the limited gene selection of Compugen probes by the microarray facility. When the mapping was performed at the exon level for both academic cDNA and Compugen platforms, there was a low percentage of matches, thus few overlaps occurred across platforms. Another possible cause of bias for the Compugen platform was the fact that these probes had been co-printed with the Operon probes on the same arrays. For Agilent, a compression of the dynamic range was observed. This may be attributed to a combination of low hybridization temperatures and insufficient washing conditions. However, the same factors may have contributed to artificially good correlations and CVs for this platform. Nevertheless, these deficiencies have been corrected in Agilent's current protocols. In sum, this emphasizes the need for careful design of cross-platform protocols and performance tuning throughout the execution of the experimental procedures.

One possible confounding factor affecting all platforms may have been the varying degrees of saturation per experiment. This could have resulted from technical choices during the scanning procedure, such as ill-chosen scanner settings, nonlinearity of photomultiplier tubes, and cross-talk of laser channels, or also larger spans in expression numbers of biological samples than the dynamic range of the individual platform could accommodate. The scanning procedures were in this case location-dependent, but we assumed that the operator was familiar with optimal technical settings for the given platform. As could be seen in signal intensity scatter plots, we observed saturation in some platforms, in a channel-dependent manner. This happened, for example, for a selection of seven genes expected to be very highly expressed in retina, and subsequently validated by QRT-PCR. Agilent and Operon were the platforms most affected by this problem. However, the microarray platforms all tended to underestimate fold changes as compared to QRT-PCR. It is difficult to assess to what extent the limitations of scanner intensity ranges influenced the comparisons reported, but as was demonstrated for highly expressed genes, correlations of percentile transformed intensities were quite good indicating that the genes are correctly identified as highly expressed.

For cross-platform comparisons, there are generally several possible strategies to match measurements. All the platforms had GenBank accession number information. Some of the platforms also provided gene designations, but rather than relying on these, we chose to use a common strategy for mapping probes to UG cluster and LL identifiers, thus avoiding problems due to annotations to different versions of the databases. However, this procedure does not eliminate biases due to wrong GenBank information. The sequence-based mappings obtained from sequence alignments using the provided probe sequence data ensured a more correct mapping from probes to genes (RS identifiers, and also exon), but at the cost of reducing the number of possible matches between platforms. Overall, the results based on different mapping strategies showed good agreement. However, the agreement on matched data tended to increase with increasing mapping specificity, that is, increasing agreement with LL versus UG, RS versus LL, and RSEXON versus RS. A possible interpretation of this is that the RS mapping eliminates biases due to splice variants, since it is on the transcript level, and that the RSEXON mapping possibly forces the probes of different platforms to be more similar, as they are confined to a limited region of each gene. This requires further systematic investigation. Exon-based mapping from original sequence information also resulted in higher correlations, indicating that the matching criterion is critical for obtaining reproducible results, and making a strong case for the disclosure of probe sequence information by the manufacturers. This criterion could be further improved by imposing requirements of probe matching based on predefined hybridization criteria, such as a nearest-neighbor melting temperature of the actual probes and potential cross-hybridization. To some extent, this would reduce the amount of compatible data between platforms, but the potential importance of this aspect requires further attention. At any rate, these considerations highlight the importance of probe design with respect to hybridization kinetics and thus, varying degrees of cross-hybridization. A systematic analysis of the effect of different oligonucleotide design biases versus platform priming strategies was not performed, and could explain parts of the variation observed.

In summary, the commercial platforms performed better than "in-house" platforms, both in internal consistency and agreement with other platforms. Although "in-house" arrays will continue to play an important role in academic laboratories, the reliability and consistency of the commercial platforms demonstrated in this study will continue to set high-performance standards for the development of gene expression technologies. Along the dye-dimension, by most measures, the one-dye platforms Affymetrix, ABI, and Amersham consistently performed among the best, as confirmed in another study [47]. The high internal consistency of Affymetrix, Amersham, and Mergen was also confirmed in the experiments conducted at a second laboratory. The observation that cross-laboratory variability using the same technology was lower than that of cross-platform variability confirmed the results of other studies [15,53].

QRT-PCR, although commonly accepted as a "gold-standard" for relative gene expression measurements [64], has technical limitations and potential biases. We used the following criteria for selecting genes for validation: (1) genes should be present in at least six platforms (see Section 12.2); (2) they should span the dynamic range; and (3) they should also include pairs with measurements that were in disagreement. We found significant correlations for genes being detected as highly expressed in most platforms, and also for some genes with lower expression values. As the expression level decreased, however, the observation discrepancy for a given gene across platforms increased. We found that for a class of genes being detected as having variable expression with different microarray platforms, there was low correlation between all platforms and QRT-PCR results. We interpret this as stochastic variation appearing at low transcript numbers in both microarrays and validation procedures, as a reflection of input amount of RNA in the procedures. If there are many genes with low transcription, the chances of stochastic variation may influence results on the QRT-PCR side, and the fluorescence sensitivity of a given platform may influence the results on the microarray side. This could to some extent be corrected in the QRT-PCR if the input RNA amount is increased. However, this will rapidly limit the number of genes due to consumption of RNA. Furthermore, it was evident that the primers mapped to lie within the same exon as that of the probe sequences (TaqMan) had higher consistency than those that did not (Universal ProbeLibrary). The Universal ProbeLibrary Kit was designed as a quick validation method, where the primers were designed using their software, to be optimal for the probes in the kit. This illustrates the importance of careful primer design for biological validations in a cross-platform study.

A possible approach for future platform comparisons could be one based on *a priori* knowledge of approximate relative expression levels of a set of genes. By selecting and analyzing the reproducibility of a gene set having high and low borderline expression, the investigator would obtain the most relevant information desired in platform comparisons, that is, the ability of a platform to consistently detect gene expression over the whole dynamic range of the platform, and that the dynamic range of the platform allows for the detection of the full biological variation of most samples.

In future studies, we will utilize other samples specifically selected to address biological and technical issues. A second pool of MR RNA was collected to examine biological variability of the same sample across the arrays [65]. Rat retina and yeast samples have been chosen to address cross-species- and cross-hybridization-related issues. The experimental design for two-dye platforms will also be extended, to allow us to evaluate the performance of the platforms in more detail. We will also analyze data for replicates of (1) dye-swaps; (2) self-hybridizations; and (3) single-sample hybridizations that will be presented elsewhere. In addition to the platforms evaluated in this manuscript, data sets have been generated but not analyzed for these platforms: Agilix [66], Febit [36], Illumina [67], NanoString (NanoString Technologies, Seattle, Washington), Phanlanx OneArray, (Phalanx Biotech, Palo Alto, California), MPSS, and SAGE. We will include these platforms in future comparisons.

The goal of this study was to illustrate a comparison framework that matched the transcripts at the sequence level. This is the first comprehensive analysis of this relatively large-scale initiative in which the sequences of all probes were known to address the question of measurement and bias in DNA microarrays [68]. The results presented here indicate that there are many available platforms that provide good quality data and that between these platforms there is generally good agreement. However, there are considerable and significant areas where the platforms disagree, despite significant developments towards standardization of gene expression profiling, and therefore many issues remain open for investigation. The ability to reliably detect low expressed genes remains a limitation of hybridization-based microarrays in general, but highly expressed genes are usually reliably measured across platforms.

ACKNOWLEDGMENTS

We would like to thank the vendors and participants for their contributions to the study. This chapter was supported in part by the Advanced Medical Counter Measures Consortium, US Department of Defense grant W81XWH.

REFERENCES

1. Fodor SP, Read JL, Pirrung MC, Stryer L, Lu AT, and Solas D. Light-directed, spatially addressable parallel chemical synthesis. *Science*, 1991; 251(4995): 767–773.
2. Schena M, Shalon D, Davis RW, and Brown PO. Quantitative monitoring of gene expression patterns with a complementary DNA microarray. *Science*, 1995; 270(5235): 467–470.
3. Velculescu VE, Zhang L, Vogelstein B, and Kinzler KW. Serial analysis of gene expression. *Science*, 1995; 270(5235): 484–487.
4. Brenner S, Johnson M, Bridgham J, Golda G, Lloyd DH, Johnson D, Luo S, et al. Gene expression analysis by massively parallel signature sequencing (MPSS) on microbead arrays. *Nat Biotechnol*, 2000;18(6): 630–634.
5. Bennett S. Solexa Ltd. *Pharmacogenomics*, 2004; 5(4): 433–438.
6. Margulies M, Egholm M, Altman WE, Attiya S, Bader JS, Bemben LA, Berka J, et al. Genome sequencing in microfabricated high-density picolitre reactors. *Nature*, 2005; 437(7057): 376–380.
7. Ball CA, Sherlock G, Parkinson H, Rocca-Sera P, Brooksbank C, Causton HC, Cavalieri D, et al. Standards for microarray data. *Science*, 2002; 298(5593): 539.
8. Shi L, Reid LH, Jones WD, Shippy R, Warrington JA, Baker SC, Collins PJ, et al. The MicroArray Quality Control (MAQC) project shows inter- and intraplatform reproducibility of gene expression measurements. *Nat Biotechnol*, 2006; 24(9): 1151–1161.
9. Edgar R, Domrachev M, and Lash AE. Gene Expression Omnibus: NCBI gene expression and hybridization array data repository. *Nucleic Acids Res*, 2002; 30(1): 207–210.
10. Brazma A, Parkinson H, Sarkans U, Shojatalab M, Vilo J, Abeygunawardena N, Holloway P, et al. ArrayExpress—a public repository for microarray gene expression data at the EBI. *Nucleic Acids Res*, 2003; 31(1): 68–71.
11. Ali-Seyed M, Laycock N, Karanam S, Xiao W, Blair ET, and Moreno CS. Cross-platform expression profiling demonstrates that SV40 small tumor antigen activates Notch, Hedgehog, and Wnt signaling in human cells. *BMC Cancer*, 2006; 6(1): 54.
12. Bammler T, Beyer RP, Bhattacharya S, Boorman GA, Boyles A, Bradford BU, Bumgarner RE, et al. Standardizing global gene expression analysis between laboratories and across platforms. *Nat Methods*, 2005; 2(5): 351–356.

13. Barczak A, Rodriguez MW, Hanspers K, Koth LL, Tai YC, Bolstad BM, Speed TP, Erle DJ. Spotted long oligonucleotide arrays for human gene expression analysis. *Genome Res*, 2003; 13(7): 1775–1785.

14. Barnes M, Freudenberg J, Thompson S, Aronow B, and Pavlidis P. Experimental comparison and cross-validation of the Affymetrix and Illumina gene expression analysis platforms. *Nucleic Acids Res*, 2005; 33(18): 5914–5923.

15. Irizarry RA, Warren D, Spencer F, Kim IF, Biswal S, Frank BC, Gabrielson E, et al. Multiple-laboratory comparison of microarray platforms. *Nat Methods*, 2005; 2(5): 345–350.

16. Kothapalli R, Yoder SJ, Mane S, and Loughran TP, Jr. Microarray results: how accurate are they? *BMC Bioinformatics*, 2002; 3(1): 22.

17. Kuo WP, Jenssen TK, Butte AJ, Ohno-Machado L, and Kohane IS. Analysis of matched mRNA measurements from two different microarray technologies. *Bioinformatics*, 2002; 18(3): 405–412.

18. Larkin JE, Frank BC, Gavras H, Sultana R, and Quackenbush J. Independence and reproducibility across microarray platforms. *Nat Methods*, 2005; 2(5): 337–344.

19. Lee JK, Bussey KJ, Gwadry FG, Reinhold W, Riddick G, Pelletier SL, Nishizuka S, et al. Comparing cDNA and oligonucleotide array data: concordance of gene expression across platforms for the NCI-60 cancer cells. *Genome Biol*, 2003; 4(12): R82.

20. Li J, Pankratz M, and Johnson JA. Differential gene expression patterns revealed by oligonucleotide versus long cDNA arrays. *Toxicol Sci*, 2002; 69(2): 383–390.

21. Mecham BH, Wetmore DZ, Szallasi Z, Sadovsky Y, Kohane I, and Mariani TJ. Increased measurement accuracy for sequence-verified microarray probes. *Physiol Genomics*, 2004; 18(3): 308–315.

22. Park PJ, Cao YA, Lee SY, Kim JW, Chang MS, Hart R, Choi S. Current issues for DNA microarrays: platform comparison, double linear amplification, and universal RNA reference. *J Biotechnol*, 2004; 112(3): 225–245.

23. Parrish ML, Wei N, Duenwald S, Tokiwa GY, Wang Y, Holder D, Dai H, et al. A microarray platform comparison for neuroscience applications. *J Neurosci Methods*, 2004; 132(1): 57–68.

24. Pylatuik JD and Fobert PR. Comparison of transcript profiling on *Arabidopsis* microarray platform technologies. *Plant Mol Biol*, 2005; 58(5): 609–624.

25. Rogojina AT, Orr WE, Song BK, and Geisert EE, Jr. Comparing the use of Affymetrix to spotted oligonucleotide microarrays using two retinal pigment epithelium cell lines. *Mol Vis*, 2003; 9: 482–496.

26. Schlingemann J, Habtemichael N, Ittrich C, Toedt G, Kramer H, Hambek M, Knecht R, Lichter P, Stauber R, and Hahn M, Patient-based cross-platform comparison of oligonucleotide microarray expression profiles. *Lab Invest*, 2005; 85(8): 1024–1039.

27. Shippy R, Sendera TJ, Lockner R, Palaniappan C, Kaysser-Kranich T, Watts G, and Alsobrook J, Performance evaluation of commercial short-oligonucleotide microarrays and the impact of noise in making cross-platform correlations. *BMC Genomics*, 2004; 5(1): 61.

28. Tan PK, Downey TJ, Spitznagel EL, Jr., Xu P, Fu D, Dimitrov DS, Lempicki RA, Raaka BM, and Cam MC, Evaluation of gene expression measurements from commercial microarray platforms. *Nucleic Acids Res*, 2003; 31(19): 5676–5684.

29. Walker SJ, Wang Y, Grant KA, Chan F, and Hellmann GM. Long versus short oligonucleotide microarrays for the study of gene expression in nonhuman primates. *J Neurosci Methods*, 2006; 152(1–2): 179–189.

30. Wang HY, Malek RL, Kwitek AE, Greene AS, Luu TV, Behbahani B, Frank B, Quackenbush J, and Lee NH. Assessing unmodified 70-mer oligonucleotide probe performance on glass-slide microarrays. *Genome Biol*, 2003; 4(1): R5.

31. Warnat P, Eils R, and Brors B. Cross-platform analysis of cancer microarray data improves gene expression based classification of phenotypes. *BMC Bioinformatics*, 2005; 6: 265.

32. Woo Y, Affourtit J, Daigle S, Viale A, Johnson K, Naggert J, and Churchill G. A comparison of cDNA, oligonucleotide, and Affymetrix GeneChip gene expression microarray platforms. *J Biomol Tech*, 2004; 15(4): 276–284.

33. Yauk CL, Berndt ML, Williams A, and Douglas GR. Comprehensive comparison of six microarray technologies. *Nucleic Acids Res*, 2004; 32(15): e124.

34. Yuen T, Wurmbach E, Pfeffer RL, Ebersole BJ, and Sealfon SC. Accuracy and calibration of commercial oligonucleotide and custom cDNA microarrays. *Nucleic Acids Res*, 2002; 30(10): e48.

35. Zhu B, Ping G, Shinohara Y, Zhang Y, and Baba Y. Comparison of gene expression measurements from cDNA and 60-mer oligonucleotide microarrays. *Genomics*, 2005; 85(6): 657–665.

36. Baum M, Bielau S, Rittner N, Schmid K, Eggelbusch K, Dahms M, Schlauersbach A, et al. Validation of a novel, fully integrated and flexible microarray benchtop facility for gene expression profiling. *Nucleic Acids Res*, 2003; 31(23): e151.

37. Jurata LW, Bukhman YV, Charles V, Capriglione F, Bullard J, Lemire AL, Mohammed A, Pham Q, Laeng P, Brockman JA, and Altar CA. Comparison of microarray-based mRNA profiling technologies for identification of psychiatric disease and drug signatures. *J Neurosci Methods*, 2004; 138(1–2): 173–188.

38. Mah N, Thelin A, Lu T, Nikolaus S, Kuhbacher T, Gurbuz Y, Eickhoff H, et al. A comparison of oligonucleotide and cDNA-based microarray systems. *Physiol Genomics*, 2004; 16(3): 361–370.

39. Jarvinen AK, Hautaniemi S, Edgren H, Auvinen P, Saarela J, Kallioniemi OP, and Monni O. Are data from different gene expression microarray platforms comparable? *Genomics*, 2004; 83(6): 1164–1168.

40. Stec J, Wang J, Coombes K, Ayers M, Hoersch S, Gold DL, Ross JS, et al. Comparison of the predictive accuracy of DNA array-based multigene classifiers across cDNA arrays and Affymetrix GeneChips. *J Mol Diagn*, 2005; 7(3): 357–367.

41. Ulrich RG, Rockett JC, Gibson GG, and Pettit SD. Overview of an interlaboratory collaboration on evaluating the effects of model hepatotoxicants on hepatic gene expression. *Environ Health Perspect*, 2004; 112(4): 423–427.

42. Waring JF, Ulrich RG, Flint N, Morfitt D, Kalkuhl A, Staedtler F, Lawton M, Beekman JM, and Suter L, Interlaboratory evaluation of rat hepatic gene expression changes induced by methapyrilene. *Environ Health Perspect*, 2004; 112(4): 439–448.

43. Dobbin KK, Beer DG, Meyerson M, Yeatman TJ, Gerald WL, Jacobson JW, Conley B, et al. Interlaboratory comparability study of cancer gene expression analysis using oligonucleotide microarrays. *Clin Cancer Res*, 2005; 11(2 Pt 1): 565–572.

44. Petersen D, Chandramouli GV, Geoghegan J, Hilburn J, Paarlberg J, Kim CH, Munroe D, et al. Three microarray platforms: an analysis of their concordance in profiling gene expression. *BMC Genomics*, 2005; 6(1): 63.

45. Severgnini M, Bicciato S, Mangano E, Scarlatti F, Mezzelani A, Mattioli M, Ghidoni R, et al. Strategies for comparing gene expression profiles from different microarray platforms: application to a case-control experiment. *Anal Biochem*, 2006; 353(1): 43–56.

46. Wang Y, Barbacioru C, Hyland F, Xiao W, Hunkapiller KL, Blake J, Chan F, Gonzalez C, Zhang L, Samaha RR, et al. Large scale real-time PCR validation on gene expression measurements from two commercial long-oligonucleotide microarrays. *BMC Genomics*, 2006; 7: 59.

47. de Reynies A, Geromin D, Cayuela JM, Petel F, Dessen P, Sigaux F, and Rickman DS, Comparison of the latest commercial short and long oligonucleotide microarray technologies. *BMC Genomics*, 2006; 7: 51.

48. Sherlock G. Of fish and chips. *Nat Methods*, 2005; 2(5): 329–330.

49. Kuo WP, Liu F, Trimarchi J, Punzo C, Lombardi M, Sarang J, Whipple ME, et al. A sequence-oriented comparison of gene expression measurements across different hybridization-based technologies. *Nat Biotechnol*, 2006; 24(7): 832–840.

50. Carter SL, Eklund AC, Mecham BH, Kohane IS, and Szallasi Z. Redefinition of Affymetrix probe sets by sequence overlap with cDNA microarray probes reduces cross-platform inconsistencies in cancer-associated gene expression measurements. *BMC Bioinformatics*, 2005; 6(1): 107.

51. Mecham BH, Klus GT, Strovel J, Augustus M, Byrne D, Bozso P, Wetmore DZ, Mariani TJ, Kohane IS, and Szallasi Z. Sequence-matched probes produce increased cross-platform consistency and more reproducible biological results in microarray-based gene expression measurements. *Nucleic Acids Res*, 2004; 32(9): e74.

52. Lee ML, Kuo FC, Whitmore GA, and Sklar J. Importance of replication in microarray gene expression studies: statistical methods and evidence from repetitive cDNA hybridizations. *Proc Natl Acad Sci USA*, 2000; 97(18): 9834–9839.

53. Chu TM, Deng S, Wolfinger R, Paules RS, and Hamadeh HK. Cross-site comparison of gene expression data reveals high similarity. *Environ Health Perspect*, 2004; 112(4): 449–455.

54. Wang H, He X, Band M, Wilson C, and Liu L. A study of inter-lab and inter-platform agreement of DNA microarray data. *BMC Genomics*, 2005; 6(1): 71.

55. Berger JA, Hautaniemi S, Jarvinen AK, Edgren H, Mitra SK, and Astola J. Optimized LOWESS normalization parameter selection for DNA microarray data. *BMC Bioinformatics*, 2004; 5(1): 194.

56. Bolstad BM, Irizarry RA, Astrand M, and Speed TP. A comparison of normalization methods for high density oligonucleotide array data based on variance and bias. *Bioinformatics*, 2003; 19(2): 185–193.

57. Workman C, Jensen LJ, Jarmer H, Berka R, Gautier L, Nielser HB, Saxild HH, Nielsen C, Brunak S, and Knudsen S. A new non-linear normalization method for reducing variability in DNA microarray experiments. *Genome Biol*, 2002; 3(9): research0048.

58. Pounds S and Cheng C. Statistical development and evaluation of microarray gene expression data filters. *J Comput Biol*, 2005; 12(4): 482–495.

59. Bussey KJ, Kane D, Sunshine M, Narasimhan S, Nishizuka S, Reinhold WC, Zeeberg B, Ajay W, and Weinstein JN. MatchMiner: a tool for batch navigation among gene and gene product identifiers. *Genome Biol*, 2003; 4(4): R27.
60. Kent WJ. BLAT—the BLAST-like alignment tool. *Genome Res*, 2002; 12(4): 656–664.
61. Gentleman RC, Carey VJ, Bates DM, Bolstad B, Dettling M, and Dudoit S. Bioconductor: open software development for computational biology and bioinformatics. *Genome Biol*, 2004; 5(10): R80.
62. Imbeaud S and Auffray C. 'The 39 steps' in gene expression profiling: critical issues and proposed best practices for microarray experiments. *Drug Discov Today*, 2005; 10(17): 1175–1182.
63. Evertsz EM, Au-Young J, Ruvolo MV, Lim AC, and Reynolds MA. Hybridization cross-reactivity within homologous gene families on glass cDNA microarrays. *Biotechniques*, 2001; 31(5): 1182, 1184, 1186 passim.
64. Qin LX, Beyer RP, Hudson FN, Linford NJ, Morris DE, and Kerr KF. Evaluation of methods for oligo-nucleotide array data via quantitative real-time PCR. *BMC Bioinformatics*, 2006; 7: 23.
65. Liu F, Jenssen TK, Trimarchi J, et al. Comparison of hybridization-based and sequencing-based gene expression technologies on biological replicates. *BMC Genomics*, 2007; 8: 153.
66. Roth ME, Feng L, McConnell KJ, Schaffer PJ, Guerra CE, Affourtit JP, and Piper KR. Expression profiling using a hexamer-based universal microarray. *Nat Biotechnol*, 2004; 22(4): 418–426.
67. Gunderson KL, Kruglyak S, Graige MS, Garcia F, Kermani BG, Zhao C, and Che D. Decoding randomly ordered DNA arrays. *Genome Res*, 2004; 14(5): 870–877.
68. Quackenbush J. Weighing our measures of gene expression. *Mol Syst Biol*, 2006; 2: 63.

13 DNA Methylation Arrays: Methods and Analysis

Natalie P. Thorne, John Carlo Marioni, Vardhman K. Rakyan,
Ashraf E.K. Ibrahim, Charles E. Massie, C. Curtis,
James D. Brenton, Adele Murrell, and Simon Tavaré

CONTENTS

13.1 INTRODUCTION

Over the last decade, microarrays have become a fundamental tool in biological research laboratories throughout the world. During this time, methods for performing microarray experiments have improved and expanded rapidly, creating an enormous demand for evaluation and comparison of emerging and existing technologies. Importantly, the responsibility for doing this lies as much with the data analyst as the data generator. Such evaluations are difficult since they are influenced by many factors, both financial and scientific. They require a good understanding of both the biological underpinnings of new array technologies and their applications, as well as the statistical issues involved when analyzing the resulting data. To date, there have been many empirical comparisons of technologies for expression array profiling, but newer applications are still lagging in this respect. With the growth in interest in applying microarrays to study a different aspect of the genome, namely the epigenome, this problem has again come to the fore. While there are many publications exploring the biology of DNA methylation and the epigenome, and a large number of articles describing the development of approaches for studying DNA methylation, there are few articles that address the analytic issues involved in these new experiments. This chapter aims to address this problem. It is aimed at the biologist who wants to understand the limitations in analyzing data obtained from different DNA methylation arrays, and the computational biologist wanting an entry point into this new and exciting area.

13.2 MAMMALIAN DNA METHYLATION

Mammalian DNA methylation describes a chemical modification that predominantly affects the cytosine base of CG dinucleotides (Figure 13.1a) [1,2], commonly represented as CpG (the p indicates the phosphodiester bond that forms the backbone of the DNA strand). A CpG found on the sense strand of the DNA duplex will have a CpG in the reverse sense on the opposite strand (Figure 13.1b). DNA methylation of a CpG covalently adds a methyl group to the 5th carbon position on the cytosine base. In lower organisms, such as plants and *Escherichia coli*, methylation can also target other bases, including adenine [3]. CpGs are statistically underrepresented in the human genome [4] and are associated with repetitive DNA sequences including centromeric repeats, retroviral elements, and retrotransposons [5,6].

Methylation in promoter regions and other regulatory sequences can prevent transcription and these regions are often heavily methylated, suggesting that CpG methylation may have evolved as a defense mechanism to silence viral DNA [2]. However, CpG-rich sequences in actively transcribed gene-rich regions are mostly unmethylated and resistant to changes in methylation [6–8]. By convention, these regions are known as CpG islands and are often associated with gene promoters and regulatory regions [6,9]. CpG islands are defined by criteria including the length of the region, GC content, and CpG density [4,10].

Methylation patterns in the genome can be maintained through cell division and replication [2,11]. However, methylation may be dynamic, and the pattern and density of methylation in areas of active transcription may change during development to control key genes in a temporal or tissue-specific manner [2]. Regions that are normally methylated and become less methylated are referred to as hypomethylated and those that become methylated are called hypermethylated. The regulation of DNA methylation is closely associated with other covalent modifications of the histone proteins on which DNA is assembled to form chromatin [2]. These protein modifications include acetylation

(a) Methylated CpG dinucleotide

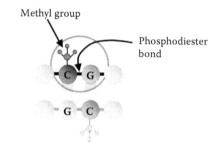

(b) Mammalian CpG methylation

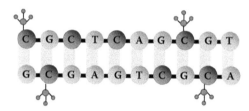

FIGURE 13.1   Illustration of (a) a methylated CpG dinucleotide. The cytosine and guanine bases are joined by a phosphodiester bond and a methyl group has been added to the cytosine. (b) gives a detailed illustration of double-stranded DNA with methylated CpGs in positive and negative strands.

(transcriptionally activating) and methylation (transcriptionally repressive). It remains unclear whether DNA methylation is a consequence or a cause of histone modification [2,11]. As there are known mechanisms for maintenance of methylated DNA during replication, it is plausible that histone methylation is maintained secondarily to DNA methylation [2].

Another generally accepted notion of DNA methylation is that it will spread locally [12,13] that is, once a region starts to become methylated, all CpGs within the region will become methylated (Figure 13.2a). This is consistent with the concept of a CpG island with boundaries defined by some signal in the DNA sequence. Using this principle, regions where a minority of CpGs are methylated would be called unmethylated since the region's methylation status is judged as a whole (see Figure 13.2b). Despite this, it is possible that small blocks of methylated (or unmethylated) regions may exist within a given CpG island. Furthermore, it is believed that certain CpGs within a region may be more important than others (i.e., some CpGs may be held under tighter evolutionary control [4]). Indeed, because mutational repair of methylated CpGs is harder than for nonmethylated CpGs, CpGs would tend to be lost through selection. This perhaps explains the lower than expected number of CpGs found throughout the mammalian genome.

Methylation of CpG islands within the promoters and body of a gene can lead to transcriptional silencing, while a lack of methylation may permit active transcription of the associated gene [14]. This regulation of gene expression is thought to occur as a result of conformational changes in the chromatin structure, altered binding capacity of transcription factors to methylated motifs in the promoter, and other effects altering regulatory elements such as enhancer and repressor sites [15].

A historical role for epigenetics has been in cancer research, especially in the search for abnormally hypomethylated oncogenes or hypermethylated tumor suppressor genes (i.e., genes promoting cancer that have become activated through hypomethylation, and genes suppressing cancer that have been deactivated through hypermethylation) [11,16,17]. Most CpG islands are usually unmethylated but, in cancer, promoter-associated CpG islands of certain genes can be hypermethylated [18]. Many of these hypermethylated genes are specific to certain cancers, suggesting that their aberrant methylation may be important [11,19]. Consequently, understanding the epigenome will

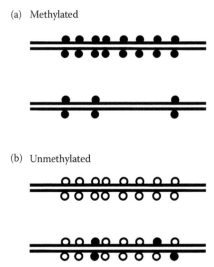

FIGURE 13.2 Two double-stranded DNA fragments that are both (a) methylated and (b) unmethylated. Filled circles represent methylated CpGs, whereas open circles represent unmethylated CpG sites. Two hemimethylated CpG sites are shown in the lower fragment in (b).

allow the control mechanisms of gene transcription to be better modeled, and, as a result, this area of research is growing rapidly. Moreover, DNA methylation is a potentially reversible modification and demethylating chemotherapies are being developed and considered for cancer treatment [6,11]. Further, sites of differential methylation between cell types are not limited to promoter regions. They have been found in exons, introns, enhancer sites, and intergenic regions—suggesting they might regulate miRNAs, reverse strand transcripts, or alternative splicing [2].

Unlike aberrant hypermethylation, cancer-related hypomethylation occurs on a more global scale [11]. Seemingly, indiscriminant hypomethylation occurs throughout the genome of cancer cells, affecting vast amounts of non-CpG island DNA that is normally methylated. Moreover, pervasive hypomethylation is also found in premalignant neoplastic cells, implying that epigenetic changes may constitute the earliest steps of tumorigenesis [19]. While the mechanism and role of global hypomethylation are not well understood, it is observed widely and undoubtedly plays a role in cancer initiation [2]. One hypothesis proposes that it unlocks normally silent repetitive elements, activating transposons that promote genomic rearrangements or interfere with normal transcriptional regulation in the tissue [20].

Of course, to study epigenetic changes in cancer, a basic understanding of DNA methylation in normal tissue is also required. It had been assumed that DNA methylation played a key role in gene switching events during development, but this view is currently being questioned and many classical notions of the epigenome are being scrutinized. Much of the controversy stems from inconsistent results from experiments involving knockout mouse models of genes that encode DNA methylation maintenance proteins [2]. Problems arise since changes in growth conditions and other environmental factors can directly alter epigenetic states [21,22]; experiments with cell lines have also encountered this limitation. Therefore, studies investigating epigenetic mechanisms must be highly controlled, carefully planned, and cautiously interpreted.

13.2.1 Measuring DNA Methylation

The ability to measure the extent of methylation at every CpG would most improve our understanding of the effects of DNA methylation. However, such precise measurements are currently possible only with low-throughput technologies and are therefore limited to small portions of the genome. Despite this, these methods have been employed extensively in molecular biology research and DNA methylation patterns associated with some genes (typically candidate genes identified by other studies) have been studied extensively.

In recent years, efforts have been made to develop high-throughput, whole-genome approaches for measuring DNA methylation [23]. These have emerged in light of the continued evolution of microarray technologies for expression [24], copy number [25], single nucleotide polymorphism (SNP), and ChIP profiling [26]. Current approaches for DNA methylation arrays rely on one of the following principles:

- *Enrichment*: Beginning with fragmented genomic DNA, the first approach enriches or separates fragments that are methylated from those that are not; one or both fractions are then hybridized to an array. Methods for enriching methylated sequences typically employ either methylation-sensitive restriction enzyme digestion or methyl-cytosine antibody precipitation.
- *Bisulfite conversion*: The second approach is much like SNP detection with microarrays. Probes are designed to discriminate target sequences containing methylated CpGs from those with unmethylated CpGs. This discrimination is possible because of the base conversion of unmethylated cytosine to uracil that occurs after bisulfite treatment of the DNA.

There are many variations on both approaches, all with specific advantages and disadvantages. Significant limitations of each approach are related to the quality and type of arrays used, including probe design and density across the genome. The data obtained inherit all biases, sources of variability, and limitations associated with a given approach. Since technologies and approaches for measuring DNA methylation are varied and still evolving, there is no consensus for statistical analyses. However, common themes arise, such as normalization issues, the effect of CpG and GC content of probes and genomic regions of interest, amplification biases, and enzyme and enrichment method efficiencies. Finally, we note that short-read resequencing of bisulfite-treated DNA provides another approach to measuring DNA methylation on a genome-wide scale (e.g., [26]).

13.2.2 CHAPTER AIMS

Lately, the field of epigenetics has been growing at a phenomenal rate due principally to advances in technology that are enabling high-resolution, high-throughput quantitation of DNA methylation; the rest of this chapter reviews the microarray approaches that have been adapted for this purpose. Emphasis is given to the design of array-based DNA methylation experiments and their ability to answer different types of epigenetic questions, as well as the normalization and analysis considerations involved. The chapter will consider the main approaches to array-based DNA methylation assays including the platform, array, and probe design choices for each. Finally, a review of methods for validating array-based results will be given.

13.3 EXPERIMENTAL DESIGN CONSIDERATIONS

When designing microarray experiments, many factors need to be considered to ensure that the biological question of interest has the greatest chance of being answered. For DNA methylation arrays, the most important factors to acknowledge are the limitations a given approach has on the user's ability to answer this question.

13.3.1 WHAT SAMPLES TO HYBRIDIZE?

DNA methylation array experiments are typically 2-color hybridizations. However, while design issues for two-color array-based experiments performed using other technologies typically revolve around which mRNA or DNA samples to compare on an array, DNA methylation experiments are much more involved. The first decision to be made is the number of samples to hybridize to an array.

For a given sample, three fractions can be obtained (Figure 13.3a). The first is simply the non-enriched *input* sample, the second is the fraction enriched for methylated sequences, and the third is the fraction enriched for unmethylated sequences. Methods of enriching for methylated or unmethylated sequences are described in Section 13.4.

For within-array comparisons between samples (Figure 13.3b), either (i) methylated fractions or (ii) unmethylated fractions can be compared (*direct comparison*). Alternatively (Figure 13.3c), methylated or unmethylated fractions from two samples may each be compared to a common reference fraction (*indirect comparison*). This design choice is used in the method of differential methylation hybridization (DMH) (see Section 13.4 [27–32]). Two limitations of this approach are that the methylation status of the reference sample is usually unknown and, when comparing methylation between samples, it is often hard to find an appropriate common reference. This makes interpretation difficult, particularly when the extent, rather than simply the direction, of change in methylation is of interest.

Single-sample approaches (Figure 13.3d) comparing the (i) methylated or (ii) unmethylated enriched fraction to the unenriched input fraction avoid the problems associated with two-sample approaches [33,34]. Additionally, the log-ratio data obtained from such experiments are potentially easier to interpret. However, it has been shown that methylated and unmethylated sequences are not equally detectable for this design (Figure 13.4b)—the dynamic range of the log-ratios is restricted, and (theoretically) positive values are not possible since there is no enrichment for both methylated and unmethylated fractions in the hybridization.

In contrast, when the methylated and unmethylated fractions from a single sample are compared within an array (Figure 13.3e), a wider range of log-ratio values is possible. In the MA-plot shown in Figure 13.4a, methylated sequences have positive (log-ratio, or M) values and unmethylated sequences have negative values. Sequences with values close to zero have ambiguous methylation status, something that can occur for a number of reasons: they may only be methylated in some of the sample (possibly due to tissue heterogeneity) or may not be fully methylated, and so can be enriched in both fractions hybridized to the array.

A limitation of this approach is that by enriching methylated and unmethylated fractions from a single sample, different systematic errors may be introduced in each channel. Such errors are difficult to identify and hard to account for in the analysis step since they are typically confounded with dye-biases and real methylation differences. Enzyme approaches are particularly susceptible to this since differing enzyme efficiencies occur. Additionally, the presence and frequency of enzyme recognition sites varies between sequences, which can introduce biases into the enrichment process. Thus, it is important to use bioinformatic methods to predict sequences that may be subject to such bias (i.e., those with few or no restriction sites for a given digestion enzyme) and to account for this in the analysis step.

13.3.2 OTHER EXPERIMENTAL DESIGN ISSUES

In addition to the classical statistical notions that have been applied to microarray experimental design [35,36], appropriate experimental planning is equally critical to ensure that high-quality experiments are achieved, designs are robust, and all aspects of the experiment are meticulously recorded for quality assessment purposes. This is particularly important for DNA methylation arrays since they tend to be more complicated (i.e., involving multiple digestion or enrichment steps, purification, and amplification), which can lead to the introduction of systematic errors. Moreover, despite their high-throughput status, microarray experiments are expensive and time-consuming—consequently, these errors can be extremely costly. Another important factor to consider relates to the acquisition of samples. In particular, since tissue samples can be extremely heterogeneous, the investigator must be aware that the resulting data are based on averaging over cells that might possess quite different levels of methylation.

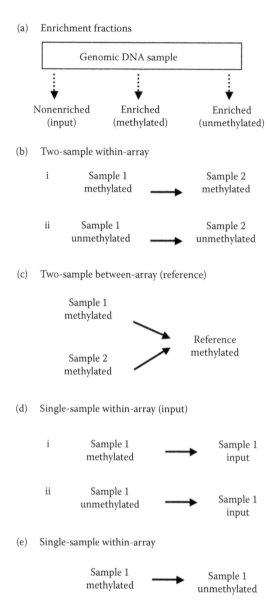

(a) Enrichment fractions

Genomic DNA sample

Nonenriched (input)   Enriched (methylated)   Enriched (unmethylated)

(b) Two-sample within-array

i   Sample 1 methylated   Sample 2 methylated

ii   Sample 1 unmethylated   Sample 2 unmethylated

(c) Two-sample between-array (reference)

Sample 1 methylated

Sample 2 methylated

Reference methylated

(d) Single-sample within-array (input)

i   Sample 1 methylated   Sample 1 input

ii   Sample 1 unmethylated   Sample 1 input

(e) Single-sample within-array

Sample 1 methylated   Sample 1 unmethylated

FIGURE 13.3 Fundamental design choices for two-color DNA methylation array experiments. Solid arrows indicate that target samples are cohybridized to an array. Shown in (a) are the three possible target fractions that can be obtained from a single genomic DNA sample; the input sample itself, a fraction enriched for methylated sequences from the input sample, or a fraction enriched for unmethylated sequences from the input sample. Given either the methylated or unmethylated fractions from two different samples, (b) illustrates the two-sample within-array designs. In (b), (i) compares methylated fractions from two samples and (ii) compares unmethylated fractions from two samples. These correspond to direct two-sample comparisons, whereas (c) is a two-sample between-array indirect comparison, commonly referred to as a reference design. Shown in (d) are single-sample within-array designs where the (i) methylated or (ii) unmethylated fraction is cohybridized with the input fraction from the same genomic sample. (e) Shows the single-sample within-array design that directly compares the methylated and unmethylated fractions from a given sample.

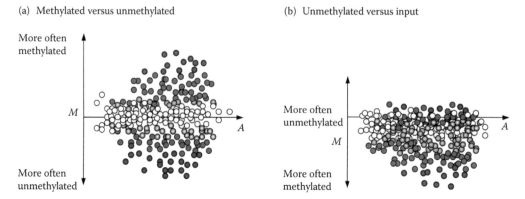

FIGURE 13.4 (**See color insert following page 138.**) Illustrations of MA-plots for DNA methylation array data for experiments where (a) the methylated and unmethylated fractions from a single sample are cohybridized and (b) the unmethylated fraction from a single sample is cohybridized with the input (unenriched) fraction. M represents the log-ratio of the Cy5 and Cy3 channel intensities and A represents the log geometric mean of the Cy5 and Cy3 channel intensities. Dots represent M and A values for probes on a microarray. Methylation status is based on an average over the genomic DNA from all cells in the sample. Blue dots correspond to sequences that are mostly methylated in the sample, whereas red dots correspond to sequences that are mostly unmethylated in the sample. Unfilled circles correspond to sequences whose methylation status is ambiguous. Methylated and unmethylated sequences are more easily distinguished in (a) than (b). Theoretically, there should be no positive M values in (b).

13.4 DNA METHYLATION ARRAY APPROACHES

As described in Section 13.2.1, there are two main approaches for detecting DNA methylation using array technology: enrichment-based methods and schemes that rely on bisulfite conversion. The following sections give an overview of various methods that exploit these techniques to measure DNA methylation.

13.4.1 Restriction Endonuclease Enzymes

Restriction endonucleases cut double-stranded DNA by utilizing specific recognition sequences in the DNA (see Figure 13.5a–d for a description of the distinct cutting properties of different enzymes). Some restriction enzymes are methylation-sensitive: if their recognition sequence contains a CpG, methylation at this site can prevent endonuclease activity [37,38]. Similarly, other enzymes are methylation-dependent (Figure 13.5d). Additionally, many methylation-sensitive restriction enzymes (e.g., *HpaII* and *MspI*) also have nonsensitive isochizomers; these cut the same sequence and are therefore useful in control experiments. As a result of their ability to detect methylated or unmethylated DNA sequences, several low- and high-throughput techniques for assessing DNA methylation are based on methylation-sensitive restriction endonucleases.*

13.4.2 Methylation-Sensitive/Dependent Digestion with PCR Enrichment

A popular enzyme-based approach for DNA methylation arrays combines the following steps (Figure 13.6). The genomic DNA sample is first digested with a frequent cutting methylation-sensitive/dependent enzyme (Figure 13.6a). Since this enzyme's recognition site is small enough to be found regularly throughout the genome and does not contain a CpG, the resulting fragments are usually small enough to be subjected to polymerase chain reaction (PCR) amplification. The digested DNA

---

* A list of the canonical recognition sequences of methylation-sensitive enzymes is available at http://rebase.neb.com/rebase/rebms.html.

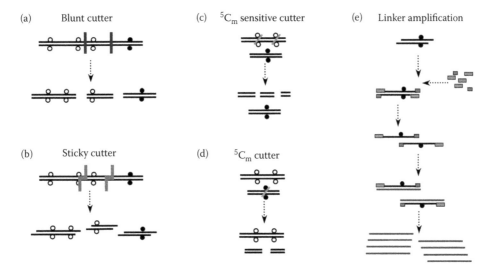

FIGURE 13.5 **(See color insert following page 138.)** Restriction endonuclease enzymes cut double-stranded DNA at given recognition sites. These enzymes have different properties and activities, including (a) blunt cutting enzymes that leave no overhanging ends; (b) sticky cutters that leave overhanging ends; (c) methylation-sensitive enzymes (that only cut if their recognition site is unmethylated); and (d) methylation-dependent enzyme cutters. Linkers can be ligated to sticky or blunt ends and used to prime PCR amplification (e).

fragments are then ligated with linkers (Figure 13.5e); subsequent steps depend on the experiment's design (see Section 13.3).

If the design in Figure 13b(i) is employed, a methylation-sensitive enzyme is used and the process is repeated independently for a second sample (Figure 13.6b). During linker-mediated PCR the uncut fragments are amplified, leading to an enrichment of methylated sequences. The enriched fractions from each sample are then cohybridized to a microarray.

This method is generally known as DMH [27–32] and was developed to compare the methylation status of CpG islands in a test and reference sample. For the initial cutting step DMH uses *MseI* digestion; *MseI*'s recognition sequence (TTAA) is found frequently within bulk DNA, but rarely in CpG islands that therefore remain intact [10] after digestion. For the digestion step, DMH uses a combination of the methylation-sensitive enzymes, *BstUI*, *HhaI*, and *HpaII*. However, these enzymes are active under different conditions, which result in two separate digestion steps.

The next two methods start by splitting the fragmented sample into two. If the design described in Figure 13.3d is used, half the sample is set aside while the other half is digested using either a methylation-sensitive or methylation-dependent enzyme (Figures 13.6b,d or Figures 13.6c,d, respectively). The two fractions are then amplified separately using linker-mediated PCR before being cohybridized. This method is a modification of DMH that allows a sample's methylation status to be measured without using reference DNA [39]. Nouzova et al. [39] applied this method, using *MseI* for the initial digestion step, before creating the unmethylated fragment by digesting half the sample with *McrBC*, an enzyme that restricts methylated sequences and has a recognition sequence that is very frequent within CpG islands.

Finally, if the design in Figure 13.3e is used, one half of the sample is digested using a methylation-sensitive restriction enzyme while the other is digested with a methylation-dependent enzyme (Figures 13.6b,c). The two fractions are then amplified (using linker-mediated PCR) before being cohybridized. This approach was adopted by [33,34] and was motivated by a desire to compare methylated and unmethylated sequences within a single sample on an array. Like DMH and the method of Nouzova et al. [39], they begin by digesting genomic DNA with a frequent cutter, before digesting unmethylated sequences using a bioinformatically derived set of methylation-sensitive

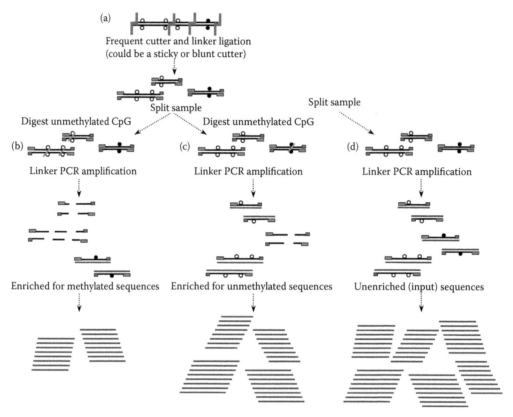

FIGURE 13.6 **(See color insert following page 138.)** Enzyme approaches for obtaining methylated (b), unmethylated (c), and input (d) fractions from a single sample. Genomic DNA is cut with a frequent sticky cutter (a), linkers are ligated and the sample is split. Then separate fractions can be subjected to (b) unmethylated sites are digested such that intact methylated sequences are amplified; (c) methylated sites are digested such that intact unmethylated sequences are amplified; and (d) the input sample (with ligated linkers) is amplified. Linker PCR amplification involves denaturation step to make single-stranded DNA that is used as a template for amplification. Any of the fractions from (b), (c) or (d) can then be cohybridized to an array.

enzymes that provide the maximum number of restriction sites within the predicted target fragments. In contrast, methylation-dependent digestion is performed using only a single enzyme.

13.4.3 STICKY ENZYME APPROACHES

Another method takes advantage of sticky cutting enzymes (Figure 13.5b) to separate methylated and unmethylated DNA; two variations on this approach are described in the following sections.

13.4.3.1 Enrichment of Consecutive Methylated Sites

Genomic DNA is digested with a methylation-sensitive blunt cutting enzyme [Figure 13.7b(i)] before a sticky cutting enzyme with the same recognition sequence is applied [Figure 13.7b(ii)]. If the recognition sequence for these enzymes is unmethylated, the DNA will be cut with blunt ends; the situation is reversed if the recognition sequence is methylated. Linkers are then ligated to the sticky ends [Figure 13.7b(iii)] and only fragments with sticky cuts at both ends are amplified [Figure 13.7b(iv)]. The amplified fragments correspond to sequences that contain consecutive methylated recognition sites.

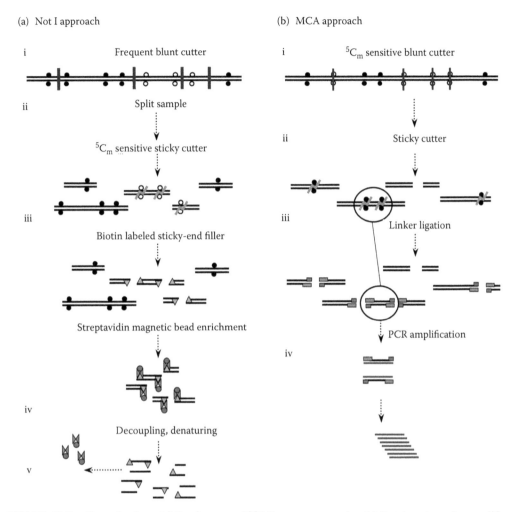

FIGURE 13.7   (See color insert following page 138.) Enzyme approaches (a) that do not require amplification and (b) that require consecutive methylated recognition site for enrichment. In (a), genomic DNA is digested with a blunt cutter (i) and the sample is split (ii). One half is further digested with a methylation-sensitive sticky cutter (iii), the sticky ends are filled with biotin-labeled nucleotides and strong binding streptavidin-coated magnetic beads are added. Sequences bound to the beads are extracted (using a magnet) (iv) and a decoupling reaction is used to release the bound sequences (v). In (b), a methylation-sensitive blunt cutter digests the genomic DNA (i) and the sample is split (ii). One half is further digested with an enzyme [with the same recognition sequence as in (i)] whose activity is not inhibited by methylation (ii). After linker ligation (iii), only sequences with both 3′ and 5′ linkers are amplified (iv).

The method of methylated CpG island amplification (MCA) [40] uses this approach. It compares two samples by applying the methylation-sensitive blunt cutting restriction enzyme *Sma*I followed by the sticky cutting enzyme *Xma*I, both of which have the same recognition sequence. Historically, MCA was combined with a low-throughput method called representational difference analysis (RDA) [41]. One disadvantage of this approach is its dependence on an amplification step to enrich for sequences with consecutively methylated recognition sites. However, its biggest limitation is its reliance on the recognition sites (of the enzymes chosen) occurring frequently throughout the genome. If the gap between recognition sites is large compared to the size of a methylated region, the methylated sequence may not be detected. Hence, this method is only useful for detecting methylation in regions with closely spaced recognition sites.

13.4.3.2   Methylation-Sensitive Sticky Cut Enrichment

For this method, genomic DNA is digested with a frequent cutting blunt DNA restriction enzyme such as *EcoRV* (Figure 13.7a), along with a methylation-sensitive restriction enzyme (e.g., *NotI* or *BssHII*) that leaves sticky ends, resulting in an overhang on one strand of DNA. Consequently, unmethylated fragments of DNA (including those contained in unmethylated CpG islands) have the characteristic overhang seen in Figure 13.5b. The cleaved ends of these fragments are then filled with modified complementary nucleic acids [Figure 13.7a(iii)] that have been altered by labeling one of the nucleotides with a chemical, such as biotin-dNTP, that is used as a reporter. Biotin compounds form a strong ionic bond with streptavidin particles and hence, by coating magnetic beads with streptavidin, unmethylated fragments can be extracted from the digested DNA. Subsequently, the unmethylated fragments are removed from the magnetic beads using affinity purification before being cohybridized with reference genomic DNA. Ching et al. [42] applied this method using the enzyme, *NotI*.

An advantage of this approach is that the extracted unmethylated DNA does not have to be amplified before hybridization. However, blunt double-stranded cutting enzymes cut at a hexamer located randomly throughout the genome and, consequently, they are unable to discriminate between methylated and unmethylated regions. This could lead to small methylated sequences being included in the (supposedly) unmethylated DNA fragment. It is possible to use bioinformatic techniques to determine exactly where the enzyme will cut and so, in theory, this situation could be modeled. Another problem is that the ionic bond formed between biotin and streptavidin is extremely strong. To break this bond and extract the unmethylated fragment, the affinity purification requires the use of a low pH, a high temperature, and a strong denaturing agent (such as formamide). The subsequent purification of the extracted DNA prior to hybridization can result in a loss of yield.

13.4.4   LIMITATIONS OF ENZYME-BASED APPROACHES

Methylation-sensitive restriction-based approaches have several advantages. In particular, no base modification is required (unlike bisulfite modification methods) and they are relatively straightforward, specific, rapid, and inexpensive (certainly compared to high performance liquid chromatography (HPLC)/mass spectrometry). Furthermore, sequence data from the Human Genome Project makes it possible to identify recognition sequences of methylation-sensitive restriction enzymes, allowing the prediction of restricted fragment sizes which enables identification of the best combination of enzymes for a particular assay. The main disadvantage of using a methylation-sensitive restriction method is the enzyme's inherent inability to digest completely the methylated sequences within the sample. For this reason, combined enzyme and antibody approaches have been suggested (Figure 13.8b). Additionally, the number of CpGs a single enzyme can assess depends upon its recognition sequence (Figure 13.9d), and the size of resulting fragments relative to the regions of methylated and unmethylated DNA (Figure 13.10b). We note that human sequence information, in conjunction with bioinformatics tools, allows the identification of restriction enzymes that can assess the methylation status of a particular CpG. Bioinformatic techniques can also be used to identify the best combination of enzymes for a particular method so that methylation levels for the largest number of loci possible can be assessed.

13.4.5   METHYL ANTIBODY APPROACH

An alternative way of measuring DNA methylation on a genome-wide basis is to use a methyl antibody approach known as methylated DNA immunoprecipitation (MeDIP) [43]. As shown in Figure 13.8a, methylation-specific antibodies are used to enrich methylated fragments of the genome [43–45] and, by cohybridizing these with a reference sample, it is possible to identify methylated regions of the genome.

The method used to enrich the methylated fragment is analogous to the immunoprecipitation step in ChIP-chip experiments. After shearing the DNA (Figure 13.9a), a mouse monoclonal antibody against methylated cytosine is used to enrich for methylated fragments. DNA that has been sheared,

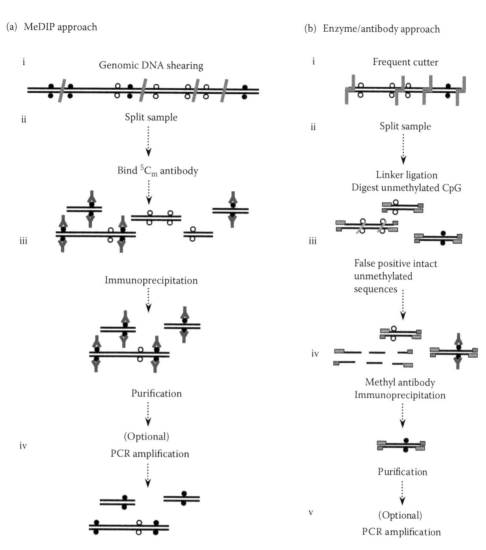

FIGURE 13.8 (See color insert following page 138.) Approaches using (a) MeDIP antibodies and (b) enzyme digestion and antibodies. In (a), genomic DNA is sheared (i) and the sample is split (ii). Antibody proteins that bind to methylated CpGs are added to one half (iii) and sequences with bound antibodies are precipitated and this enriched sample is purified. PCR amplification may be used (iv) depending on the amount of sample required for hybridization. In (b), genomic DNA is digested with a sticky cutter (i), linkers are ligated and the sample is split (ii). One half is further digested with a methylation-sensitive enzyme leaving methylated sequences intact. Any unmethylated sequences that remain intact (escaping digestion) are subsequently removed using the MeDIP appraoch (iv). Finally, the sample is purified and (v) amplification may be used.

but not treated with the antibody, is used as the reference sample. After labeling the two samples with fluorescent dyes, they are hybridized to an array, and the data generated can be used to find methylated regions. The principle difference between ChIP-chip and MeDIP is that, in ChIP-chip experiments, the regions enriched for a particular protein tend to be symmetrical (Figure 13.10b) but, when MeDIP is performed, the level of methylation across a region can vary nonsymmetrically (Figure 13.10c). For example, a CpG island might display more methylation at the 5′ end relative to the level observed across the rest of the island. This means that analysis methods for ChIP-chip experiments may not be directly applied to data obtained using MeDIP; for more discussion of this, see Section 13.6.

(a) Genomic DNA shearing

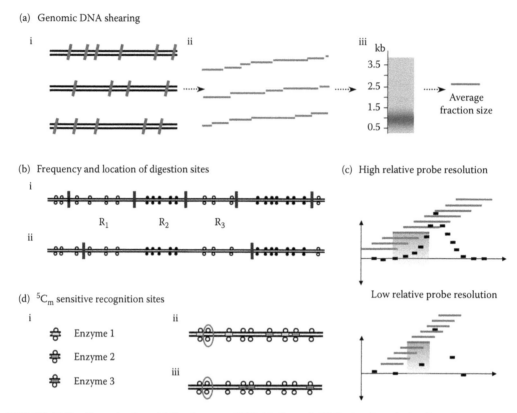

FIGURE 13.9 **(See color insert following page 138.)** (a) Genomic DNA shearing; depicted are three genomes (i) which (after shearing) results in random overlapping fragments (cyan segments) (ii) of varying size depending on the amount of shearing applied. The overall size of the fragments should be confirmed by running the sample on an electrophoresis gel (iii). Bearing in mind the overlapping fragments resulting from genomic shearing, data obtained from consecutive probes along the genome (c) will be correlated. The extent of correlation between neighboring probes depends on the size of the average target fragments (cyan segments) relative to the probe resolution (spacing). In (c) where there is a high relative probe resolution, a single fragment may potentially bind to five neighboring probes, but where there is a low relative probe resolution, a given fragment may only bind to one probe. (b) If enzyme digestion is used to fragment the genome, infrequent cutting (ii) results in regions of altered methylation status (R1–R3) with ambiguous methylation. In contrast, where restriction sites are just as frequent as regions with altered methylation status (i), target fragments will be informative for methylation. (d) Often a combination of enzymes (i) is required for effective and informative digestion of DNA. For a given fragment, enzyme 1 has three recognition sites, (ii), enzyme 2 has two recognition sites (ii) and enzyme 3 has three recognition sites (iii). The combination of enzymes 1 and 2 results in five of the nine CpGs being included in a recognition site (ii).

Antibody enrichment is known to be inefficient but importantly (and unlike restriction enzyme-based approaches), it is very specific. This means there is limited bias in the MeDIP enrichment but, if the amount of starting material is small, amplification may be required. Moreover, it is dose-dependent—the level of enrichment is positively correlated with the number of methylated cytosines (Figure 13.10c). However, there are also a number of drawbacks. In particular, the dose-dependency means that the CpG density of a region has to be considered in the analysis step to avoid the methylation of regions with low CpG content being underestimated. Additionally, it has been shown empirically that regions with a CpG density of <2% are not enriched efficiently. Notwithstanding these problems, MeDIP is a promising strategy for genome-wide methylation analysis as evidenced by a recent publication that used it to elucidate the genome-wide DNA methylation profile of *Arabidopsis*, the first high-density methylation profile of any genome [46].

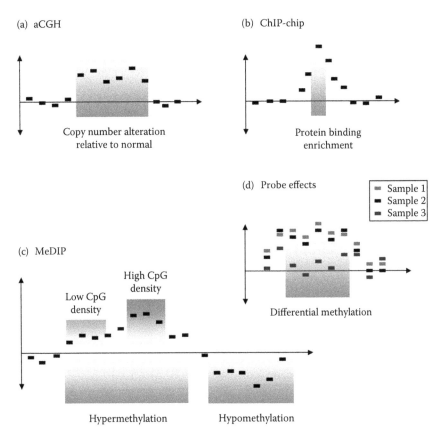

FIGURE 13.10 **(See color insert following page 138.)** Log-ratio data for probes along the genome illustrating typical changes that are observed in (a) array CGH experiments, (b) ChIP-chip experiments, and (c) MeDIP methylation array experiments. Changes in log-ratio data are expected to appear like (a) gained or lost segments, (b) positive symmetric peaks, and (c) variably peaked regions (positive or negative, after normalization). In (c), the variable peak heights of a methylated region depend on the density of methylated CpGs. In addition to CpG density effects in MeDIP data, there will be (d) probe effects. Real changes in methylation (sample 3, blue data points) can be distinguished from changes due to probe effects when there are multiple samples. Similar probe effects are observed in samples 1, 2, and 3. However, sample 3 is differentially methylated across a five-probe region compared to the other samples.

13.4.6 DETECTING METHYLATED DNA BY METHYL-CPG BINDING DOMAIN PROTEINS AFFINITY PURIFICATION

As described in Section 13.2, the methylation of CpG islands associated with genes can have a direct effect on gene expression by inhibiting the recruitment of certain transcription factors that are essential for gene expression [47]. However, accumulating evidence suggests that DNA methylation changes are associated with alterations in chromatin structure through post-translational modification of histones, or physical remodeling of chromatin structure [48,49]. Moreover, a functional link has been established between DNA methylation and chromatin structure, including a cumulative effect of both DNA methylation and chromatin modifications (e.g., histone deacetylation) on gene expression, implying a cooperative function [48]. A further link between DNA methylation and functional histone modifications is the family of DNA methyl-CpG binding domain proteins (MBD or MeCP) [31]. Certain members of this protein family bind specifically to symmetrically methylated CpGs and interact with large transcriptional repressor complexes that switch off gene expression

[49,50]. Therefore, the MBD family of proteins is thought to provide a functional link between DNA methylation and chromatin modifications that results in altered gene expression.

The specificity of certain MBD family members for methylated CpGs has been utilized in an alternative approach for identifying methylated regions. This technique, recently termed methylated-CpG island recovery assay (MIRA), involves the use of an immobilized MBD protein on a solid matrix over which the fragmented DNA sample of interest is passed, resulting in the affinity purification of methylated regions and the elution of unmethylated regions [10,51–54]. The isolated methylated DNA fragments can then be applied to genomic microarrays or analyzed by quantitative PCR. Several variations of this approach have been published using either the full-length MeCP2 protein, full-length MBD2, the MBD2/MBD3L1 complex, or the core MBD domain of MBD2 for affinity purification of methylated DNA [10,51–54]. Since full-length MBD proteins may have preference for certain sequences [55], it is likely that the use of the core MBD domain of MBD2 would provide the least biased approach. An alternative reagent has been recently described which may provide increased sensitivity and specificity of MIRA-like approaches, using an engineered poly-MBD protein [56].

MIRA has several advantages over other global methylation screening techniques. The MBD domains utilized in MIRA bind specifically to heavily methylated DNA, compared to restriction enzyme or methyl-cytosine antibody techniques which may also identify sites of single methylated CpGs. Further, the MBD proteins link DNA methylation, histone modification, and gene expression. Therefore, MBD purification approaches may isolate functionally relevant regions of the genome that are associated with the control of gene expression. However, these advantages also limit the types of sequences that can be isolated by MIRA—if a sequence is not sufficiently methylated, it may be missed by this technique.

A recent approach (COMPARE-MS) combined a restriction enzyme-based approach and MIRA to isolate methylated sequences [57]. This was reported to have increased sensitivity and to be applicable for high-throughput screening, suggesting that it may allow DNA methylation screening in the clinical setting.

13.4.7 ARRAYS FOR BISULFITE-TREATED DNA

Most of the high-throughput methods discussed thus far have used enrichment-based approaches. An alternative is to apply bisulfite conversion in conjunction with arrays that look for converted unmethylated loci and unconverted methylated loci. Previously, this approach was limited by difficulties in designing probes for converted DNA where there were effectively only three bases—this could lead to nonspecific hybridization. However, an SNP detection approach using beadarrays has been developed recently to detect specifically bisulfite converted loci [58].

For a particular locus of interest, four oligonucleotides (two allele-specific and two locus-specific) are used. The 3' end of the allele-specific oligos are designed to hybridize with the bisulfite-modified DNA—one oligo will hybridize if the cytosine base has been converted to uracil, whereas the other will hybridize if no conversion has occurred. Moreover, a different PCR priming site is attached to the 5' end of each oligo and these sites are fluorescently labeled with different dyes. The locus-specific oligos are designed so that the 5' end has a locus-specific sequence, the 3' end is a universal PCR priming site, and in the middle is an address that identifies the oligo with a genomic location. The sequences differ between the locus-specific oligos; for one oligo it is assumed that the CpG of interest is methylated and consequently all CpGs in the locus-specific sequence will also be methylated and vice versa. Subsequently, where there is allele-specific hybridization, a one-step primer extension is performed to ensure that perfect matching has occurred at the allele-specific methylation site. Next, the locus-specific oligos are ligated to their appropriate partner and the subsequent products amplified using PCR before being hybridized to a beadarray using standard techniques [59]. The methylation level can be measured by observing the amount of fluorescence emitted by the dyes attached to the primers of the allele-specific oligos.

The principal advantage and drawback of this method are related to the (very strong) assumption that all CpGs around the locus of interest share the same methylation status. While this means it is possible to have high confidence that any observed differences in methylation will be genuine, it is also known that (even locally) CpGs may not have the same methylation status. Consequently, if only one of the CpGs around the locus of interest has a different methylation status, the methylation status of the CpG of interest cannot be determined since neither allele-specific fragment can be amplified.

13.5   ARRAY CHOICES

Designing a microarray to investigate DNA methylation on a genome-wide level is a challenging problem. The investigator has to decide whether to use bacterial artificial chromosomes (BACs), CpG island clone libraries, or oligonucleotide (oligo) probes and, given this, what resolution (coverage) the probes should have and where they should be located on the genome.

One method is to use whole-genome BAC arrays originally designed for array comparative genomic hybridization (array CGH) studies [43]. BACs (~120 Kb long) are tiled across the genome and so all CpG islands (or other methylated regions) should be contained within, or straddle two or more, BACs depending upon the relative sizes of a given CpG island and the BACs covering that region. An advantage of this method is that BAC arrays already exist for many organisms and could be adapted easily for methylation studies. BACs also provide coverage in regions of the genome where it is difficult to design smaller probes due to repetitive sequence content. Additionally, because of the length of the probes, the noise associated with BAC array data is low relative to other microarray technologies. However, the length of a BAC relative to a CpG island means these arrays may not be sensitive enough to detect either small changes in methylation or small regions of methylation. In particular, many CpG islands (or other methylated regions) could be contained within a single BAC, resulting in data that are difficult to interpret. Further, even if only one methylated region is contained within a BAC, the large disparity in size between this region and the BAC may affect the hybridization and result in data where it is difficult to distinguish between noise and truly methylated regions.

An alternative approach is to use CpG island arrays where the probes are taken from CpG island libraries [10]. In this case, there is no ambiguity about the location of the methylated regions—this is the principal benefit of this method. However, there are a number of problems; for example, it is assumed that all methylation of interest occurs in CpG islands—this is not necessarily so. Moreover, CpG island arrays often include probes that are not CpG islands or that are made up of repetitive genomic sequences; this can lead to an increase in the level of background noise due to cross-hybridization. Consequently, the downstream analysis of such arrays requires expert bioinformatic support.

Another method is to use oligo arrays. This area is developing rapidly but, as yet, no publications have described its application to genome-wide DNA methylation studies. However, it is likely that such publications will arise in the near future. Unlike BACs, oligos are small (generally between 25 and 70 bp in length) and are therefore not practical to generate an array where they are tiled across the whole genome. Consequently, when using oligo arrays to investigate DNA methylation, it is necessary to choose between a number of different layouts. One option is to space the oligos (approximately) evenly across the genome. Alternatively, promoter arrays or CpG island oligo arrays, where oligos are tiled within gene promoter regions or CpG islands, may be used [60]. Both of these methods have the advantage of providing more sensitive coverage than a BAC array [61]. One obvious advantage of promoter or CpG island arrays is their extremely high coverage in genomic regions where methylation changes might be expected to occur—depending upon the resolution, it may even be possible to discriminate between different levels of methylation within the same CpG island or promoter region. Of course, any of these methods run the risk of differential methylation occurring in regions where there are no probes. However, as technology develops, it ought to become possible to design longer oligos (up to 200 bp in length) than are available at present [62]. Notwithstanding technical problems caused by repeat sequences, this should allow the design of whole-genome tiled oligo arrays that would be ideal for studying DNA methylation.

However, irrespective of the chosen layout, oligo arrays do have some drawbacks. In particular, since oligos are much shorter than BACs, the processed signal tends to be noisier. Indeed, when oligo arrays are used for array CGH experiments, it is generally necessary to average over a window of 3–5 consecutive probes to reduce the variability. Nevertheless, the effective resolution of oligo arrays may still be greater than BAC arrays, depending upon the density of oligos on a given array [61].

An additional difficulty when designing oligo arrays is the GC content of individual probes. While GC content (and correspondingly the probes' melting temperature) has been shown not to affect the assignation of probes as outliers [63], this is quite different from more subtle changes in the observed intensities that may be caused by GC-dependent hybridization efficiency biases. This effect would clearly be aggravated if a probe was located within a CpG island. Moreover, if a DNA amplification step is used, this may increase the GC-dependent bias depending upon the protocol used for target preparation (the amplification step is less commonly used in array CGH experiments, so this problem has not yet been thoroughly examined). It is worth noting that GC-dependent biases may be less of a problem if large profiling studies are being conducted, since cross-sample information can be used when calling differentially methylated regions (DMRs) in the analysis step. Of course, this assumes that much of the observed variation is systematic (i.e., the variation tends to be similar across arrays). Finally, we note that a probe's GC content is less of a problem in BAC arrays (in terms of hybridization bias) because of the probe's length; however it has still been observed to affect the quality of the generated data.

One other important problem, irrespective of the array design, is how target fractions are fragmented. In particular, the relationship between target fragments and the probes they hybridize to affect the interpretation of methylation measurements. A summary of the issues involved is given in Figure 13.11.

In summary, despite the problems mentioned above, it is likely that (as is already happening with array CGH experiments [61]), oligo arrays will supersede BAC arrays in DNA methylation studies. This is due principally to their superior resolution and their ability to better target small regions of the genome, such as CpG islands and other regulatory regions where methylation is likely to occur.

13.6   DATA ANALYSIS ISSUES

Two major problems with the analysis of DNA methylation array data are (i) normalization and (ii) calling of methylation status at given genomic loci. Owing to the nature of methylation in the mammalian genome, global levels of methylation can differ radically between samples. Therefore, normalization of arrays within an experiment can be difficult—real differences might be normalized away. Calling methylation levels across the genome is also challenging but, given accurate normalization, differential methylation measures can be obtained. Finally, many data analysis issues are dictated by the experimental approach and the array and probe design.

13.6.1   NORMALIZATION ISSUES

Regardless of the approach, most microarray-based DNA methylation assays result in log-ratio data that are characteristically asymmetric. The skewness of the log-ratio distribution arises from a fundamental imbalance in methylation levels throughout the genome. In normal cells, there are generally more methylated than unmethylated sequences whereas, in cancer or diseased cells, the opposite situation can occur. The extent of the skewness is determined largely by the global levels of methylation in the samples studied. In addition to this skewness (which is specific to DNA methylation experiments), it is also recognized widely that dye-bias is a common problem in two-color microarray experiments that must be corrected.

Many predicted target fragments ⊂ one probe	Predicted target fragments probe	Ambiguous interpretation · Intensity at probes may be due to multiple predicted target fragments · gives information on methylation within and around probe
Predicted target fragments = probes	Predicted target fragments probes	Clear interpretation · each probe hybridize single matching target fragment · gives information on methylation within each probe
One predicted target fragment > many probes	Predicted target fragment probes	Ambiguous interpretation · multiple probes hybridize single predicted target fragment · gives information on methylation within and around probes
One predicted target fragment > one probe	Predicted target fragment probe	Clear interpretation · single predicted target fragment hybridizes to single (smaller) probe · gives information on methylation at and around probe
Many random target fragments > many probes	Random target fragment probes	Clear interpretation · random (overlapping) target fragments hybridize to multiple probes · gives (dependent) information on methylation within and around probes

FIGURE 13.11 **(See color insert following page 138.)** Relationship between target fragments and probes in the interpretation of methylation measurements across probes. Fractionation of the genome by enzyme digestion results in predicted target fragments. Fractionation by shearing yields random (i.e., overlapping) target fragments. Target fragments that hybridize to a given probe determine the amount of methylation measured at that probe.

Unfortunately, real differences in methylation between samples can be removed through inappropriate use of common normalization procedures. This fundamental problem has received little attention in the literature despite the fact that it can have a dramatic impact on the results. We now consider an example (Figure 13.12) of the type of problem that might arise if an inappropriate between-array method is used to remove the skewness. Two arrays are used: in the first, methylated sequences enriched from a normal sample are compared to unenriched sequences from the same sample, whereas, in the second, methylated sequences enriched from tumor tissue are compared to the corresponding unenriched sequences. The aim of the experiment is to compare DNA methylation in the tumor and normal samples. In array 2, many more negative log-ratios are obtained (corresponding to a large number of hypomethylated sequences), whereas in array 1 (the normal tissue array) more positive log-ratios are observed (corresponding to more methylated than unmethylated sequences). Assume for the moment that there is no dye-bias present in either of these arrays so that no correction for this is required (Figure 13.12a). Clearly, the median log-ratio for the first array is less than zero, whereas for the second array it is greater than zero. If a (between-array) median normalization were performed to make the two arrays "comparable" (i.e., both centred at zero), real differences in methylation between the tumor and normal array would be removed, as shown in Figure 13.12b. Consequently, even in the absence of dye-bias, performing between array normalization can be dangerous and should be done only with great care.

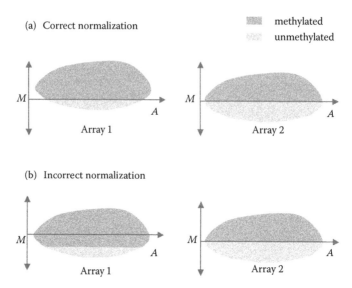

FIGURE 13.12   (See color insert following page 138.) Diagram of MA-plots for two arrays after (a) correct normalization and (b) incorrect normalization. Shaded regions in the plots represent the areas of the plot where data would be observed. Blue areas represent data from truly methylated sequences. Orange areas represent data from truly unmethylated sequences. The sample hybridized to array 1 has considerably more methylated sequences than the sample hybridized to array 2. After correct normalization (a), methylated and unmethylated regions coincide with the $M = 0$ line for both arrays; positive M-values correspond to true methylation and negative M-values correspond to true unmethylation. After incorrect normalization, array 2 (which has been globally centered) has some of the methylated region below the $M = 0$ line, falsely represented as unmethylated.

13.6.2   NORMALIZATION OPTIONS

In an ideal experiment with no dye-bias (or other biases due, for example, to amplification or labeling), any global shift or skewness in log-ratios could be interpreted as a real difference in the proportion of methylated and unmethylated sequences present. However, dye-biases do occur and this variation is confounded with global changes in methylation. Consequently, it is difficult to assess whether a shift in log-ratios is due to dye-bias or real differences in global methylation. This is a particular problem in experiments that involve DNA methylation array profiling of samples with very different methylation profiles (e.g., experiments involving normal and tumor samples are very difficult to normalize). The one situation where dye-biases may be effectively normalized is where appropriate and reliable exogenous controls are available. Otherwise, it may be preferable to avoid dye-bias normalization. Alternatively, the investigator might normalize any obvious intensity-dependent effects without shifting the overall location of the log-ratios to zero. This might be achieved by applying a Loess normalization followed by a global correction of the log-ratios back to their orginal median value. However, this approach does not seem very satisfactory. Methods such as VSN [64] or other affine transformations [65] may provide a normalization that is more robust to the asymmetry.

The practicality of using exogenous spiked controls depends on two main factors. The first is whether the intensity values of the controls cover sufficiently the range of possible intensities. This determines if an intensity-dependent normalization based on these controls is possible. The availability of such controls depends entirely on the number of different control probes on the array and whether the experiment is designed to use them in this way. The second, and probably more important, factor is whether the spiked material has been added in precisely the same amounts in both channels of the experiment—if not, a spike bias will be introduced into the normalization.

Spike biases can occur due to pipetting errors arising from trying to aliquot very small quantities of material—typically the amount of spiked material is very small relative to the amount of target material. To account for this problem, spiked material is generally diluted so that greater volumes are added, minimizing the chance of pipetting-based errors. However, there may be a constraint on the total volume of target material allowed (e.g., when a hybridization chamber is used). Additionally, it is possible that to cover the full range of intensities, spike controls at high concentrations may dominate the amplification step.

In expression array experiments, titration controls have also been used [36]. Here, all probes on the array are pooled, and a serial dilution of the pool is spotted onto the array, assuming that both samples hybridize equally to this pool, these controls can be used for normalization. A disadvantage of this approach is its presumption that these custom spots can be printed on the array. Additionally, the assumption that both target fractions will hybridize equally to the titration pool may not be valid for all samples and approaches.

For enzyme-based DNA methylation array approaches there exists a class of probes that is potentially useful for normalization. These so-called *uninformative probes* do not have restriction sites for the enzymes used to enrich for methylated or unmethylated fractions. The presence of such probes depends on the type of array being used. Probes on CpG island arrays are based on sequence libraries obtained by cutting the genome with *MseI*. Therefore, if *MseI* genomic digestion is used to fragment the genome, the probes will correspond to predicted target sequences. Target sequences without internal digestion sites should hybridize equally (to their corresponding probes) in both channels. If an oligo-based array is used, such uninformative probes may not be available. However, if the oligo probes are designed to be contained within predicted target sequences, there will be uninformative probes that may be used for normalization. Obviously, if the genomic DNA is sheared, rather than fragmented into predicted sequences using enzyme digestion, uninformative probes will not exist. Uninformative probes are also useful for quality assessment—the variability in their log-ratios reflect the inherent quality of the experiment and the log-intensities indicate the level at which single-copy sequences are detected (after amplification).

In experiments incorporating technical dye-swaps for every array, dye-bias self-correction may be sufficient. However, care must be taken to check the basic assumption that the dye-bias is the same between the dye-swap pair. If labeling occurs during the amplification step, dye-swaps are not true technical replicates since different amplification steps would have been used. In this case, dye-bias self-correction may not be appropriate since differences in amplification biases between arrays are confounded with the dye effect. It might be hoped that normalization of a large reference design experiment would be helped by utilizing intelligently the fact that the reference channel is common to all arrays. However, as with using dye-swaps for self-normalization, it is important that the reference channel is indeed a technical replicate (i.e., same fragmentation, enzyme digestion or enrichment, amplification, and labeling steps) in each array.

DNA methylation experiments involving samples with similar global levels of methylation are easier to normalize. While the log-ratios for each array may still be considerably skewed, the important assumption is that the extent of skewness is roughly the same for all arrays within an experiment. Consequently, within-array intensity-dependent dye-bias normalization would modify the overall log-ratios for each array by the same amount. In other words, the arrays would be comparable to each other, but perhaps not to external experiments that include samples with significantly different global methylation levels. This approach to normalization and subsequent analysis of DNA methylation array data is thus restricted to homogeneous collections of samples.

13.6.3 QUALITY ASSESSMENT OF DNA METHYLATION ARRAYS

Quality assessment is a critical step in the analysis of any microarray experiment. As for expression arrays, it is crucial to identify features that determine whether an experiment is of good quality. For example, one feature that might indicate a problem in the experiment is obvious dye-bias. One way of exploring if this occurs is to compare plots of the red and green channel log-intensity distribution within and between arrays, as well as between different experimental batches. Other features of concern are spatial effects across the array or a low overall foreground-to-background signal intensity ratio.

In addition to examining these problems, another crucial quality control step is to generate MA-plots for every array and to compare the scatter before and after processing steps such as background correction and normalization. Caricatures of typical MA-plots observed for methylation experiments produced using different experimental designs and protocols are shown in Figure 13.4.

Given the extremely complicated experimental protocols described in Section 13.4 to effectively perform the quality assessment steps described above, it is crucial to have a sufficient number of replicate arrays, particularly technical replicates and dye-swaps. Besides being a very useful tool for determining the reliability of the data, replicate arrays also (for example) allow the investigator to discern what sort of MA-plot is associated with good quality data for a particular approach. Additionally, as more and more experiments are performed, the characteristics that are important in determining whether a dataset is of good quality will become apparent.

13.6.4   ANALYSIS OF METHYLATION DATA

A key motivation for using DNA methylation arrays is to answer the question: how and where in the genome is my sample methylated? This involves trying to infer *absolute methylation* levels across the genome for a given sample. Alternatively, where a sample is compared to another (within an array), an equivalent aim would be to call changes in *relative methylation* levels across the genome. While single arrays can be used to determine relative methylation levels, multiple arrays are required to find DMRs of the genome (i.e., regions that are consistently differentially methylated).

In Section 13.4.2, we describe methods for finding DMR (i.e., regions of the genome that are differentially methylated *between* samples—arrays), while in the following section we concentrate on methods for calling methylation *within* an array.

13.6.4.1   Calling Methylation within an Array

Low-throughput approaches (e.g., those based on bisulfite sequencing) often summarize results at individual CpGs as a percentage methylation measurement. Regardless of whether this is entirely meaningful (even from low-throughput data inferences) it is unlikely to be a feasible objective for array experiments.

Even ignoring problems caused by sample heterogeneity, a number of experimental factors make determining absolute methylation levels extremely difficult. For example, if the MeDIP approach has been used (Section 13.4.5), the number of binding sites of the methyl antibody in a particular region clearly impacts upon the amount of DNA amplified and this will affect the intensity (and hence the amount of methylation) observed. In particular, if two regions are fully methylated, but one has more binding sites, it will be enriched more efficiently and appear to be more methylated (Figure 13.10c). Consequently, it may not be biologically meaningful to state that methylation in one genomic region is greater than another. Similarly, for methylation-sensitive restriction enzyme-based approaches (including the NotI approach) the location of the restriction sites will determine how much sample DNA can be extracted and subsequently hybridized. Thus, if two regions are fully methylated but one contains a larger number of restriction sites, it might be more efficiently enriched and thus appear to be more methylated. One way to resolve this problem could be to locate an enzyme's restriction sites and factor this information into the analysis. However, given that hybridization biases will also occur, this is likely to be difficult. While bioinformatic approaches may be able to tackle this problem, it is not clear that they will be able to model other sources of bias that can impact upon the investigator's ability to measure absolute methylation. For example, will it be possible to model probe effects, enzyme or antibody efficiencies, restriction site or binding site frequencies, and amplification biases? These biases will undoubtedly be protocol (and even lab) specific and very difficult to generalize. Moreover, very large experiments, involving extensive replication would probably be required to measure them.

Even if sample homogeneity could be assumed and the biases mentioned above were either non-existent or removed, it is unlikely that regions 80% methylated could be distinguished with confidence from regions 90% methylated in the same sample—there would be considerable overlap in the data from both groups. However, it is realistic to expect (barring uninformative probes) that regions that are 10% methylated can be confidently distinguished from those that are 90% methylated. Despite appearing rather trivial, this represents a very meaningful biological difference, and is arguably more important than a measured (but not necessarily real) difference of 10% in methylation. To make this process more straightforward, probes in uninformative regions (i.e., having few restriction or binding sites, for example) can be identified bioinformatically and removed or downweighted. Moreover, probes that bind to fragments containing repetitive elements can also be considered for downweighting in statistical analyses.

Despite this, analyses that aim to attach an absolute methylation score (such as a percentage value) should proceed with caution for the reasons described above. However, calling changes in methylation levels across multiple probes is somewhat easier, particularly if the probe and target fragment resolution are high enough (Figure 13.9c). Analyses that average adjacent probes' log-ratios might allow better calling of true methylation changes in that region since many of the biases discussed above can essentially be averaged out. This may be acheived using a method as sophisticated as a hidden Markov model (HMM) or as simple as a sliding window averaging technique. It is likely that HMM approaches might be problematic, not least because (i) many arrays will not have enough probes (relative to the size of methylated regions); and (ii) probes will probably be unequally spaced across the genome (i.e., there may be large gaps between probed promoter or CpG island regions).

Instead of attempting to find absolute levels of methylation for each probe on an array, an alternative is to flag probes or regions where there is a difference in the relative methylation of the test sample to the reference sample; a number of analytic methods have been proposed for finding such regions.

Where microarrays have been used to investigate methylation at a genome-wide level [42,43], threshold-based approaches have been employed. This flags probes if their log-ratio is above or below a threshold generally derived from the standard or median absolute deviation of the log-ratios from all of the probes on the array. While this approach is simple to understand and implement, it fails to take account of all the information provided. It does not utilize the spatial dependency (i.e., probes that are genomically adjacent to each other are more likely to have the same methylation status than probes which are further apart) inherent in the data (Figure 13.9c). As the resolution of the arrays used for methylation analysis increases, it will become more important to take this into account; this will necessitate the development of model-based approaches.

Many analysis methods for finding copy number changes using array CGH experiments or regulatory elements/transcription factors using ChIP-chip techniques take spatial features into account [66–69] and, at first glance, some of these methods can be easily modified and applied to the problem at hand. However, there are a number of difficulties. The principle problem when modifying methods designed for analyzing array CGH data is that methylation status can change gradually over a region of the genome, whereas for array CGH data, it is assumed that copy number changes occur in steps (Figure 13.10a). Consequently, such methods might lead to an underestimation of the size of methylated regions (Figure 13.10c). (This will also be a problem for threshold-based approaches.) For ChIP-chip experiments on the other hand, analysis methods are designed to look for "bumps" (Figure 13.10b) which, superficially, seems more desirable. However, ChIP-chip analysis methods often assume that the bump is symmetrical and they are generally only interested in finding its centre—the peripheral region is of less interest. Additionally, ChIP-chip analysis techniques often rely on there being a large number of replica experiments [66]; this is unlikely to be the case for DNA methylation arrays due to the often large amounts of DNA required, the cost of such experiments, and the fact that this has not yet become a requirement in the literature.

Despite this, it seems likely that model-based approaches for tackling this problem will be developed; these approaches will depend on the technology used and the possibilities afforded by more repeated design experiments. However, it is difficult to speculate about the form this method will take. Moreover, until large datasets exist where the methylation status of the whole (or at least large parts of the) genome have been confirmed, it will not be possible to assess the efficacy of different methods.

13.6.4.2 Differential Methylation

All the methods described in the previous section can be thought of as "within-array" analysis. Another problem is how to combine information across arrays to explore whether regions of the genome are differentially methylated between samples. Methods for finding DMRs are heavily dependent upon the experimental design. For example, if the design illustrated in Figure 13.3c is employed, the same common reference sample must be used, otherwise cross-array inferences to find DMRs are more difficult. One of the major advantages of finding DMRs (rather than determining absolute or relative methylation levels within an array) is that a lot of the probe effects and other technical problems will be neutralized if we assume that the effects are the same for each array (Figure 13.10d). If we make this assumption, we can find DMRs by simply comparing the log-ratios from one array to another.

Of course, this also assumes that all of the arrays have been properly normalized. As discussed earlier, this is difficult; consequently, it may be hard to compare different arrays. In particular, even a slight difference in amplification efficiencies could result in subtle differences in the log-ratios which could lead to problems. Thus, developing methods for finding DMRs is more complicated than is apparent at a first glance. Because of this, and the lack of datasets where such methods can be tested, this remains an open and interesting research question.

13.7   ENZYME APPROACHES AND GENOMIC COPY NUMBER EFFECT

When using microarrays (or other technologies) to examine the methylation status of genomic regions, it is important to consider the number of copies of the genome present since this could affect the amount of methylation observed. For example, suppose we are interested in the methylation status of the same genomic region in two individuals, one of whom has two copies of the region and the other has three (or more) copies. We also assume that each strand of DNA in this region is methylated to the same extent for both patients. In this case, when the methylation status of this region is measured, it will appear as if more methylation occurs in the second individual relative to the first due to the additional number of copies of the genome. We note that this confounding will occur only if one of the experimental designs illustrated in Figures 13.3b, c, or e are used. By hybridizing the test fraction alongside the input fraction, as shown in Figure 13.3d, the effect of copy number is neutralized. However, when the experimental designs described in Figures 13.3b, c, or d are used to properly assess methylation, it is also necessary to have a good understanding of the number of copies of the genome that might be present. This will be the case when methylation is examined using either one- or two-channel arrays.

One way of determining copy number is to hybridize DNA sheared using *XbaI* (or another digestion enzyme) to the same array used to measure DNA methylation. This has the advantage that the use of array CGH for determining copy number is well known and the protocols/analysis techniques are well established. Additionally, it ensures that copy number can be detected for every probe on an array.

Alternatively, technology is being developed that enables the measurement of copy number, loss of heterozygosity (LOH), and DNA methylation using the same array [70]. After using *XbaI* to shear the DNA and a methylation restriction enzyme (such as *HpaII*) to enrich for methylated fragments, this method (called MSNP) separates SNPs into three groups depending upon whether an *XbaI* (DNA) fragment contains a *HpaII* binding site and whether this binding site might have been eliminated/caused by the presence of a polymorphism. Subsequently, *XbaI* fragments not containing a

HpaII binding site are used to measure copy number, and fragments containing a *HpaII* binding site are used to measure methylation.

While this approach has the advantage that information about a number of different genetic features can be determined from the same array, it also has a number of drawbacks. In particular, it has been observed [33] that multiple methylation restriction enzymes have to be used for DNA methylation to be measured at a sufficient number of locations across the genome—this will significantly reduce the number of SNPs that can be used to determine copy number. Additionally, it is not possible to detect copy number at the same SNPs where methylation has been measured, which means that the resolution may be insufficient to analyse both methylation and copy number separately. Consequently, perhaps the best way of confidently determining copy number and DNA methylation level is to carry out two different hybridizations, one using DNA sheared using only *XbaI* (or a similar enzyme) and the other where the DNA has been treated in such a way that methylation can be detected.

13.8   VALIDATION CHOICES

Validating data obtained from microarray experiments is essential. To confirm that genes with known methylation levels are correctly identified, it is necessary to select a number of genes and compare the results obtained in the array experiment with methylation analyses performed using a different technology. To analyze a particular gene, knowledge of its structure and sequence is required since (in most cases) its methylation status will be examined in only a small region, usually in CpG islands or CpG-rich regions near the gene promoter. The examination of such regions is typically performed using methods based upon bisulfite conversion. Bisulfite sequence analysis is performed by treating DNA with sodium bisulfite which results in the deamination of unmethylated cytosine to uracil, while leaving methylated cytosine residues unchanged [71]. DNA sequencing can then be used to identify methylated cytosines with the exercise being reduced to differentiating between SNPs (cytosines vs thymines).

Variations of bisulfite sequence analyses offer the opportunity to examine a number of CpGs simultaneously and can be scaled up to assay multiple sample sets. In this section, we give a brief description of some analysis methods that use bisulfite conversion in conjunction with PCR to interrogate the methylation status of a small genomic region. For a more detailed review see [72].

A number of commercially available kits for bisulfite conversion are available and are continually being improved with regard to the yield and stability of bisulfite converted DNA, enabling longer amplicons (typically up to 700 bp) to be obtained from small amounts of starting material. However, the limiting step in bisulfite sequence analyses is the conversion process itself, which results in significant DNA degradation such that 84–96% of DNA is affected [73]. While numerous attempts have been made to optimize bisulfite treatment by striking a balance between achieving complete cytosine conversion and minimal DNA degradation [73–75], degradation remains an issue. Thus, it is important to determine the amount of degradation that occurs during specific reaction conditions and to consider the effect on the amplicon of interest, and a recent method towards this end has been described [76]. Fragmentation not only sets an empirical upper size limit on the PCR amplicon (~400–500 bp), but the longer the amplicon, the fewer intact templates there will likely be. In addition, bisulfite treatment results in reduced sequence diversity, generating AT-rich regions and long stretches of thymines, which can be difficult for polymerases to read. Thus, PCR amplification of bisulfite-treated DNA can be challenging and requires careful primer design to avoid mispriming and primer–dimer formation.

13.8.1   Bisulfite Sequencing

Conventional bisulfite sequencing consists of amplifying a specific region of interest and then sequencing the PCR products. To simplify the sequence analysis, PCR products are cloned into bacterial plasmids, and single clones, each representing one PCR amplicon, are sequenced. This approach enables a number of adjacent CpGs (up to 50) to be analyzed on a single amplicon and, by analyzing multiple clones of the same sample, sample-specific profiles can be generated. For an application of bisulfite sequencing to the study of cell division, see Ref. [77]. This approach could be made allele-specific if combined with SNP detection (provided that the SNP is not masked by the bisulfite conversion), and is still the method of choice for examining imprinted genes. The disadvantage of this method is that cloning PCR products is time-consuming, and the cost of sequencing multiple clones of the same PCR reaction can be high.

Bioinformatics software that enables direct quantitative analyses from sequence traces is emerging [78]. These programs allow for direct sequencing of complex mixes of amplicons generated from a single PCR. An obvious advantage of sequencing over other methods is that single CpG profiles can be generated for a locus of interest. In addition, sequencing can be outsourced and thus expensive equipment need not be purchased.

13.8.2 METHYLATION-SPECIFIC PCR/QUANTITATIVE METHYLATION-DEPENDENT PCR (METHYLIGHT)

Methylation-specific PCR (MSP), employs methylation-specific primers that exploit the sequence differences in methylated versus unmethylated bisulfite-treated DNA at a particular locus [79]. Parallel amplification reactions using unmethylation-specific primers should also be performed for each DNA sample. Thus, methylation is determined by the ability of specific primers to allow for amplification. The PCR products can be examined following nondenaturing polyacrylamide gel electrophoresis and ethidium bromide staining such that the presence of a band of the appropriate molecular weight indicates the methylation status of the allele in the original sample. Such products may be compared but, due to variations in PCR efficiency with different primers, quantitative interpretation should be cautious. Specificity can be improved by designing primers that anneal to multiple CpG sites or with the CpG dinucleotide at the 3′ end of the primer. This approach is most useful for querying densely methylated CpG islands.

A variation on this approach allows for more quantitative measures of methylation by employing real-time quantitative PCR and is referred to as the MethyLight method. As in MSP, methylation-specific primers are employed, but a methylation-specific fluorescence reporter probe that anneals to the amplified region of interest is also incorporated. Quantification is performed based on methylated reference sequences that are included on each plate to control for plate-to-plate variations and involves several optimization steps such as the generation of standard curves. Although these assays can be scaled up to quite high throughput, the number of CpGs that can be assayed depends on the probe and generally these are designed based on the assumption that all CpGs within the region queried share the same methylation status.

13.8.3 COMBINED BISULFITE RESTRICTION ANALYSIS

Combined bisulfite restriction analysis (COBRA) is a low-throughput method that determines the level of methylation at specific genomic loci. After DNA has been bisulfite-treated, the region of interest is amplified using primers that do not span CpG sites to generate an amplicon that contains a CG recognizing restriction site. Digestion of the PCR product with an appropriate restriction enzyme results in the digestion of only those products that have unmodified (i.e., methylated) cytosine in the CpG. For example, BstUI's recognition sequence, CGCG, if methylated, would be unchanged after bisulfite modification. But if the recognition sequence were unmethylated, bisulfite modification would change it into TGTG and the PCR product would not be restricted [80].

A drawback of this method is that since a single restriction site is analyzed, if more than one CpG is present in the amplicon, these may or may not be identically methylated. Consequently, the PCR product could be a complex mixture of various amplicons which could impact upon the efficiency of the digestion step. Additionally, during the melting and annealing steps of PCR, heteroduplexes of methylated and unmethylated PCR strands that are resistant to restriction digestion may form, and therefore give false-negative results. False-positive results can occur when there is incomplete bisulfite conversion.

13.8.4 METHYLATION-SENSITIVE SINGLE NUCLEOTIDE PRIMER EXTENSION

Methylation-sensitive single nucleotide primer extension (Ms-SNuPE) is an adaptation of a technique originally designed for the analysis of SNPs in the context of mutation detection [81] and for the quantification of allele-specific expression [82,83]. Essentially, this approach employs paired primer extensions such that the Ms-SNuPE primers anneal to the PCR-generated template and subsequently terminate 5′ of the cytosine residue to be queried [84,85]. In this application, bisulfite-treated DNA is PCR amplified and the gel excised PCR product is incubated with the appropriate Ms-SNuPE primers, polymerase, and radiolabeled dNTPs. The incorporation of [^{32}P]dCTP or [^{32}P] dTTP is then assessed following denaturing polyacrylamide gel-electrophoresis and phosphorimaging, and is used to determine the relative amounts of methylated (C) versus unmethylated (T) cytosines at the original CpG site. Similarly, the complementary strand can be queried using primers designed to incorporate either [^{32}P]dATP or [^{32}P]dGTP. Nonradioactive fluorescent labeling and quantification schemes can also be adapted to this assay. This approach allows for the simultaneous analysis of several CpG dinucleotides in a single reaction and provides a quantitative readout of the ratio of methylated to unmethylated cytosines at a particular CpG site.

Adaptations of this approach employ matrix-assisted laser desorption ionization/time-of-flight (MALDI-TOF) mass spectrometry to discriminate between the two primer extension products based on the GOOD assay for SNP analysis.

13.8.5 PYROSEQUENCING

Pyrosequencing is a sequencing by synthesis method that offers rapid and accurate quantification of CpG methylation sites [86]. After bisulfite conversion and PCR amplification, a sequence-specific primer is hybridized to the strand to be interrogated. The nucleotides are dispensed sequentially according to the predicted sequence which is programed into the pyrosequencer (Biotage). Each time a nucleotide is incorporated into a sequence, pyrophosphate (PPi) is released, and this energy is used for the enzymatic conversion of luciferin to oxyluciferin. This generates light in proportion to the released PPi, which is captured on a charge-coupled device (CCD) camera and recorded as a peak. Before the addition of the next nucleotide in the sequence, a nucleotide-degrading enzyme or apyrase such as uracil N-glycosylase (UNG) is employed to remove excess nucleotides. The result is synchronized nucleotide addition or real-time quantitative sequencing such that the amount of cytosine and thymine incorporation during extension can be used to quantitatively determine the C/T ratio at specific loci. At present this technology is expensive, but the technology enables direct read out of C/T ratios. The advantages of this technology are its accuracy, ease of use, high-throughput nature, and the minimization of variation between experiments.

13.9 CONCLUSIONS

Using high-throughput microarrays to interrogate DNA methylation on a genome-wide level is an exciting research area that could yield important insights into many biological problems. However, as described in this chapter, there are many outstanding technological and analytical issues that have to be resolved so that the investigator can have more confidence that the data obtained are of "good" quality. Determining data quality might be made easier by the Human Epigenome Project [22] which uses low-throughput technology to assess the methylation status of each base of the genome for a number of individuals—this will provide a rich test dataset on which the efficacy of technologies and analysis methods might be assessed. Besides testing the performance of these methods, another important consideration will be how to improve the effective resolution of DNA methylation microarrays. This will be crucial if we want to combine data from DNA methylation arrays with other microarray technologies (e.g., expression, array CGH, ChIP-chip, or micro RNA), since such a comparison will be limited by the effective resolutions of the data obtained. Such combined analysis will be essential to improve our understanding of how different genetic phenomena interact and contribute to (for example) biological development and tumorigenesis.

REFERENCES

1. Doerfler, W. 1983. DNA methylation and gene activity. *Annu Rev Biochem*, 52, 93–124.
2. Bird, A. 2002. DNA methylation patterns and epigenetic memory. *Genes Dev*, 16, 6–21.
3. Finnegan, E. J., Peacock, W. J., and Dennis, E. S. 2000. DNA methylation, a key regulator of plant development and other processes. *Curr Opin Genet Dev*, 10, 217–223.
4. Gardiner-Garden, M. and Frommer, M. 1987. CpG islands in vertebrate genomes. *J Mol Biol*, 196, 261–282.
5. Yoder, J. A., Walsh, C. P., and Bestor, T. H. 1997. Cytosine methylation and the ecology of intragenomic parasites. *Trends Genet*, 13, 335–340.
6. Baylin, S. B. 2005. DNA methylation and gene silencing in cancer. *Nat Clin Pract Oncol*, 2, S4–S11.
7. Bird, A. P. 1986. CpG-rich islands and the function of DNA methylation. *Nature*, 321, 209–213.
8. Song, F., Smith, J. F., Kimura, M. T., Morrow, A. D., Matsuyama, T., Nagase, H., and Held, W. A. 2005. Association of tissue-specific differentially methylated regions (tdms) with differential gene expression. *Proc Natl Acad Sci U S A*, 102, 3336–3341.
9. Larsen, F., Gundersen, G., Lopez, R., and Prydz, H. 1992. CpG islands as gene markers in the human genome. *Genomics*, 13, 1095–1107.
10. Cross, S. H., Charlton, J. A., Nan, X., and Bird, A. P. 1994. Purification of CpG islands using a methylated DNA binding column. *Nat Genet*, 6, 236–244.
11. Laird, P. W. 2005. Cancer epigenetics. *Hum Mol Genet*, 14 Spec No 1, R65–R76.
12. Bestor, T. H. 1992. Activation of mammalian DNA methyltransferase by cleavage of a Zn binding regulatory domain. *EMBO J*, 11, 2611–2617.
13. Pradhan, S., Bacolla, A., Wells, R. D., and Roberts, R. J. 1999. Recombinant human DNA (cytosine-5) methyltransferase. I. Expression, purification, and comparison of de novo and maintenance methylation. *J Biol Chem*, 274, 33002–33010.
14. Robertson, K. D. 2005. DNA methylation and human disease. *Nat Rev Genet*, 6, 597–610.
15. Robertson, K. D. 2002. DNA methylation and chromatin—unraveling the tangled web. *Oncogene*, 21, 5361–5379.
16. Costello, J. F. and Plass, C. 2001. Methylation matters. *J Med Genet*, 38, 285–303.
17. Esteller, M. 2007. Cancer epigenomics: DNA methylomes and histone-modification maps. *Nat Rev Genet*, 8, 286–298.
18. Esteller, M., Corn, P. G., Baylin, S. B., and Herman, J. G. 2001. A gene hypermethylation profile of human cancer. *Cancer Res*, 61, 3225–3229.
19. Feinberg, A. P., Ohlsson, R., and Henikoff, S. 2006. The epigenetic progenitor origin of human cancer. *Nat Rev Genet*, 7, 21–33.
20. Feinberg, A. P. and Vogelstein, B. 1983. Hypomethylation distinguishes genes of some human cancers from their normal counterparts. *Nature*, 301, 89–92.
21. Paz, M. F., Fraga, M. F., Avila, S., Guo, M., Pollan, M., Herman, J. G., and Esteller, M. 2003. A systematic profile of DNA methylation in human cancer cell lines. *Cancer Res*, 63, 1114–1121.
22. Murrell, A., Rakyan, V. K., and Beck, S. 2005. From genome to epigenome. *Hum Mol Genet*, 14 Spec No 1, R3–R10.
23. Callinan, P. A. and Feinberg, A. P. 2006. The emerging science of epigenomics. *Hum Mol Genet*, 15 Spec No 1, R95–R101.
24. Allison, D. B., Cui, X., Page, G. P., and Sabripour, M. 2006. Microarray data analysis: From disarray to consolidation and consensus. *Nat Rev Genet*, 7, 55–65.
25. Pinkel, D. and Albertson, D. G. 2005. Comparative genomic hybridization. *Annu Rev Genomics Hum Genet*, 6, 331–354.

26. Hoheisel, J. D. 2006. Microarray technology: Beyond transcript profiling and genotype analysis. *Nat Rev Genet*, 7, 200–210.

27. Taylor, K., Kramer, R., Davis, J., Guo, J., Duff, D., Xu, D., Caldwell, C., and Shi, H. 2007. Ultradeep bisulfite sequencing analysis of DNA methylation patterns in multiple gene promoters by 454 sequencing. *Cancer Res*, 67, 8511–8518.

28. Huang, T. H., Perry, M. R., and Laux, D. E. 1999. Methylation profiling of CpG islands in human breast cancer cells. *Hum Mol Genet*, 8, 459–470.

29. Wei, S. H., Chen, C.-M., Strathdee, G., Harnsomburana, J., Shyu, C.-R., Rahmatpanah, F., Shi, H., et al. 2002. Methylation microarray analysis of late-stage ovarian carcinomas distinguishes progression-free survival in patients and identifies candidate epigenetic markers. *Clin Cancer Res*, 8, 2246–2252.

30. Yan, P. S., Perry, M. R., Laux, D. E., Asare, A. L., Caldwell, C. W., and Huang, T. H. 2000. CpG island arrays: An application toward deciphering epigenetic signatures of breast cancer. *Clin Cancer Res*, 6, 1432–1438.

31. Yan, P. S., Chen, C. M., Shi, H., Rahmatpanah, F., Wei, S. H., Caldwell, C. W., and Huang, T. H. 2001. Dissecting complex epigenetic alterations in breast cancer using CpG island microarrays. *Cancer Res*, 61, 8375–8380.

32. Yan, P. S., Chen, C.-M., Shi, H., Rahmatpanah, F., Wei, S. H., and Huang, T. H.-M. 2002. Applications of CpG island microarrays for high-throughput analysis of DNA methylation. *J Nutr*, 132, 2430S–2434S.

33. Yan, P. S., Efferth, T., Chen, H.-L., Lin, J., Rodel, F., Fuzesi, L., and Huang, T. H.-M. 2002. Use of CpG island microarrays to identify colorectal tumors with a high degree of concurrent methylation. *Methods*, 27, 162–169.

34. Schumacher, A., Kapranov, P., Kaminsky, Z., Flanagan, J., Assadzadeh, A., Yau, P., Virtanen, C., et al. 2006. Microarray-based DNA methylation profiling: Technology and applications. *Nucleic Acids Res*, 34, 528–542.

35. Ibrahim, A. E., Thorne, N. P., Baird, K., Barbosa-Morais, N. L., Tavaré, S., Collins, V. P., Wyllie, A. H., Arends, M. J., and Brenton, J. D. 2006. MMASS: An optimized array-based method for assessing CpG island methylation. *Nucleic Acids Res*, 34(20), e136.

36. Glonok, G. F. and Solomon, P. J. 2004. Factorial and time course designs for cDNA microarray experiments. *Biostatistics*, 5, 89–111.

37. Yang, Y. H., Dudoit, S., Luu, P., Lin, D. M., Peng, V., Ngai, J., and Speed, T. P. 2002. Normalization for cDNA microarray data: A robust composite method addressing single and multiple slide systematic variation. *Nucleic Acids Res*, 4, e15.

38. McClelland, M., Nelson, M., and Raschke, E. 1994. Effect of site-specific modification on restriction endonucleases and DNA modification methyltransferases. *Nucleic Acids Res*, 22(17), 3640–3659.

39. McClelland, M. 1983. The effect of site specific methylation on restriction endonuclease cleavage (update). *Nucleic Acids Res*, 11, r169–r173.

40. Nouzova, M., Holtan, N., Oshiro, M. M., Isett, R. B., Munoz-Rodriguez, J. L., List, A. F., Narro, M. L., Miller, S. J., Merchant, N. C., and Futscher, B. W. 2004. Epigenomic changes during leukemia cell differentiation: Analysis of histone acetylation and cytosine methylation using CpG island microarrays. *J Pharmacol Exp Ther*, 311, 968–981.

41. Toyota, M., Ho, C., Ahuja, N., Jair, K. W., Li, Q., Ohe-Toyota, M., Baylin, S. B., and Issa, J. P. 1999. Identification of differentially methylated sequences in colorectal cancer by methylated CpG island amplification. *Cancer Res*, 59, 2307–2312.

42. Lisitsyn, N., Lisitsyn, N., and Wigler, M. 1993. Cloning the differences between two complex genomes. *Science*, 259, 946–951.

43. Ching, T. T., Maunakea, A. K., Jun, P., Hong, C., Zardo, G., Pinkel, D., Albertson, D. G., et al. 2005. Epigenome analyses using BAC microarrays identify evolutionary conservation of tissue-specific methylation of SHANK3. *Nat Genet*, 37, 645–651.

44. Weber, M., Davies, J. J., Wittig, D., Oakeley, E. J., Haase, M., Lam, W. L., and Schübeler, D. 2005. Chromosome-wide and promoter-specific analyses identify sites of differential DNA methylation in normal and transformed human cells. *Nat Genet*, 37, 853–862.

45. Mukhopadhyay, R., Yu, W., Whitehead, J., Xu, J., Lezcano, M., Pack, S., Kanduri, C., et al. 2004. The binding sites for the chromatin insulator protein CTCF map to DNA methylation-free domains genome-wide. *Genome Res*, 14, 1594–1602.

46. Keshet, I., Schlesinger, Y., Farkash, S., Rand, E., Hecht, M., Segal, E., Pikarski, E., et al. 2006. Evidence for an instructive mechanism of de novo methylation in cancer cells. *Nat Genet*, 38, 149–153.

47. Zhang, X., Yazaki, J., Sundaresan, A., Cokus, S., Chan, S. W., Chen, H., Henderson, I. R., et al. 2006. Genome-wide high-resolution mapping and functional analysis of DNA methylation in arabidopsis. *Cell*, 126, 1189–1201.

48. Inamdar, N. M., Ehrlich, K. C., and Ehrlich, M. 1991. CpG methylation inhibits binding of several sequence-specific DNA-binding proteins from pea, wheat, soybean and cauliflower. *Plant Mol Biol*, 17, 111–123.

49. Cameron, E. E., Bachman, K. E., Myohanen, S., Herman, J. G., and Baylin, S. B. 1999. Synergy of demethylation and histone deacetylase inhibition in the re-expression of genes silenced in cancer. *Nat Genet*, 21, 103–107.

50. Hendrich, B. and Bird, A. 1998. Identification and characterization of a family of mammalian methyl-CpG binding proteins. *Mol Cell Biol*, 18, 6538–6547.

51. Zhang, Y., Ng, H. H., Erdjument-Bromage, H., Tempst, P., Bird, A., and Reinberg, D. 1999. Analysis of the NuRD subunits reveals a histone deacetylase core complex and a connection with DNA methylation. *Genes Dev*, 13, 1924–1935.

52. Shiraishi, M., Chuu, Y. H., and Sekiya, T. 1999. Isolation of DNA fragments associated with methylated CpG islands in human adenocarcinomas of the lung using a methylated DNA binding column and denaturing gradient gel electrophoresis. *Proc Natl Acad Sci U S A*, 96, 2913–2918.

53. Brock, G. J., Huang, T. H., Chen, C. M., and Johnson, K. J. 2001. A novel technique for the identification of CpG islands exhibiting altered methylation patterns (ICEAMP). *Nucleic Acids Res*, 29, E123.

54. Rauch, T. and Pfeifer, G. P. 2005. Methylated-CpG island recovery assay: A new technique for the rapid detection of methylated-CpG islands in cancer. *Lab Invest*, 85, 1172–1180.

55. Rauch, T., Li, H.,Wu, X., and Pfeifer, G. P. 2006. MIRA-assisted microarray analysis, a new technology for the determination of DNA methylation patterns, identifies frequent methylation of homeodomaincontaining genes in lung cancer cells. *Cancer Res*, 66, 7939–7947.

56. Klose, R. J., Sarraf, S. A., Schmiedeberg, L., McDermott, S. M., Stancheva, I., and Bird, A. P. 2005. DNA binding selectivity of MeCP2 due to a requirement for A/T sequences adjacent to methyl-CpG. *Mol Cell*, 19, 667–678.

57. Jorgensen, H. F., Adie, K., Chaubert, P., and Bird, A. P. 2006. Engineering a high-affinity methyl-CpG-binding protein. *Nucleic Acids Res*, 34, e96.

58. Yegnasubramanian, S., Lin, X., Haffner, M. C., DeMarzo, A. M., and Nelson, W. G. 2006. Combination of methylated-DNA precipitation and methylation-sensitive restriction enzymes (COMPARE-MS) for the rapid, sensitive and quantitative detection of DNA methylation. *Nucleic Acids Res*, 34, e19.

59. Bibikova, M., Lin, Z., Zhou, L., Chudin, E., Garcia, E. W., Wu, B., Doucet, D., et al. 2006. High-throughput DNA methylation profiling using universal bead arrays. *Genome Res*, 16, 383–393.

60. Gunderson, K. L., Kruglyak, S., Graige, M. S., Garcia, F., Kermani, B. G., Zhao, C., Che, D., et al. 2004. Decoding randomly ordered DNA arrays. *Genome Res*, 14, 870–877.

61. Weber, M., Hellmann, I., Stadler, M., Ramos, L., Päabo, S., Rebhan, M., and Schübeler, D. 2007. Distribution, silencing potential and evolutionary impact of promoter DNA methylation in the human genome. *Nat Genet*, 39, 457–466.

62. Ylstra, B., van den Ijessel, P., Carvalho, B., Brakenhoff, R. H., and Meijer, G. A. 2006. BAC to the future! or oligonucleotides: A perspective for micro array comparative genomic hybridisation (array CGH). *Nucleic Acids Res*, 34, 445–450.

63. Egeland, R. D. and Southern, E. M. 2005. Electrochemically directed synthesis of oligonucleotides for DNA microarray fabrication. *Nucleic Acids Res*, 33, e125.

64. Leiske, D. L., Karimpour-Fard, A., Hume, P. S., Fairbanks, B. D., and Gill, R. T. 2006. A comparison of alternative 60-mer probe designs in an in-situ synthesized oligonucleotide microarray. *BMC Genomics*, 7, 72.

65. Huber, W., von Heydebreck, A., Sultmann, H., Poustka, A., and Vingron, M. 2002. Variance stabilization applied to micrarray data calibration and to the quantification of differential expression. *Bioinformatics*, 18, S96–S104.

66. Bengtsson, H. and Hossjer, O. 2006. Methodological study of affine transformations of gene expression data with proposed robust non-parametric multi-dimensional normalization method. *BMC Bioinformatics*, 7, 100.

67. Buck, M. J. and Lieb, J. D. 2004. ChIP-chip: Considerations for the design, analysis and application of genome-wide chromatin immunoprecipitation experiments. *Genomics*, 83, 349–360.

68. Marioni, J. C., Thorne, N. P., and Tavaré, S. 2006. BioHMM: A heterogeneous hidden Markov model for segmenting array CGH data. *Bioinformatics*, 22, 1144–1146.

69. Olshen, A. B., Venkatraman, E. S., Lucito, R., and Wigler, M. 2004. Circular binary segmentation for the analysis of array-based DNA copy number data. *Biostatistics*, 5, 557–572.

70. Qi, A., Rolfe, P. A., MacIsaac, K., Gerber, G. K., Pokholok, D., Zeitlinger, J., Danford, T., et al. 2006. High-resolution computational models of genome binding events. *Nat Biotechnol*, 24, 963–970.

71. Yuan, E., Haghighi, F., White, S., Costa, R., McMinn, J., Chun, K., Minden, M., and Tycko, B. 2006. A single nucleotide polymorphism chip-based method for combined genetic and epigenetic profiling: Validation in decitabine therapy and tumor/normal comparisons. *Cancer Res*, 66, 3443–3451.

72. Frommer, M., McDonald, L., Millar, D., Collis, C., Watt, F., Grigg, G., Molloy, P., and Paul, C. 1992. A genomic sequencing protocol that yields a positive display of 5-methylcytosine residues in individual DNA strands. *Proc Natl Acad Sci U S A*, 89, 1827–1831.

73. Fraga, M. F. and Esteller, M. 2002. DNA methylation: A profile of methods and applications. *Biotechniques*, 33, 632–649.

74. Grunau, C., Clark, S. J., and Rosenthal, A. 2001. Bisulfite genomic sequencing: Systematic investigation of critical experimental parameters. *Nucleic Acids Res*, 29(13), 65.

75. Raizis, A., Schmitt, F., and Jost, J. 1995. A bisulfite method of 5-methylcytosine mapping that minimizes template degradation. *Anal Biochem*, 226, 161–166.

76. Olek, A., Oswald, J., and Walter, J. 1996. A modified and improved method for bisulphite based cytosine methylation analysis. *Nucleic Acids Res*, 24, 5064–5066.

77. Ehrich, M., Zoll, S., Sur, S., and van den Boom, D. 2007. A new method for accurate assessment of DNA quality after bisulfite treatment. *Nucleic Acids Res*, 35, e29.

78. Shibata, D. and Tavaré, S. 2006. Counting divisions in a human somatic cell tree: How, what and why? *Cell Cycle*, 5, 610–614.

79. Lewin, J., Schmitt, A. O., Adorjan, P., Hildmann, T., and Piepenbrock, C. 2004. Quantitiative DNA methylation analysis based on four-dye trace data from direct sequencing of PCR amplificates. *Bioinformatics*, 20, 3005–3012.

80. Herman, J., Graff, J., Myöhänen, S., Nelkin, B., and Baylin, S. 1996. Methylation-specific PCR: A novel PCR assay for methylation status of CpG islands. *Proc Natl Acad Sci U S A*, 93, 9821–9826.

81. Xiong, Z. and Laird, P. W. (1997). COBRA: A sensitive and quantitative DNA methylation assay. *Nucleic Acids Res*, 25, 2532–2534.

82. Kuppuswamy, M., Hoffmann, J., Kasper, C., Spitzer, S., Groce, S., and Bajaj, S. 1991. Single nucleotide primer extension to detect genetic diseases: Experimental application to hemophilia B (factor IX) and cystic fibrosis genes. *Proc Natl Acad Sci U S A*, 88, 1143–1147.

83. Singer-Sam, J., LeBon, J., Dai, A., and Riggs, A. 1992. A sensitive, quantitative assay for measurement of allele-specific transcripts differing by a single nucleotide. *PCR Methods Appl*, 1, 160–163.

84. Szabó, P. and Mann, J. 1995. Allele-specific expression and total expression levels of imprinted genes during early mouse development: Implications for imprinting mechanisms. *Genes Dev*, 9, 3097–3108.

85. Gonzalgo, M. and Jones, P. 1997. Rapid quantitation of methylation differences at specific sites using methylation-sensitive single nucleotide primer extension (Ms-SNuPE). *Nucleic Acids Res*, 25, 2529–2531.

86. Gonzalgo, M. and Liang, G. 2007. Methylation-sensitive single-nucleotide primer extension (Ms-SNuPE) for quantitative measurement of DNA methylation. *Nat Protoc*, 2, 1931–1936.

87. Tost, J. and Gut, I. 2007. DNA methylation analysis by pyrosequencing. *Nat Protoc*, 2, 2265–2275.

14 Electronic Microarrays: Progress toward DNA Diagnostics

Howard Reese, Dan Smolko, Paul Swanson, and Dalibor Hodko

CONTENTS

14.1 INTRODUCTION

Nearly a decade ago it was once quipped that deoxyribonucleic acid (DNA) diagnostics was a technology looking for a market. It was implied that DNA diagnostics was too expensive to compete with existing technology. Today, DNA diagnostics is starting to penetrate the diagnostics market, but the penetration is still very small. Nevertheless, despite the challenging market hurdles, DNA diagnostics has enormous potential if the technical issues can meet the market challenges. Many biotech companies continue to be long-term players in the DNA diagnostics arena despite the fact that the road to profits has been long and rough. Owing to the enormous potential of DNA diagnostics, however, it is not a question of "if," but only a question of "when" the technology will own a substantial share of the diagnostics market.

To meet the market challenges scientists and engineers will have to develop assays and instrumentation that are faster, cheaper, more accurate, and more sensitive. Although there is currently no single technology that meets all these requirements, there are technologies that satisfy one or two of these requirements. For example, real-time polymerase chain reaction (PCR) is sensitive and moderately fast, but the ability to multiplex many assays or samples is limited. If one compares the cost of analyzing a single sample using DNA diagnostics versus conventional technology enzyme-linked immunosorbent assay (ELISA, high-performance liquid chromatography) HPLC, colorimetric, and so forth, the conventional technology is generally less expensive. The lesser expense is due in large part to the technical maturity of conventional technology.

Microarrays, however, offer the advantage that many assays can be performed on a single device. In marketing terminology the economy of scale is functional at the microarray level. Multiplexing can be performed not only by analyzing a large number of genes or mutations in a sample, but also by testing multiple patient samples. For example, Nanogen's 400-site microarray is capable of analyzing 23 genetic mutations for each of 64 samples on a single disposable microarray. Thus, costs are significantly reduced because many samples/assays are run on a single disposable microarray.

DNA diagnostics offers more than just a replacement for conventional technology. The development of pharmacogenomics is a new field that has the potential to improve patient care by providing patient-specific drug dosing [1]. For example, patients whose genetic makeup indicates that they are fast metabolizers of a specific drug need a different drug dosage than patients who are diagnosed as slow metabolizers. Gene expression is another new field that has the potential to significantly improve patient care [2]. Gene expression patterns are able to more accurately and rapidly determine the cause of a patient's illness compared to a battery of conventional tests, particularly for diseases that are difficult to diagnose. For example, cancer gene expression data can accurately confirm a pathologist's finding from conventional microscopic examinations of tissue and/or tumor cells separated by laser capture microdissection [3].

In this chapter, we will discuss electronic DNA microarrays and their progress in meeting the market challenges. Electronic microarrays are defined as microarrays whose individual sites are electrodes used for electrophoretic transport to perform a variety of assay tasks including capture loading, target addressing, hybridization, stringency, and DNA amplification. The technology will be discussed in detail and examples of applications will be presented.

14.2   ELECTRONIC MICROARRAY TECHNOLOGY

14.2.1   ELECTROPHORESIS ON THE MICROSCALE

Figure 14.1 shows a 10×10 microarray of working electrodes surrounded by 20 larger counter electrodes. A detail of the working electrode is shown in Figure 14.2. The 80-μm diameter platinum working electrodes are coated with a hydrogel permeation layer. During a typical assay the working electrode is positively biased at 1.9 V (versus Ag/AgCl ref) to electrophoretically transport capture or target DNA to the working electrode. Owing to the orientation of the field lines emanating from the positive electrode, the majority of the DNA in the bulk solution will migrate through the bulk solution until it is over the biased electrode where it starts to penetrate the hydrogel permeation layer. As most of the DNA transport occurs in solution (free-field conditions), the transport is extremely rapid.

The arrangement of the working electrodes with respect to the counter electrodes has a significant impact on the accumulation of DNA. Figure 14.3 shows the results of a finite-element analysis of a model depicting a single working electrode (40 μm radius) surrounded by a ring counter electrode [4]. It can be seen that the electric field strength falls off quite rapidly with distance from the working electrode. In fact, at 5 radii distant from the center of the working electrode (200 μm) the

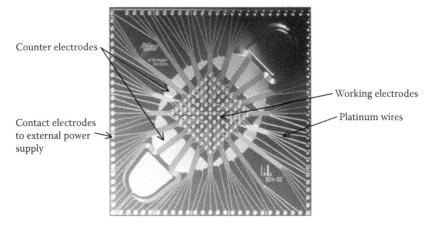

FIGURE 14.1 Image of 100-site microarray. Platinum working electrodes in a 10×10 array are connected by platinum wires to contact pads located on the periphery of the chip. Surrounding the 10×10 array are 20 platinum counter electrodes. A SiO_2 dielectric layer overlaying the chip is etched away to expose the platinum working and counter electrodes and the contact pads.

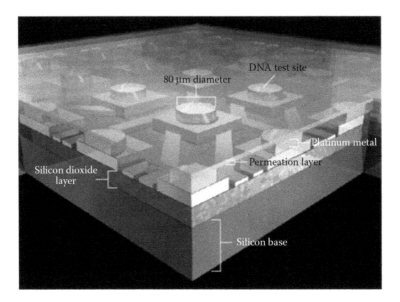

FIGURE 14.2 Illustration of the details of the construction of the 100-site electronic microarray.

field strength is almost negligible. On the 100-site microarray, the electrodes are spaced on $200\,\mu m$ centers. Thus, when a specific electrode is biased the effect on neighboring electrodes should be minimal. In practice, it is often observed that nearest-neighbor electrodes are affected when an electrode is biased, but second nearest-neighbor electrodes are not affected. As a consequence of this nearest-neighbor electrode effect, assays are designed to prevent the simultaneous biasing of neighboring electrodes.

The effect of a rapidly declining electric field strength means that most of the DNA accumulated on the working electrode originates from within ~5 radii of the biased electrode. Although it would

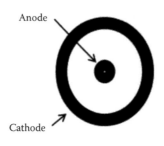

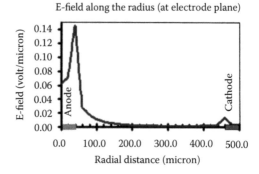

E-field along the radius (at electrode plane)

FIGURE 14.3 Finite-element analysis model of working electrode with ring cathode (counter electrode). Working electrode has a 40-μm radius. The model results shown on the right indicate that the electric field drops very rapidly with distance and is very small at 200 μm. The working electrodes on the 100-site microarray are located on 200 μm centers; consequently, the neighboring array sites should be minimally disturbed when a specific array site is biased. In practice, small neighboring site effects are observed, but compensation is made for this effect by not simultaneously biasing neighboring sites.

be desirable to accumulate DNA from a larger volume of the overlying bulk solution, it has rarely been necessary to do so because the DNA accumulation over the working electrode represents up to a 1000-fold enhancement of the DNA concentration compared to the concentration of the bulk solution. Furthermore, and very important, this concentration enhancement occurs in <2 min.

The situation where multiple electrodes are simultaneously biased has also been modeled by finite-element analysis [4]. Model results agree qualitatively with experimental data: As more electrodes are biased, a greater proportion of the DNA in the overlying bulk solution is accumulated at the biased electrodes.

14.2.2 ELECTROCHEMISTRY ON THE WORKING ELECTRODE

During electrophoretic transport of DNA, the potential is so high that water oxidation and reduction occur at the working and counter electrodes, respectively (Equations 14.1 and 14.2).

Oxidation: $$H_2O \rightarrow 2H^+ + 1/2\ O_2 + 2e^- \qquad (14.1)$$

Reduction: $$2e^- + 2H_2O \rightarrow 2OH^- + H_2 \qquad (14.2)$$

At the surface of the working electrode (positively biased), the extremely low pH minimizes or prevents DNA hybridization. Furthermore, the DNA itself may become oxidized should it make contact with the working electrode. Electro-oxidation of adenine and guanine has been used as a potential direct detection method for DNA; however, the signals were not satisfactory for detection of small amounts of DNA present in patient samples. Finally, there is a concern that the DNA may stick to the electrode and not be capable of hybridizing to its complementary strand. In short, the surface of the working electrode is to be avoided.

Two strategies are used to circumvent these problems: First, a 10-μm thick hydrogel permeation layer is molded on top of the electrode (Figure 14.2). The hydrogel contains covalently bound streptavidin for binding biotin-labeled oligonucleotides. When the oligonucleotides accumulate at the working electrode, they bind with the streptavidin and most are prevented from approaching the surface of the platinum working electrode. Second, a buffer is used to mitigate the extreme pH near the electrode. In most applications the buffer is 50 mM histidine. The imidazole ring of the histidine has a secondary amine with a buffering capacity near pH 6.0. The counteracting effects of H^+ ion generation and buffering by histidine cause a pH gradient at the working electrode with increasing pH with distance from the working electrode surface. This raises the pH within the hydrogel sufficiently high to permit DNA hybridization.

There are two additional benefits to using histidine as the buffer. Amino acids near their pI are zwitterionic. As zwitterions they have little tendency to migrate in an electric field. Consequently, most of the charge-carrying species in histidine buffer are the other added electrolytes. In electronic microarray assays, the added electrolyte is DNA. Thus, the use of zwitterionic histidine allows a greater amount of DNA to be electrophoretically transported. In contrast, if 5 mM NaCl is added to the histidine buffer, the negatively charged chloride ions become the major charge-carrying species. (The fraction of total current carried by each ion type is directly related to the mobility of the ion [5].) As a consequence, the negatively charged DNA will become a minor charge carrier and the resulting DNA accumulation at the working electrode will decrease (Figure 14.4). In electrochemical terminology, the transport number of DNA has been reduced by adding NaCl.

The second added benefit of using a histidine buffer occurs when the pH of the solution over the working electrode drops below the pI of the imidazole nitrogen in histidine. Under these circumstances, histidine becomes protonated and has a net +1 charge. Apart from the protonated histidine being transported away from the positively biased working electrode, the protonated histidine will also function to shield the negative charge of single-stranded DNA. As a consequence, the charge repulsion between two complementary strands of DNA is reduced, thereby allowing hybridization to occur. This phenomenon is more familiar in hybridization assays where, instead, high salt conditions are used to facilitate DNA hybridization. Thus, the use of a histidine buffer with electro-generated protons mimics the effects of high salt without reducing the electrophoretic mobility of DNA.

The phenomenon whereby target DNA molecules are electrophoretically transported to the working electrode and hybridize to bound capture oligonucleotides is coined "electronic hybridization." Electronic hybridization is schematically illustrated in Figure 14.5. A key feature unique to electronic microarrays is that electronic hybridization occurs within 1–2 min compared to 1 or 2 h for passive hybridization at the same initial concentration [6,7]. The speed of electronic hybridization is due to the accumulation of a high concentration of target DNA that drives the equilibrium toward hybridization.

Although histidine is the main component of the electrophoresis buffer, α-thioglycerol is also sometimes added to help reduce bubbling on the working electrode. In Equation 14.1, it can be seen that oxygen is also a product of water oxidation. At the electrode where oxidation is occurring,

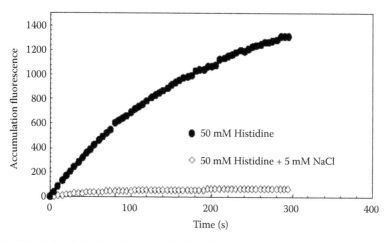

FIGURE 14.4 Effect of added salt on the accumulation of fluorescently labeled DNA at a working electrode.

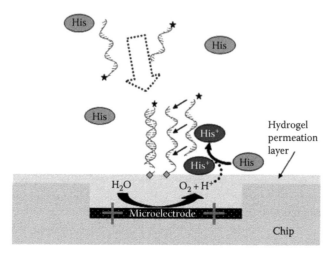

FIGURE 14.5 **(See color insert following page 138.)** Illustration demonstrating the interactions between the electrochemistry at the electrode and DNA hybridization. Water oxidation at the positive electrode generates H^+ ions which protonate histidine. The protonated histidine acts to shield the negative charge on DNA thereby allowing the two single strands of DNA to approach and hybridize.

atomic oxygen is first created, which then combines to form molecular oxygen, O_2. As the concentration of O_2 increases eventually the saturation limit is reached and a bubble is nucleated. When these bubbles become large enough or large enough in quantity, they interfere with electrophoretic transport. The addition of α-thioglycerol slows the development of molecular oxygen by electro-oxidation of α-thioglycerol and/or reaction of the thiol moiety of α-thioglycerol with atomic oxygen at the electrode surface, thereby delaying the formation of molecular oxygen and, hence, bubbles.

14.2.3 FUNCTIONAL HYDROGEL PERMEATION LAYERS

The hydrogel permeation layer serves three very important functions. First, streptavidin is covalently bound within the hydrogel permeation layer to provide attachment chemistry for biotin-labeled oligonucleotides. Second, the thickness of the permeation layer (~10 μm) moves the oligonucleotides away from the harsh environment at the surface of the platinum working electrode. Third, the hydrogel permeation layer provides a three-dimensional attachment site that is capable of binding more DNA than a planar, two-dimensional surface, thereby facilitating fluorescence detection.

Current hydrogels are formulated with acrylamide, a crosslinker, a porogen, a photoinitiator, and streptavidin that has been modified such that it will copolymerize with acrylamide. A small drop of the formulation is placed on a bare microarray followed by placement of a micromachined mold that defines the thickness of the hydrogel. With the mold in place, the formulation is exposed to ultraviolet (UV) light to rapidly polymerize the formulation. Figure 14.6 shows an example of a hydrogel permeation layer molded on a 100-site electronic microarray.

The performance of an assay is very dependent on the porosity of the hydrogel permeation layer. If the pores are too small, the oligonucleotides will have a hard time penetrating the gel. Consequently, most of the binding will be confined to the surface of the hydrogel. However, if the pores are too large, the oligonucleotides can easily reach the harsh electrochemical environment at the platinum electrode surface. At an intermediate pore size, oligonucleotides are capable of entering the hydrogel permeation layer, but their mobility is significantly reduced.

The factors that control the pore size include the concentrations of all the above reagents, the intensity of the UV light, and the temperature [8]. In practice, the parameters that are easily adjusted

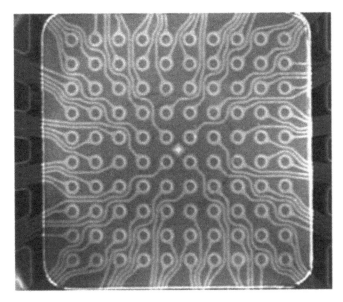

FIGURE 14.6 Acrylamide hydrogel permeation layer molded over 100-site array. The mold is micro-machined such that the thickness of the hydrated permeation layer is 10 μm thick. The hazy appearance of the permeation layer is due to light scattering from micropores that form within the hydrogel during the UV curing process.

and cause the largest changes are the concentrations of the crosslinker and the porogen and the total concentration of solids in the formulation. For example, Figure 14.7 shows the effect of the Brij 700 porogen on hydrogel porosity.

We have learned that porosity can have a strong effect on assay performance. This is illustrated in Figure 14.8 where assay performance is compared on four hydrogels with different porosity. The assay run for this comparison was an on-chip strand displacement amplification (SDA) reaction. (Details of on-chip amplification are given in Section 14.5.3.)

Although hydrogel formulations can be optimized for every assay, we have found that it is not necessary to do so. For example, hydrogel B2 shown in Figure 14.8 works well for assays involving on-chip amplification of bacterial target DNA. In contrast, assays to detect single nucleotide polymorphisms (SNPs) use a different hydrogel formulation that works well for all SNP detection assays. In practice, we have found that two hydrogel formulations are capable of giving good assay

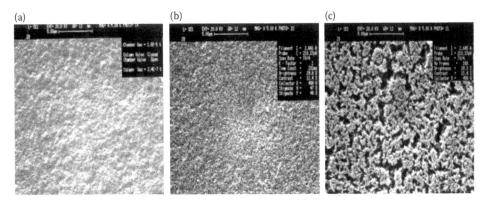

FIGURE 14.7 Hydrogel permeations layers of differing porosity. The porosity was controlled by the amount of added porogen, Brij 700: (a) 0 mg/mL; (b) 11 mg/mL; and (c) 18 mg/mL.

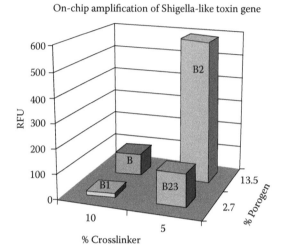

On-chip amplification of Shigella-like toxin gene

FIGURE 14.8   **(See color insert following page 138.)** Comparison of assay results on four different hydrogel permeation layers. The hydrogels were formulated to optimize the performance of an on-chip SDA reaction. In this experiment, the effects of the acrylamide crosslinker and the porogen were examined. Formulation B2 gave the best performance with this type of assay.

performance for all of our applications. Recently, a single hydrogel formulation was developed that satisfies all current applications.

14.2.4   HARDWARE

The microarrays shown in Figures 14.1 and 14.6 are 100-site arrays where the platinum electrodes are connected individually or as user-defined groups to a single power supply. Although these microarrays work very well for most applications, they have limitations when a group of electrodes are selected for biasing. For example, when simultaneously biasing more than one electrode, the total current can be regulated, but the currents at individual electrodes will vary depending on their location within the array and the cleanliness of the electrodes. Thus, the combined current for two electrodes might be 800 nA, but one electrode might be drawing 375 nA while the other will be drawing 425 nA. Consequently, there will be some electrode-to-electrode variability.

To overcome the electrode-to-electrode variability a new "smart" chip was designed using complementary metal–oxide–semiconductor (CMOS) technology (Figure 14.9). These chips have overall dimensions smaller than the 100-site microarray. They have 400 working electrodes, and most importantly have on-board circuitry to individually regulate each electrode for current or potential [9]. In essence, each electrode has its own potentiostat (or galvanostat). Consequently, the electrode-to-electrode variability is significantly reduced. Furthermore, the 400-site chip has on-board memory for storing data and a p-n junction thermal sensor for temperature regulation. The 400-site chip is mounted on a small cartridge with fluidic and electrical connections. The insert in Figure 14.10 shows the 400-site cartridge.

To run an assay the cartridge is inserted in the front of the fully automated NanoChip® 400 System shown in Figure 14.10. This system uniquely combines robotized fluidics manipulation for dispensing microliter volumes of reagents or samples to the cartridge and precise electrophoretic transport for further directing of the DNA molecules toward array sites. The electrophoretic transport occurs within a microscale cell (ca. 30 μL of fluid) that covers the electronic array. The user loads his samples on a 96-well or 384-well microtiter plate, which is mounted on the platen below

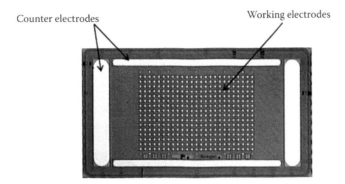

FIGURE 14.9 The 400-site smart chip built with CMOS technology. All working electrodes and counter electrodes are independently regulated for potential and current. The platinum working electrodes are 50 μm in diameter and are spaced on center-to-center distances of 150 μm. The chip has an on-board thermistor for temperature control and on-board memory for storing information.

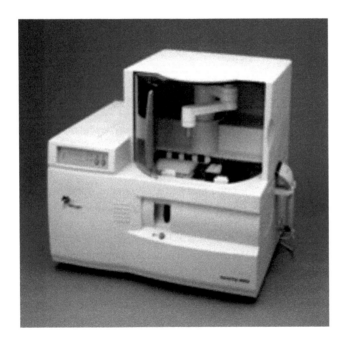

FIGURE 14.10 NanoChip 400 System is an instrument for complete automated analysis of DNA samples. The samples are loaded in one or two microtiter plates (96 well or 384 well), which are mounted on a platen just below the robot arm. Reagent packs designed for specific assays are also mounted on the platen. The 400-site cartridge is inserted in the front of the instrument. All operations including sample transfer, cartridge washing, electronic biasing, and image processing are computer controlled and require no user intervention once the assay is started.

the robotic arm. Reagent packs for specific assays are also mounted on the platen. When the assay is started the robotic arm transfers reagents and samples to the 400-site cartridge. Electronic biasing, fluidic washes, temperature control, and fluorescence imaging are all automatically controlled by software. The imaging system uses a charge-coupled device (CCD) camera with red and green filters to measure the fluorescence intensity of red and green reporter dyes. Focusing, pad finding, image acquisition, and image processing are all automated in software. When the assay is finished

FIGURE 14.11 Low-cost instrument originally developed with a Dual Use in Science and Technology Government Contract. The instrument uses the 400-site cartridge and has all the functionality of the NanoChip 400 System; however, samples are manually loaded and the assays require user intervention. A simple scripting language permits assays to be run where the user is prompted to load samples or reagents. The instrument is best used for quick development of simple assays, verification of assays, or for educational purposes.

the analyzed results are stored on the computer and are available to the user. No user intervention is required during the assay.

In contrast to the NanoChip 400 System, Figure 14.11 shows a very simple, low-cost, prototype instrument that was developed under a Dual Use in Science and Technology federal contract. This instrument was designed to be a light-weight, portable field instrument for the detection of biowarfare agents [10]. It uses the same cartridge that is used on the NanoChip 400 System. Thus, each array site is still capable of being individually regulated for current or potential. The main difference between the NanoChip 400 System and the low-cost prototype instrument is that the low-cost instrument lacks a robot for total automation. There is, however, a simple scripting language that will run multiple tasks, but it will prompt the user whenever reagents or samples need to be loaded. Although the instrument lacks full automation, it has proved to be useful for rapid development and validation of simple assays. The instrument is also being considered as an educational aid because it allows simple assay design and direct user intervention. (The results shown in Figure 14.12 were obtained on this instrument.)

14.3   OPEN PLATFORM ELECTRONIC MICROARRAYS

One of the unique features of electronic microarrays is that the end user can decide which assays or combination of assays to run on the microarray—electronic microarrays are not preloaded with capture probes. Furthermore, unlike most other microarrays, the electronic microarray can run multiple samples on a single microarray. In addition, after the capture probes are loaded, the microarray can be reused up to 10 times to detect the same analyte from as many as 10 different samples. Figure 14.13 shows the results for 10 reuses of the same microarray to detect 8 analytes.

SNP/Mutation	No Signal	Correct Call
Taq1D	2/72	70/70
Taq1A	3/72	69/69
−141delC	1/72	71/71
C957T	1/72	71/71
S311C	1/72	71/71

FIGURE 14.12 Results of study to detect DRD2 mutants in 72 samples. The no-signal samples were later rereported on a new chip and correctly called.

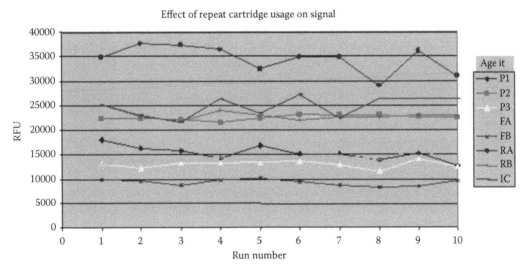

FIGURE 14.13 The signal strength of 8 different agents was measured 10 times by stripping the reagent and reloading it for each measurement. The signal strength remains relatively constant during the 10 measurements.

14.4 FUNDAMENTAL ASSAY PROCEDURES ON ELECTRONIC MICROARRAY

There are four fundamental assay procedures that are unique to electronic microarrays: electronic capture loading, electronic hybridization, electronic stringency, and target preparation. These four fundamental procedures are used to prepare the microarray for an assay, reduce the assay time, and to improve the signal-to-noise ratio.

14.4.1 ELECTRONIC CAPTURE LOADING

Aside from electronic microarrays, all other microarrays have their capture oligonucleotides deposited by some form of spotting technology or the oligonucleotides are synthesized on the substrate.

In contrast, electronic microarrays currently use electrophoretic transport to load capture probes. Typically, a 100-nM biotinylated oligonucleotide solution in 50 mM histidine is loaded in the cartridge containing the microarray. Selected array sites are then biased positive at 350 nA for 30 s causing the capture probes to concentrate and bind at the selected array sites. The array is then washed and the next set of capture probes is loaded.

14.4.2 ELECTRONIC HYBRIDIZATION

As previously mentioned, electronic hybridization refers to the process where target DNA molecules are electrophoretically addressed to specific arrays sites and hybridize to previously loaded capture probes anchored at those sites [11]. The developing high concentration of target DNA drives the equilibrium toward hybridization. Furthermore, the protonated histidine generated in the vicinity of the array site acts to minimize charge repulsion between the hybridizing strands of DNA. Most importantly, electronic hybridization is extremely rapid with hybridization that completes in <2 min. After hybridization, the microarray is washed with buffer to remove excess reagent.

14.4.3 ELECTRONIC STRINGENCY

After electronic hybridization, a protocol is needed that will easily discriminate single base pair mismatches from fully complementary DNA. Generally, this discrimination is based on the fact that there is usually a two or three degree difference in the thermal stability of complementary double-stranded DNA versus double-stranded DNA with a single base pair mismatch [12]. The preferred method for removing nonspecifically bound DNA is to use thermal stringency. Here the temperature is slowly increased or decreased to remove the nonspecifically bound DNA.

Electronic microarrays can also perform electronic stringency. In this protocol, the working electrode is negatively biased. Water reduction occurs at the negative electrode and generates H_2 and OH^- (Equation 14.2). The hydroxyl ions destabilize the DNA with the single base pair mismatch more than the fully complementary DNA. In addition, noncomplementary DNA is electrophoretically forced away from the electrode site. After negative biasing for a specific time and magnitude, the array is washed to remove the mismatched single strand DNA.

One advantage of electronic stringency over thermal stringency is that electronic stringency can be run on selected array sites instead of the entire array. Furthermore, the stringency conditions can be different at each array site.

14.4.4 TARGET PREPARATION

The purpose of target preparation is to reduce the background signal caused by target DNA that might have been electrophoretically driven into the hydrogel permeation layer, but not specifically bound to a capture oligonucleotide. By reversing the bias at the array site, the nonspecifically bound DNA is electrophoretically transported out of the hydrogel permeation layer where it is subsequently washed away.

14.5 EXAMPLES OF SPECIFIC ASSAYS

14.5.1 TARGET DETECTION: CAP/SEPSIS ASSAY

The community acquired pneumonia (CAP)/sepsis assay is an example of a relatively complex assay. CAP is a serious and relatively common respiratory tract infection [13]. If left untreated, pneumonia can lead to sepsis (blood poisoning) and eventually death. Approximately 4 million cases of CAP are reported each year within the United States with 20% of those cases requiring hospitalization. *Streptococcus pneumoniae* is the bacterial pathogen responsible for the majority of

CAP cases. The other cases are caused by atypical pathogens such as *Mycoplasma pneumoniae*, *Chlamydia pneumoniae*, *Legionella* species, and other species.

The CAP/sepsis assay is designed to detect 11 pathogens representing both bacterial and viral targets. The viral ribonucleic acid (RNA) from patient samples is extracted and then amplified by a multiplex reverse transcriptase PCR reaction. Two target-specific capture probes complementary to sequences within the generated amplicons are electronically loaded on each of five array sites, for a total of 10 target-specific probes, Figure 14.14 [the respiratory syncytial virus (RSV)

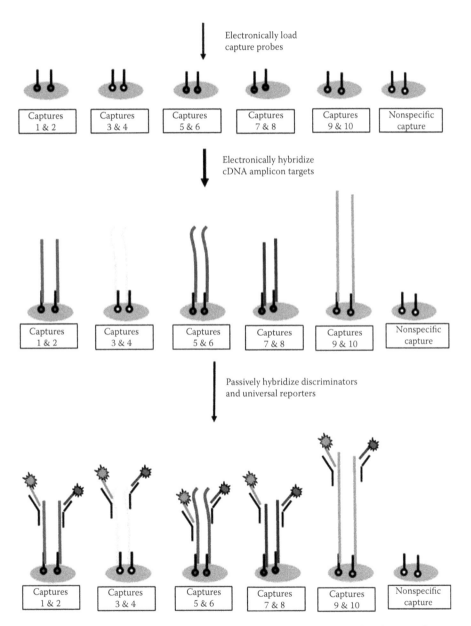

FIGURE 14.14 **(See color insert following page 138.)** Assay procedure for CAP/sepsis assay. Capture probes are loaded two at a time on five array sites. The sixth array site is addressed with a nonspecific capture and used as a background control. After capture addressing, all of the cDNA amplicon targets are simultaneously hybridized to all six array sites. Next, discriminator probes and universal reporter probes are passively hybridized to the capture-amplicon complex. Finally, the red/green fluorescence signals are measured and the data analyzed.

System	Input templates	Signal	SBR
Staphylococcus aureus	1.2	647	0.9
		613	0.7
	12	715	1.1
		667	1.1
	120	5911	8.3
		8403	8.5
	1200	15956	20.2
		22438	23.6
Streptococcus pneumoniae	1.2	92	0.3
		327	1.0
	12	550	1.6
		785	2.4
	120	3569	10.7
		3652	10.9
	1200	7928	23.7
		11362	34.0

FIGURE 14.15 Partial results of CAP/sepsis assay. A positive result requires at least 1000 relative fluorescence units and a signal-to-background ratio of 3 : 1.

probe captures both RSV A and RSV B). A sixth array site has a capture oligonucleotide not complementary to any of the amplicons addressed to it, and is used as a background control. Next, amplification reactions are addressed to a set of six pads representing all of the capture oligonucleotides. Under the electronic conditions used, any amplicon present in the reaction hybridizes to its complementary capture oligonucleotide. This process is repeated until all samples have been addressed. After electronic hybridization, a reporter mix containing 500 nM discriminators complementary to each of the amplicons and 750 nM reporters is allowed to passively hybridize to the capture-amplicon complex. The discriminator probe is an oligonucleotide whose sequence contains a region specific to the amplicon and another region specific to the universal reporter. Fluorescence signal intensities are then measured on the red and green channels for the entire array. (Because there are two analytes on each array site, one analyte is detected using red fluorescence and the other using green fluorescence.)

Results for two of the pathogens are shown in Figure 14.15. Based on the criteria for a positive call, *Streptococcus pneumoniae* and *Staphylococcus aureus* are detectible at 120 copies.

14.5.2 Pharmacogenomics: Research Assay for DRD2 Mutants

The dopamine receptor D2 (DRD2) research assay is an example of a protocol for the detection of SNPs. Four SNPs (Taq1A, Taq1D, S311C, and C957T) and a deletion mutant (-141delC) are analyzed on a 100-site microarray (the assay has also been transferred to the 400-site microarray).

Medications that block the dopamine receptor are often used for the treatment of schizophrenia, alcohol addiction, and psychiatric disorders. Mutations within the receptor site may inhibit the effects of these medications [14,15]. In particular, the Taq1A SNP and the -141delC deletion mutation are correlated with the inhibition of these medications [16]. Consequently, patients with these mutations are less responsive to their medication and should be treated with a higher dosage or an alternative medication.

The assay for the DRD2 mutants begins with DNA extraction from a patient sample and multiplex PCR amplification of the five amplicons. Capture probes for the five amplicons are loaded on

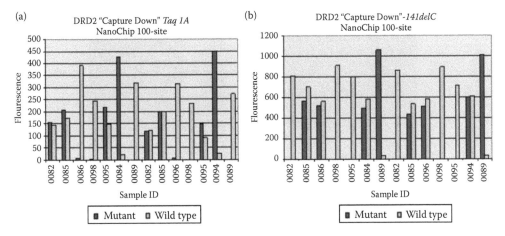

FIGURE 14.16 Simultaneous detection of the DRD2 mutants for 14 patient samples. Data for two of the mutants, Taq1A and -141delC, are shown in panels (a) and (b), respectively. (The other three mutants were also detected but the data are not shown.) Samples with nearly equal wild type and mutant signals are heterozygous mutant strains.

five separate array sites. All five amplicons are then electronically hybridized to the capture probes in a single biasing step. Next, discriminator probes and universal reporter probes are passively hybridized to the amplicons. In the final step of the assay, the hybridized probes undergo a combination of thermal stringency and salt washes. The fluorescence intensity of the array sites is then measured in the red and green channels. The discriminator probes are designed such that the green signal represents the wild-type allele and a red signal represents the mutant allele. Because the capture probes capture both wild-type and mutant alleles, each array site can fluoresce red, green, or both colors. The latter case represents a heterozygous mutant sample.

Figure 14.16 shows a subset of the results in a study to detect the DRD2 mutants in 72 samples using a 100-site microarray. All five mutants were detected in a single reporting step. A summary of the results is shown in Figure 14.12.

14.5.3 ON-CHIP AMPLIFICATION: AMPLIFICATION AND DETECTION OF STX

Aside from providing attachment chemistry for oligonucleotides, the hydrogel permeation layer can also be considered as a solid-phase support for chemical reactions. In this example, the hydrogel permeation layer is used to anchor primers for SDA. SDA is an isothermal amplification method that is more complicated than PCR amplification, but has the advantage of not requiring a thermal cycler and is capable of very rapid amplification [17,18]. Aside from the normal forward and reverse primers and a polymerase, SDA also requires forward and reverse bumper primers (for strand displacement) and a nicking enzyme. Initially, the double-stranded template DNA is thermally denatured at 95°C. This denaturation step is done off-chip, and is the only thermal denaturation step in the entire protocol.

For on-chip SDA, forward and reverse primers (which are biotinylated) are electronically addressed to specific array sites. The denatured template DNA is then electronically hybridized to the anchored primers. Bumper primers (not biotinylated) and enzymes are loaded on the cartridge followed by incubation for 30 min to 1 h at 50–55°C. The cartridge is then washed and reported with fluorescently labeled reporter probes complementary to the amplicon.

Figure 14.17 shows the amplification results for a gene fragment of the Shigella-like toxin, Shigella-like toxin (STX). In this experiment the template DNA is titrated to lower concentrations at different array sites. Eventually, a concentration limit is reached where the number of template molecules within the strong electric field of the biased electrode was almost zero. It then becomes a

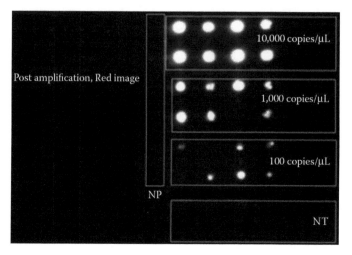

FIGURE 14.17 SDA of a Shigella-like toxin gene fragment; NP, no primer control; and NT, no template control. Assay was run on a 400-site cartridge using the low-cost, prototype instrument shown in Figure 14.11. Sample size was 10 mL.

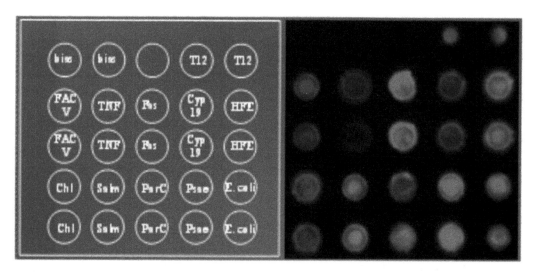

FIGURE 14.18 Five human gene fragments and five bacterial gene fragments relevant to infectious diseases/biological warfare were simultaneously amplified and detected on a 25-site electronic microarray using anchored SDA.

matter of statistical probability of whether or not a template molecule is within the strong electric field of the array site chosen for the assay. In Figure 14.17, it can be seen that the limit is being approached at 100 copies/µL. (Other factors, such as gel entanglement and hybridization efficiency also affect the lower limit where amplification can proceed.)

Figure 14.18 shows another example of on-chip amplification [19]. Here, 10 different gene fragments were simultaneously amplified using on-chip SDA.

14.6 SUMMARY

The electronic microarray technology offers an unprecedented advantage of extremely rapid and controlled DNA hybridization that occurs within seconds compared to hours in passive microarrays.

This yields an instrument that offers the possibility of an open platform where the users can prepare their own array and perform the assay within the same or shorter amount of time it takes for the assay performance with the competitive array technologies. A wide variety of assay designs are available to the user in this open platform including capture down, amplicon down, or sandwich-type assays. The platform uniquely offers a convenient fluidics manipulation using common microtiter plates as input and robotized fluid transport as well as a high-precision electrophoretic addressing and accumulation of probes and amplicons to the arrays sites. The latter is considered to be a unique feature of this technology because it enables rapid array printing that becomes a part of the assay.

In the introduction it was stated that the technology for DNA diagnostics needs to be faster, cheaper, more accurate, and more sensitive in order to gain wider acceptance in the diagnostics market. What was not mentioned, yet is extremely important, is the need for the integration of sample preparation and detection. In the last example above, amplification and detection were integrated on a single microarray. Direct detection has been proposed (and demonstrated in the research setting) to circumvent the current need for amplification [20]. Although this is a step in the right direction, the major bottleneck to integration is the extraction of DNA from patient samples. The extraction methodology must be capable of handling different sample sources (blood, nasopharyngeal, or sputum), and the extraction efficiency should be high particularly for infectious diseases that might contain gram-positive spores (i.e., water, food, or agricultural samples, and so forth).

In the past, electronic microarrays have been integrated with sample preparation methods, but the criteria of faster, cheaper, more accurate, and more sensitive still need to be met. Electronic microarrays offer the advantages of an open platform (multiplex samples/assays) and rapid target hybridization. Other technologies also offer significant advantages for DNA diagnostics. To the extent that these advantages can be incorporated in a fully integrated sample-to-answer device, DNA diagnostics will become a major player in the diagnostics market and the health care needs of patients will be improved.

ACKNOWLEDGMENTS

The authors would like to acknowledge Dr. David Cantor and Dr. Mathew Harris for reviewing the CAP/sepsis and DRD2 sections of the manuscript and Dr. Ray Radtkey for providing additional information concerning the DRD2 research assay.

REFERENCES

1. Evans WE and Relling MV. Pharmacogenomics: Translating functional genomics into rational therapeutics. *Science* 1999; 286: 487–491.
2. Young RA. Biomedical discovery with DNA arrays. *Cell* 2000; 102: 9–16.
3. Stuart RO, Wachsman W, Berry CC, Wang-Rodriguez J, Wasserman L, Klacansky I, Masys D, et al. *In-silico* dissection of cell-type-associated patterns of gene expression in prostate cancer. *Proc Natl Acad Sci USA* 2004; 101(2): 615–620.
4. Kassagne SK, Reese H, Hodko H, Yang JM, Sarkar K, Smolko D, Swanson P, Raymond D, Heller MJ, and Madou MJ. Numerical modeling of transport and accumulation of DNA on electronically active biochips. *Sens Actuators B Chem* 2003; 94: 81–98.
5. Atkins PW. *Physical Chemistry*, 1978, p. 829. Great Britain: Oxford University Press.
6. Chee M, Yang R, Hubbell E, Berno A, Huang XC, Stern D, Winkler J, Lockhart DJ, Morris MS, and Fodor SPA. Accessing genetic information with high-density DNA arrays. *Science* 1996; 274: 610–614.
7. Taton TA, Mirkin CA, and Letsinger RL. Scanometric DNA array detection with nanoparticle probes. *Science* 2000; 289: 1757–1760.
8. Okay O. Macroporous copolymer networks. *Prog Polym Sci* 2000; 25: 711–779.
9. Swanson P, Gelbart R, Atlas E, Yang L, Grogan T, Butler WF, Ackley DE, and Sheldon E. A fully multiplexed CMOS biochip for DNA analysis. *Sens Actuators B Chem* 2000; 64: 22–30.
10. Huang Y, Hodko D, Smolko D, and Lidgard G. Electronic microarray technology and applications in genomics and proteomics. In: *BioMEMS and Biomedical Nanotechnology: Micro and Nano-Technologies*

for Genomics and Proteomics, M Ozkan and MJ Heller (eds), 2005, vol. 2, pp. 3–21. Springer Science+Business Media, Inc.

11. Edman CF, Raymond DE, Wu DJ, Tu E, Sosnowski RG, Butler WF, Nerenberg M, and Heller MJ. Electric field directed nucleic acid hybridization on microchips. *Nucleic Acids Res* 1997; 25(24): 4907–4914.

12. Cantor CR and Smith CL. *Genomics: The Science and Technology behind the Human Genome Project*, 1999, p. 77. New York: Wiley.

13. Bartlett JG, Breiman RF, Mandell LA, and File TM. Community-acquired pneumonia in adults: Guidelines for management. *Clin Infect Dis* 1998; 26: 811–838.

14. Noble EP. D2 dopamine receptor gene in psychiatric and neurologic disorders and its phenotypes. *Am J Med Genet Part B Neuropsychiatr Genet* 2003; 116B: 103–125.

15. Arinami T, Gao M, Hamaguchi H, and Toru M. A functional polymorphism in the promoter region of the dopamine D2 receptor gene is associated with schizophrenia. *Hum Mol Genet* 1997; 6(4): 577–582.

16. Samochowiec J, Ladehoff M, Pelz J, Smolka M, Schmidt L, Rommelspacher H, and Finckh U. Predominant influence of the 3′-region of dopamine D2 receptor gene (DRD2) on the clinical phenotype in German alcoholics. *Pharmacogenetics* 2000; 10: 471–475.

17. Walker GT, Nadeau JG, Spears PA, Schram JL, Nycz CM, and Shank DD. Multiplex strand displacement amplification (SDA) and detection of DNA sequences from *Mycobacterium tuberculosis* and other mycobacteria. *Nucleic Acids Res* 1994; 22(13): 2670–2677.

18. Spargo CA, Fraiser MS, Van Cleve M, Wright DJ, Nycz CM, Spears PA, and Walker GT. Detection of *M. tuberculosis* DNA using thermophilic strand displacement amplification. *Mol Cell Probes* 1996; 10: 247–256.

19. Westin L, Xu X, Miller C, Wang L, Edman CF, and Nerenberg M. Anchored multiplex amplification on a microelectronic chip array. *Nat Biotechnol* 2000; 18: 199–204.

20. Park S, Taton TA, and Mirkin CA. Array-based electrical detection of DNA with nanoparticle probes. *Science* 2002; 295: 1503–1506.

15 New Applications for Microarrays

Mark Andersen, Steve Warrick, and Christopher Adams

CONTENTS

15.1 INTRODUCTION

Genomic alterations are one of several key factors involved in carcinogenesis. Over the years, much focus has been dedicated to the study of genomic copy number changes using fluorescent *in situ* hybridization (FISH) techniques such as comparative genomic hybridization (CGH) and spectral karyotyping. Mapping copy number gains and losses has led to potentially valuable diagnostic and prognostic markers for cancer and prenatal defects. In recent years, researchers have applied the concept of CGH to high-density deoxyribonucleic acid (DNA) microarrays (Figure 15.1) (array CGH), enabling unprecedented characterization of genomic aberrations on a genome-wide level [1].

It has been shown that genomic copy changes impact messenger ribonucleic acid (mRNA) expression levels in tumors, suggesting that deregulation of mRNA expression of tumor-suppressor genes and oncogenes may be involved in tumorigenesis [2]. However, mRNA is not the only RNA transcript that is impacted by genomic gains and losses. MicroRNAs (miRNAs) are an endogenous class of non-coding RNAs that play a significant role in gene regulation by acting as posttranscriptional inhibitors of gene expression. A recent analysis of the genes corresponding to 186 human miRNA suggested that miRNA genes are often associated with regions of the genome that are prone to copy

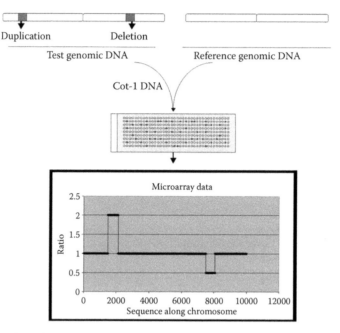

FIGURE 15.1   Overview of array CGH.

gains and losses [3]. In a parallel analysis of genomic aberrations and miRNA expression, 283 human miRNA genes were analyzed by array CGH. A significant portion of these genes exhibited copy number alterations in ovarian, breast, and skin cancer, and corresponding changes in miRNA expression. In addition to impacting miRNA expression by altering the copy number of miRNA genes, genomic aberrations were also found in the genes of *Dicer1*, *Argonaute 2*, proteins that play essential roles in miRNA biogenesis [4].

Microarray-based CGH (array CGH) and miRNA profiling are two of several emerging applications that enable effective, global interrogations of the genomic alterations, gene expression, and epigenetic mechanisms that underlie complex diseases. Here we describe these methods in more detail.

15.2 USING MICROARRAYS TO COMPARE GENOMIC DNA SAMPLES

Alterations in chromosomes that lead to deletions and amplifications are being identified as critical components of tumorigenesis. Moreover, inherited and noninherited DNA copy number variations have also been directly correlated with changes in mRNA levels, indicating that underlying genetic imbalances significantly impact the development and progression of many tumor-specific expression profiles [5,6]. CGH has become a common method of whole-genome analysis used to detect and map widespread amplifications and deletions of DNA sequences.

Clinical protocols for CGH have involved probing metaphase chromosome spreads with two labeled DNA samples: a test and a reference sample. This type of analysis can identify gross structural abnormalities in the DNA, such as large insertions, deletions, or rearrangements. Upon identification, more focused approaches such as FISH have proved useful in characterizing specific deletions and amplifications correlated with certain disorders.

Recent advances in microarray methods—including the use of genomic bacterial artificial chromosome (BAC) vector clones, complementary DNA (cDNA) clones, and oligonucleotides—have circumvented some of the limitations of conventional CGH analytical methods [1,7–9]. Array-based CGH methods allow researchers to perform high-resolution genome scans of test samples with very high accuracy and throughput. Thus, genome-wide array CGH offers improved resolution and sensitivity over *in situ* methods, and has become an important tool in detecting chromosomal imbalances in cancer and medical genetic studies [8,10].

Like standard CGH, array CGH compares two genomic DNA samples: a test sample and a reference sample, which is usually composed of normal genomic DNA. The test and reference DNA are labeled with different fluorescent dyes and simultaneously hybridized to a microarray with unlabeled Cot-1 DNA® to block the hybridization of repetitive sequences. The relative fluorescent intensity of each dye is correlated to the amount of genetic material from each sample at sequences queried by each array feature. The ratio of dye signals represents the copy number of sequences in each DNA sample, and thereby identifies amplifications and deletions at specific loci along the genome.

Array CGH offers two dramatic improvements over FISH-based CGH analysis. First, it provides the ability to specifically interrogate the genome with much finer resolution. The resolution, or ability to detect smaller and smaller differences between the test and reference samples, can be readily controlled by the length and overlap of features. This is particularly true with oligo arrays and tiling approaches. Second, FISH has a limited ability to multiplex. Therefore, in cases where abnormal regions are not known a priori, the traditional FISH approach may require multiple, often iterative, interrogation of different sites. Conversely, high-density array CGH can interrogate the entire genome in a single experiment.

15.2.1 SAMPLE PREPARATION

Reliable genomic DNA purification, quantification, and labeling approaches are key factors in obtaining accurate array CGH results. They are particularly important for cancer studies, where typical test samples are derived from tumor tissue that can be heavily vascularized, fibrotic, or otherwise refractory to effective cellular disruption and nucleic acid extraction. Accurate quantification of isolated genomic DNA is important because the methods employed in array CGH require precise input of equal and often small amount of precious starting material (e.g., from biopsy samples) into the array CGH dye-labeling reaction in order to reliably resolve twofold gene copy number differences. Similarly, the integrity and purity of isolated genomic DNA are important for efficient and balanced dye labeling and detection. Finally, retrospective array CGH studies leveraging the large number of banked fixed and/or embedded clinical samples present an increased level of difficulty, as genomic DNA integrity, efficient recovery, and effective dye labeling are often more challenging when using these samples.

15.2.2 DNA LABELING

The two primary methods employed to label genomic DNA for CGH analysis are nick translation and random priming. The nick translation approach utilizes a limited DNase I treatment to nick double-stranded DNA followed by treatment with *Escherichia coli* DNA polymerase I holoenzyme. The 3′ ends of the nicks are polymerization start points, and as the 5′ → 3′ nuclease activity degrades the unlabeled starting material, it is replaced by a new strand with dye-labeled nucleotides. In this method, nicks move along the sites of new DNA synthesis, but the starting material does not undergo amplification.

A more popular approach for DNA labeling and amplification is based on random priming. In this method, DNA samples are first fragmented by physical or enzymatic methods. After denaturation, high concentrations of short random primers are annealed to the target sample, and *E. coli* DNA polymerase I Klenow fragment is used to extend the primers. Since Klenow lacks 5′ → 3′ nuclease activity, target samples can be amplified by at least 10-fold using this method. Fluorescent dyes can be incorporated into the genomic DNA directly during the Klenow reaction, or indirectly in a separate step.

Using the indirect method, an amino-derivatized nucleotide is incorporated during primer extension and is later conjugated with the ester form of the fluorescent dye. This two-step approach to labeling provides additional flexibility with dye selection and the ability to cost-effectively perform hybridizations with more than the standard two colors. However, it is also more time-consuming than direct labeling. With direct labeling, dye-labeled nucleotides are directly incorporated into the primer extension products.

The BioPrime® Total Genomic Labeling System provides a direct labeling method using Alexa Fluor®-labeled nucleotides and mutant form of the Klenow fragment described below [11]. Dye-labeled nucleotides, primers, enzymes, buffers, and purification columns are all included within the system (Figure 15.2).

15.2.2.1 Polymerases

The native Klenow fragment has been used in several DNA labeling protocols. However, native Klenow fragment tends to produce low DNA yields, probably due to its $3' \rightarrow 5'$ exonuclease activity degrading primers, templates, and labeled probes during the course of the reaction.

Systems that use $3' \rightarrow 5'$ exonuclease-minus Klenow (exo-Klenow) have been shown to improve labeling efficiency and DNA yield. Exo-Klenow is unique in that it lacks both the proofreading activity ($3' \rightarrow 5'$ exonuclease) and $5' \rightarrow 3'$ nuclease activity, while maintaining high levels of polymerase and strand displacement activity [12]. Since DNA is not degraded by exo-Klenow during the course of the reaction, higher concentrations of labeled probes are generated with improved reproducibility.

Labeled DNA probes created with exo-Klenow yield higher normalized median signal than standard nick translation and Klenow-based random labeling methods. In addition, exo-Klenow labeled probes yield greater signal-to-noise ratios than the other two methods. This results in an increased number of positive features and higher levels of accuracy (Figure 15.3).

15.2.2.2 Fluorescent Dyes

Polymethine or cyanine dyes are the most commonly used fluorophores in array applications. They are closely aligned spectrally with the excitation sources used in most array scanners and have high extinction coefficients which, when coupled to their quantum yield, enable detection of low concentrations of labeled material.

Alexa Fluor dyes used in current array applications have many of the same features as cyanine dyes. In addition, Alexa Fluor dyes have an advantage over cyanine dyes in terms of their resistance to ozone degradation. This improved stability is derived from modifications made to the core structure, inhibiting breakdown of the polymethine backbone of the fluorophore. This is particularly true for Alexa Fluor 647, which has demonstrated improved performance over other red, cyanine-based structures. Furthermore, the presence of increased negatively charged pendant groups on the Alexa Fluor dye structure also inhibits aggregation of the dyes, limiting quenching of fluorescence from labeled biomolecules and improving the labeling "sweet spot" of the Alexa Fluor dye series.

A recent improvement in array CGH involves the development of a new dye-labeled nucleotide pairing. A combination of new nucleotide linker, dye structure, and DNA polymerase has been designed for optimal, balanced incorporation of Alexa Fluor 3- and Alexa Fluor 5-labeled nucleotides for use in array CGH applications. The formulation is designed to simplify the labeling workflow by creating a master mix solution that has the dye-labeled nucleotides present at optimum concentration. The new dyes have identical spectra to the original Alexa Fluor 555 and Alexa Fluor 647 dye pair, but have been modified to increase efficient incorporation using exo-Klenow. The formulation balances the amount of signal in red and green channels from array experiments, simplifying data analysis, and increasing signal-to-noise ratios, in particular, in the red channel.

15.2.2.3 Purification and Assessing Labeling Efficiency

Efficient purification of DNA labeled for array CGH is essential to reduce background signals due to the presence of unincorporated dye-labeled nucleotides, unlabeled nucleotides, and random primers. To correctly identify genomic alterations, the signals between the test sample and the reference sample on an array feature must be well balanced. High signal background in either channel can

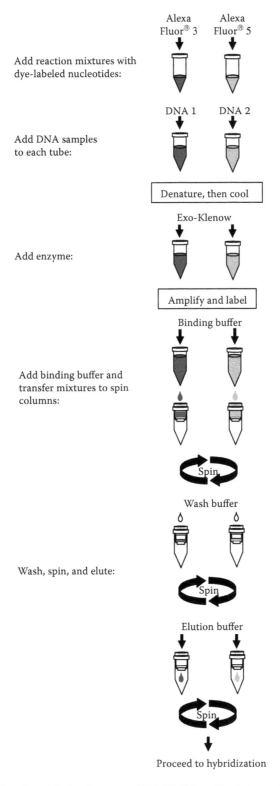

Add reaction mixtures with dye-labeled nucleotides:

Alexa Fluor® 3   Alexa Fluor® 5

Add DNA samples to each tube:

DNA 1   DNA 2

Denature, then cool

Add enzyme:

Exo-Klenow

Amplify and label

Add binding buffer and transfer mixtures to spin columns:

Binding buffer

Spin

Wash, spin, and elute:

Wash buffer

Spin

Elution buffer

Spin

Proceed to hybridization

FIGURE 15.2 **(See color insert following page 138.)** BioPrime Total Array CGH Genomic Labeling Protocol.

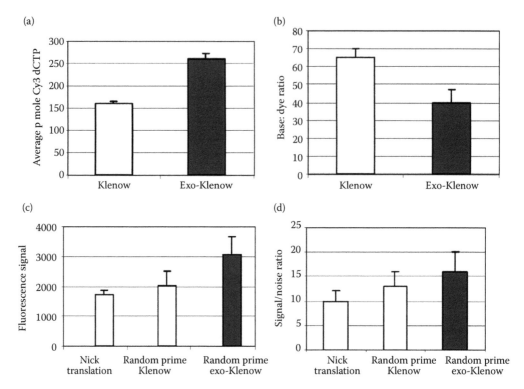

FIGURE 15.3 Comparison of exo-Klenow, Klenow, and nick translation labeled templates. (a) Spectrophotometric quantifications of dye-deoxycytidine triphosphate (dCTP) incorporation starting from 100 ng of genomic DNA labeled with 40 units of Klenow or exo-Klenow at 37°C for 4 h ($n = 3$). (b) Calculated base-to-dye ratios from Klenow and exo-Klenow labeled templates. (c) Normalized median fluorescence signal quantification from scanned images of 3000 element BAC arrays hybridized to 2 µg of genomic DNA labeled by exo-Klenow, Klenow, and nick translation methods ($n = 3$). (d) Signal-to-noise ratios calculated from the scanned images of BAC arrays hybridized with exo-Klenow, Klenow, and nick translation labeled templates. Data shown are from a representative experiment replicated a minimum of three independent times.

skew the signal ratios and conceal copy number changes, or conversely result in false copy number changes. Removing the unincorporated material reduces contaminating background signals.

Additionally, many researchers find it useful to determine the yield and dye incorporation of their labeled samples prior to committing them to an array. These values can be determined by measuring the absorbance of each sample at 260 nm for DNA yield and at the appropriate wavelength for the incorporated dye. Inefficient purification will result in an overestimation of yield due to the presence of unincorporated nucleotides and random primers (which are usually present in large excess). The amount of dye incorporation will also be overestimated due to contaminating unincorporated dye-labeled nucleotides.

Sample cleanup can be performed in a number of ways, ranging from simple ethanol precipitation to spin column purification. Ethanol precipitation has the advantage of being inexpensive and simple to perform, and if done carefully, it will remove the majority of the contaminating nucleotides and random primers. However, to ensure good recovery, it is advisable to precipitate the DNA for an extended period of time, which adds to the overall length of the protocol. In addition, care must be taken to avoid losing the DNA pellet during subsequent washes.

In contrast, spin column purification offers a quick and simple alternative. The proliferation of silica-based nucleic acid purification methods has resulted in easy-to-perform column purifications that allow the purification of several samples at the same time. The BioPrime Total Array CGH Genomic Labeling Systems include spin columns and purification buffers in the kit.

15.2.3 Array Blocking and Hybridization

A key component of successful genomic microarray experiments is the blocking of genomic repetitive elements in microarray probes to prohibit undesired hybridization between probe targets. Human Cot-1 DNA is placental DNA that is predominantly 50–300 base pairs (bp) in size and enriched for repetitive DNA sequences such as the *Alu* and *Kpn* family members [13,14]. Human Cot-1 DNA is commonly used to block nonspecific hybridization in microarray screening. It can also be used to suppress repetitive DNA sequences for the direct mapping of human DNA or mapping genomic clones to panels of somatic-cell hybrids for chromosome localization by Southern blotting. Human Cot-1 DNA is effective as a library-screening probe for somatic-cell hybrid libraries and flow-sorted chromosome libraries made from somatic-cell hybrids [15–17].

15.2.4 The Future of Array CGH

In some aspects, array CGH is more technically challenging than standard microarray expression profiling. Genomic DNA is more complex than RNA or cDNA, and genomic DNA profiling must detect small ratio deviations reflecting a single-copy loss of a tumor-suppressor gene from a mixed cell population. To detect twofold differences in copy number between samples, array CGH must involve techniques to reduce variability across the entire workflow. Labeling systems that produce high predictable yields with the least number of steps can reduce human error and at least partially compensate for imprecise DNA sample quantification. In addition, kits that that have been optimized with specific dyes to yield balanced signals on arrays can further increase the robustness and accuracy of analysis.

As with microarray expression analysis, there are a wide variety of methods and a lack of standardization across array CGH platforms [8]. As microarray gene expression profiling has matured, there have been studies comparing platforms and analysis methods [18]. Such studies are the impetus behind standardization and the introduction of universal controls, developments that will make expression analysis more suitable for clinical settings. Array CGH, while still a young method compared to microarray expression analysis, is poised to enter clinical settings. Similar developments in standardization and external controls will hasten the adoption of array CGH in regulated settings, where high reproducibility and accuracy are requirements.

15.3 USING MICROARRAYS FOR GLOBAL EXPRESSION PROFILING OF miRNAs

miRNAs are small (18–25 nucleotide) noncoding RNA molecules that have regulatory functions across a wide range of cell types and species. They are thought to regulate gene expression by binding to partially complementary sites in the 3′-untranslated region (UTR) of mRNAs and either inducing degradation of the mRNA or blocking its translation [19,20]. miRNAs were first discovered in mutants of the nematode *Caenorhabditis elegans* that lacked the ability to control the timing of specific cell fate switches during development [19,21]. Since then, hundreds of miRNAs in *C. elegans*, *Drosophila melanogaster*, plants, and mammals have been identified through computational and cloning approaches [22,23]. miRNAs are now understood to be one of the most highly abundant gene regulators in higher eukaryotes, and have been implicated in the regulation of developmental timing and pattern formation [24], restriction of differentiation potential [25], regulation of insulin secretion [26], and genomic rearrangements [27].

As the number of identified miRNAs has grown, so has the need for tools to analyze global differences in their expression levels between specific tissues, disease states, or developmental states. Several unique physical attributes of miRNAs—including their small size, lack of polyadenylated tails, and tendency to bind their mRNA targets with imperfect sequence homology—have made them elusive and challenging to study. Strong conservation between miRNA family members means that any profiling technology must be able to distinguish between ~20 base sequences that differ by only 1–2 nucleotides. Despite these obstacles, recent advances in microarray labeling and detection have enabled the use of this high-throughout technology for miRNA screening. In this section, we describe a novel method for labeling miRNAs with highly fluorescent branched dye molecules and hybridizing them to spotted oligonucleotide microarrays for subsequent detection and analysis.

This method relies on the enrichment of small RNAs from a total RNA population, followed by the polyadenylation, tagging, and fluorescent labeling of all the miRNAs in the sample. Depending on the amount of starting material, the miRNAs may also be amplified prior to labeling in a reaction that preserves the relative abundance of the different sequences in the original sample. Herein, we also describe the design and development of highly specific microarray oligonucleotide probes from validated miRNA sequences in the Sanger miRBase Sequence Database.

15.3.1 Purification of miRNAs

Because miRNAs represent only a small fraction of the RNA in a total RNA sample, using total RNA to profile miRNA expression can limit sensitivity. Therefore, it sometimes is advisable to enrich the sample for small RNAs prior to miRNA profiling. Traditional column-based RNA isolation methods are optimized to purify higher molecular weight RNA, resulting in minimal small RNA enrichment, whereas gel-based extraction methods are laborious and inefficient. In recent years, several column-based purification methods for the specific isolation of small RNAs have become available, including the PureLink™ miRNA Isolation Kit (Invitrogen) and Centricon YM-100 columns (Millipore).

The PureLink system utilizes two silica-based columns to purify small RNAs <200 nucleotides in length [including transfer RNA (tRNA), 5S ribosomal RNA (rRNA), 5.8S rRNA, miRNA, and short interfering RNA] from various sample sources. From sample lysates, RNAs greater than ~200 nucleotides are bound to the first column, allowing small RNA molecules to flow through. This flow-through fraction is mixed with a higher concentration of ethanol to increase its binding affinity to the glass fiber membranes in the second column. Bound small RNAs are then eluted free of larger RNA molecules such as 28S rRNA, 18S rRNA, and mRNA that can interfere with expression analysis of the small RNAs. The purified small RNA is suitable for most downstream applications such as microarray analysis, northern blotting, and reverse transcriptase PCR (RT-PCR) (Figure 15.4).

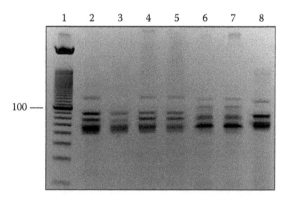

FIGURE 15.4 Small molecule purification from various sample sources. A variety of small RNA molecules were efficiently purified with the PureLink miRNA Kit and run on a NuPAGE® Novex 12% Bis-Tris Gel. Lane 1: 10 bp ladder; Lane 2: 2×10^6 HeLa cells; Lane 3: 2×10^6 293F cells; Lane 4: 5.5 mg rat spleen; Lane 5: 5 mg mouse liver; Lane 6: 300 mg spinach; Lane 7: 1×10^7 yeast cells; and Lane 8: 2×10^9 bacteria cells.

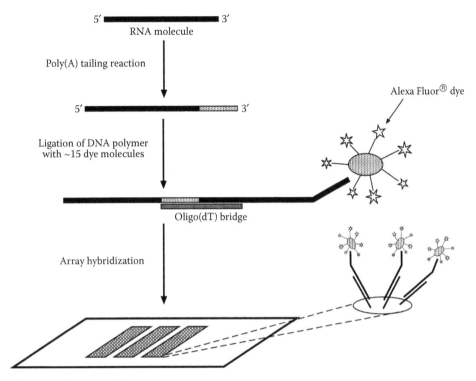

FIGURE 15.5 NCode miRNA Labeling System workflow.

15.3.2 LABELING miRNAs

Given their small size and lack of poly A tails, miRNAs are not effectively labeled by traditional procedures such as random-primed or oligo(dT)-primed cDNA synthesis. Alternative methods involving the direct addition of modified nucleotides to endogenous miRNAs through poly A "tailing" or direct ligation have been reported. However, these methods suffer from low and/or variable analyte incorporation and labeling. The NCode™ Rapid miRNA Labeling System (Invitrogen) provides a method for labeling endogenous miRNAs with fluorescent tags that results in sensitive profiling of expression patterns from small amounts of starting material (Figure 15.5).

This system can be used to label any RNA sample, including total RNA, enriched low-molecular-weight RNA, or amplified senseRNA. Using the system, a poly(A) tail is added to the RNA using poly A polymerase and an optimized reaction buffer. Then a DNA polymer labeled with multiple dye molecules is ligated to each tailed RNA using a bridging oligo. This polymer is composed of branched DNA "dendrimers," each with a core that consists of a matrix of double-stranded DNA and an outer surface composed of singled-stranded "arms." The surface arms carry ~15 Alexa Fluor 3 or Alexa Fluor 5 dye molecules.

The miRNAs in the labeled sample are hybridized to a microarray spotted with species-specific antisense miRNA probes, and then scanned using a standard microarray scanner. The high specificity of the probes and high fluorescence of the dye molecules result in strong signal-to-background ratios and signal correlations (Figure 15.6).

15.3.3 MICROARRAY PROBE CONTENT

Strong conservation between miRNA family members makes it difficult to design oligo probes that are specific at the level of 1–2 nucleotides out of an 18–25-nucleotide sequence. Researchers at

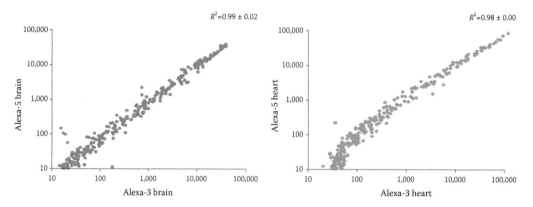

FIGURE 15.6 Reproducibility of labeling. Linear correlation of mouse brain and heart homotypic hybridizations. Data points represent mean normalized data of triplicate arrays. Mouse heart and brain miRNA were enriched using the PureLink miRNA Isolation Kit and labeled with the NCode miRNA Labeling System. Mouse miRNAs were detected on the NCode miRNA array. The mean R^2 value is ≥0.98 for both data sets, showing a high correlation between channels with minimal variability across arrays.

Rutgers University have developed a probe sequence design algorithm that balances normalized melting temperatures (T_m) for uniform hybridization with high specificity for discerning between closely related miRNAs [28]. First, mature miRNA sequences from the miRBase Sequence Database (http://microrna.sanger.ac.uk) are used to generate reverse-complement probe sequences. These are then trimmed to balance the melting temperatures of the probes, using an algorithm that calculates which end of the miRNA should be truncated for the most precise adjustment of the T_m. Because

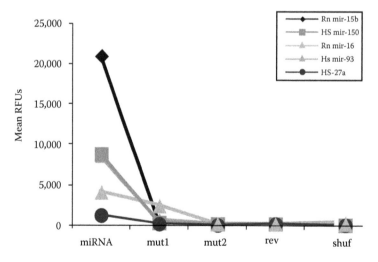

FIGURE 15.7 Specificity of probe design. Discrimination between perfect match and mismatch probes. Data points represent the mean relative fluorescence units (RFUs) ($n = 6$) from NCode miRNA arrays hybridized with mouse heart miRNA labeled with the NCode miRNA Labeling System. For a subset of miRNAs, probes were designed to evaluate the specificity of the NCode miRNA Detection System. These include probes with a single (mut1) or double (mut2) mismatch, the reverse complement (rev), and a probe in which the sequence was randomly shuffled (shuf). The data illustrate that the system is consistently able to discriminate between sequences with a double mismatch, and often between the perfect match and probes with a single mismatch, across a broad dynamic range of perfect match intensities. The reverse and shuffle probes are always negative.

studies have shown that the 5′ region of miRNAs are conserved among miRNA family members, truncation from the 5′ end is favored to preserve the more variable 3′ end of the sequence.

In analyzing the performance of the probes on microarrays, multiple negative control probes were created for each species, with a 1-nucleotide mismatch, 2-nucleotide mismatch, random sequence, shuffled sequence, and monomer generated for each selected control (Figure 15.7). The probes were found to be highly specific, even distinguishing between single-mismatch sequences. Finally, monomer, dimer, and trimer probe sequences were tested for their affinity to epoxy groups on glass slides, and dimers were selected as the final format.

The probe designs described by Goff et al. are available commercially as part of the NCode family of products (Invitrogen). They are preprinted on the NCode Multi-Species miRNA Microarray, a Corning® Epoxide-Coated Glass Slide printed with all the unique probes from miRBase for human, mouse, rat, *D. melanogaster*, *C. elegans*, and Zebrafish miRNAs, as well as predicted human miRNA sequences. NCode Mammalian and Non-Mammalian miRNA Microarray Probe Sets provide the same oligo probes in 384-well plates for self-printing on standard DNA microarray surfaces.

The NCode microarray platform is sensitive to sub-femtamolar levels of miRNAs using conventional static hybridization methods. Sensitivity can be increased to attamolar levels using active hybridization methods like the MAUI® Hybridization System (BioMicro Systems). Figure 15.8 demonstrates the overall performance of the platform using a universal RNA reference standard.

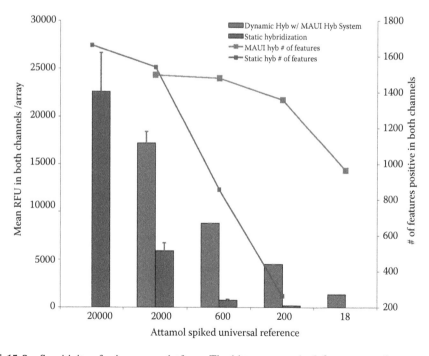

FIGURE 15.8 Sensitivity of microarray platform. The histogram to the left compares the mean signal in both channels (A3 and A5) for all mammalian features from duplicate samples spiked with decreasing quantities of a universal reference RNA. Error bars represent one standard deviation between mean signal intensities for both channels across replicate arrays. Additionally, the figure compares the number of features positive in both A3 and A5 channels on both arrays versus the spike quantity. Universal reference samples were diluted from 20 fmol to 18 amol and labeled with the NCode miRNA Labeling System. Samples were hybridized to an NCode Multi-Species miRNA Microarray. Data were background corrected. Mean intensities were averaged for all mammalian features. The number of positive features (second Y-axis) was defined as the number of features with intensities greater than 5× median background intensity in both channels.

Using a static hybridization method as described in the NCode miRNA Labeling System manual, >50% of the universal reference miRNA molecules were detectable (>5X median background) at concentrations as low as 0.6 fmol per reaction, which corresponds to 100 copies per cell for a given miRNA. Using the MAUI Hybridization System, >90% of the universal reference miRNAs were detected from the same spike quantity, whereas >50% were detectable when spiked at concentrations as low as 0.018 fmol per reaction.

Figure 15.9 demonstrates the reproducibility of the microarray platform using 17 replicate homotypic arrays with 10 µg of mouse heart RNA in each channel. The $\log_2$ %cv for both channels was ~7.7% using static hybridization. The array-to-array reproducibility improved to 3.5% using active hybridization.

15.3.4 Amplification of miRNAs in Small Samples

The biggest limitation of most microarray platforms is the quantity of starting material, typically 1–10 µg of total RNA. In many cases, it is difficult or impossible to obtain this amount of RNA, especially from laser-capture microdissections, fine-needle aspirates, and cell-sorted samples. Here, we describe a linear miRNA amplification system for amplifying "sense RNA" molecules from minute quantities of purified miRNA to generate sufficient amounts of material for downstream research.

The NCode miRNA Amplification System (Invitrogen) employs modifications to previously documented mRNA amplification methods, enabling microarray profiling from <50 ng of total RNA [29]. The resulting amplified miRNA is in the sense orientation, for direct compatibility with miRNA probe sequences on microarrays. Data indicate that the amplified sense RNA preserves the relative abundance of the different miRNA sequences in the original sample, allowing for the comparison of relative quantities across experiments.

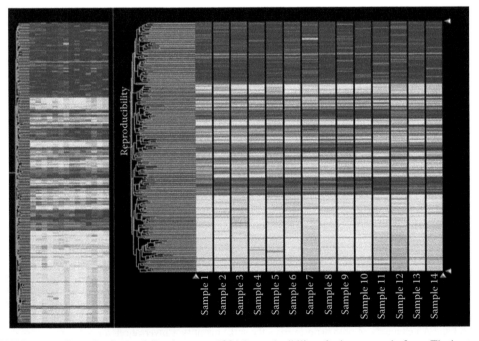

FIGURE 15.9 **(See color insert following page 138.)** Reproducibility of microarray platform. The heat map and table to the right demonstrate the precision of the NCode miRNA analysis platform comparing 18 replicate homotypic arrays of 10 µg of pooled human RNA. Data were background corrected and normalized. The %cv calculated from mean $\log_2$ expression values from one array to another.

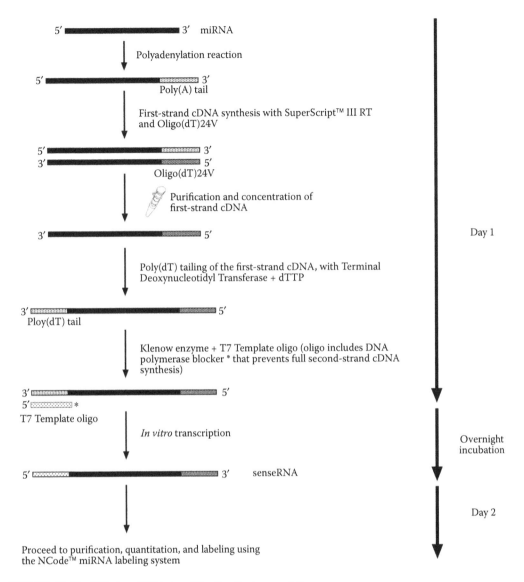

FIGURE 15.10 NCode miRNA Amplification System workflow.

Using the amplification system, small RNA enriched from 300 to 500 ng of total RNA is poly-adenylated using poly A polymerase, and first-strand cDNA is synthesized from the tailed miRNAs using SuperScript™ III Reverse Transcriptase (RT). A poly(dT) tail is added to the 3′ end of the first-strand product using terminal deoxynucleotidyl transferase, and a T7 promoter is synthesized and annealed on the tailed cDNA using Klenow enzyme and a specially designed T7 template oligo. An *in vitro* transcription reaction with an overnight incubation is performed to generate the amplified senseRNA, which is ready for microarray labeling and detection (Figure 15.10).

The amplification procedure routinely yields ~30 μg of amplified sense RNA from as little as 15 ng of enriched small RNA. The NCode microarray platform requires 1.5 μg of amplified miRNA, allowing for many technical replicates from a single amplification reaction. To measure the accuracy of amplification, small quantities of miRNA have been amplified and then labeled and hybridized on microarrays alongside much larger quantities of unamplified miRNA from the same sample.

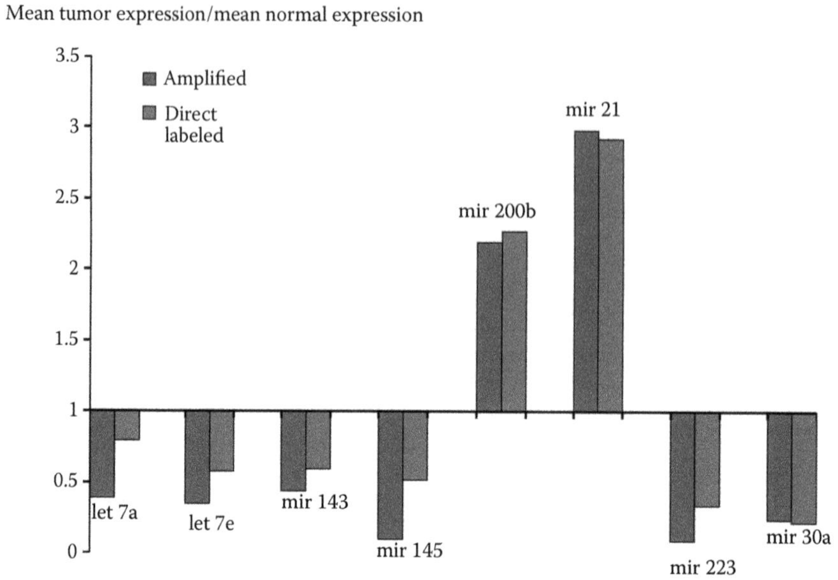

Mean tumor expression/mean normal expression

FIGURE 15.11 Relative abundance of specific amplified and unamplified miRNAs. Total RNA from large cell carcinoma and adjacent normal tissue from three patients was enriched for miRNA using the PureLink miRNA Isolation Kit. Enriched miRNA equivalent to 300 ng of total RNA from both tumor and normal samples was amplified using the NCode miRNA Amplification System, and amplified samples were labeled for array analysis using the NCode miRNA Labeling System. Additionally, enriched miRNA equivalent to 10 μg of total RNA was labeled in the same manner. Tumor samples and corresponding adjacent normal tissue were analyzed on the NCode Multi-Species miRNA Microarray, and the array data were background corrected, normalized, and averaged across patients. The ratios of all known human miRNAs in the two types of tissues were calculated, and statistically significant differentially expressed miRNAs were validated by qRT-PCR.

The r^2 value of spots from the two volumes is typically >0.8, demonstrating a high level of amplification accuracy.

Figure 15.11 demonstrates the efficiency of the amplification system. Three hundred nanograms of total RNA from large-cell carcinoma tissue and adjacent normal tissue from three separate patients was enriched and amplified. The amplified samples were compared to 10 μg of direct-labeled total RNA from the same samples by microarray analysis. Several differentially expressed miRNAs associated with the disease state were identified in both amplified and direct-labeled samples.

15.4 VALIDATION OF MICROARRAY RESULTS

Quantitative RT-PCR (qRT-PCR) has become the standard for microarray data validation, as well as a tool for profiling subsets of miRNAs with high sensitivity. Most commercially available qRT-PCR systems for miRNA analysis use proprietary, predesigned miRNA-specific primer sets for reverse transcription. These systems require a publicly available miRNA sequence and a commercial qRT-PCR assay that has been developed for that specific sequence, which limits their use for rare or recently discovered miRNAs. The NCode miRNA SYBR® Green qRT-PCR Kit (Invitrogen) combines poly A tailing of the miRNAs as described previously with a "universal" first-strand cDNA synthesis reaction that does not require a proprietary miRNA-specific primer set.

Following purification and polyadenylation of a miRNA population, SuperScript III RT and a universal RT primer are used to synthesize cDNA from the tailed miRNA. The miRNA-specific

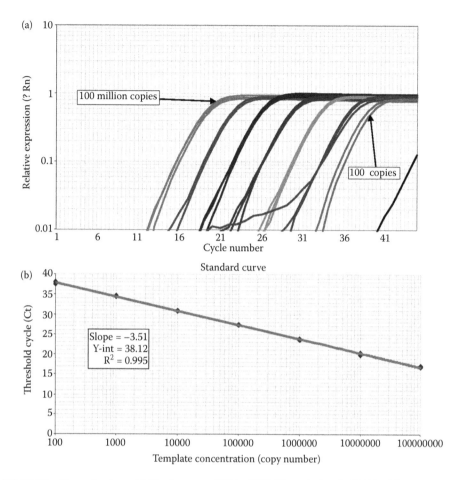

FIGURE 15.12 **(See color insert following page 138.)** qRT-PCR analysis of miRNAs. The cDNA synthesis module of the NCode SYBR Green miRNA qRT-PCR Kit was used to polyadenylate a synthetic miRNA oligo and generate cDNA from the tailed miRNA. The cDNA was then amplified in duplicate qPCR reactions on an ABI PRISM® 7700 using SYBR Green I dye, a universal qPCR primer (provided in the kit), and a forward primer specific for the oligo sequence. The amount of starting material ranged from 1×10^8 to 1×10^2 copies. The assay demonstrated a wide dynamic range of seven orders of magnitude and sensitivity down to 100 copies of template.

amplification occurs during the quantitative polymerase chain reaction (qPCR) reaction, using a PCR primer that can be ordered for any sequence of interest. The miRNA-specific forward primer is paired with a universal reverse primer in a qPCR reaction with Platinum® SYBR Green qPCR SuperMix-UDG, which includes Platinum *Taq* DNA polymerase and SYBR Green fluorescent dye.

The NCode miRNA SYBR Green qRT-PCR Kit has demonstrated a dynamic range of seven logs and sensitivity down to 100 copies per reaction (Figure 15.12). Additionally, the system is able to provide single nucleotide discrimination between closely related family members. It provides an effective validation mechanism for microarray analyses of miRNAs.

ACKNOWLEDGMENTS

The following individuals contributed to the writing and/or data generation for this manuscript: Ranjan Perera, Mark Landers, Sam An, and Mason Brooks.

REFERENCES

1. Pinkel, D., Segraves, R., Sudar, D., Clark, S., Poole, I., Kowbel, D., Collins, C., et al., 1998. High-resolution analysis of DNA copy number variation using comparative genomic hybridization to microarrays. *Nat Genet*, 20, 207–211.
2. Pollack, J. R., Perou, C. M., Alizadeh, A. A., Eisen, M. B., Pergamenschikov, A., Williams, C. F., Jeffrey, S. S., Botstein, D., and Brown, P. O., 1999. Genome-wide analysis of DNA copy-number changes using cDNA microarrays. *Nat Genet*, 23, 41–46.
3. Calin, G. A., Sevignani, C., Dumitru, C. A., Hyslop, T., Noch, E., Yendamuri, S., Shimizu, M., et al., 2004. Human microRNA genes are frequently located at fragile sites and genomic regions involved in cancers. *Proc Natl Acad Sci USA*, 101, 2999–3004.
4. Zhang, L., Huang, J., Yang, N., Greshock, J., Megraw, M., Giannakakis, A., Liang, S., et al., 2006. MicroRNAs exhibit high frequency genomic alterations in human cancer. *Proc Natl Acad Sci USA*, 103, 9136–9141.
5. Beheshti, B., Braude, I., Marrano, P., Thorner, P., Zielenska, M., and Squire, J. A., 2003. Chromosomal localization of DNA amplifications in neuroblastoma tumors using cDNA microarray comparative genomic hybridization. *Neoplasia*, 5, 53–62.
6. Hodgson, G., Hager, J. H., Volik, S., Hariono, S., Wernick, M., Moore, D., Nowak, N., et al., 2001. Genome scanning with array CGH delineates regional alterations in mouse islet carcinomas. *Nat Genet*, 29, 459–464.
7. Cai, W. W., Mao, J. H., Chow, C. W., Damani, S., Balmain, A., and Bradley, A., 2002. Genome-wide detection of chromosomal imbalances in tumors using BAC microarrays. *Nat Biotechnol*, 20, 393–396.
8. Pinkel, D. and Albertson, D. G. (2006). Comparative genomic hybridization. *Annu Rev Genomics Hum Genet*, 6, 331–354.
9. Pollack, J. R. and Iyer, V. R., 2002. Characterizing the physical genome. *Nat Genet*, 32(Suppl.), 515–521.
10. Snijders, A. M., Nowak, N., Segraves, R., Blackwood, S., Brown, N., Conroy, J., Hamilton, G., et al., 2001. Assembly of microarrays for genome-wide measurement of DNA copy number. *Nat Genet*, 29, 263–264.
11. Lieu, P. T., Jozsi, P., Gilles, P., and Peterson, T., 2005. Development of a DNA-labeling system for array-based comparative genomic hybridization. *J Biomol Tech*, 16, 104–111.
12. Derbyshire, V., Freemont, P. S., Sanderson, M. R., Beese, L., Friedman, J. M., Joyce, C. M., and Steitz, T. A., 1988. Genetic and crystallographic studies of the 3′, 5′-exonucleolytic site of DNA polymerase I. *Science*, 240, 199–201.
13. Britten, R. J., Graham, D. E., and Neufeld, B. R., 1974. Analysis of repeating DNA sequences by reassociation. *Methods Enzymol*, 29, 363–418.
14. Weiner, A. M., Deininger, P. L., and Efstratiadis, A., 1986. Nonviral retroposons: genes, pseudogenes, and transposable elements generated by the reverse flow of genetic information. *Annu Rev Biochem*, 55, 631–661.
15. Landegent, J. E., Jansen in de Wal, N., Dirks, R. W., Baao, F., and van der Ploeg, M., 1987. Use of whole cosmid cloned genomic sequences for chromosomal localization by non-radioactive in situ hybridization. *Hum Genet*, 77, 366–370.
16. Lengauer, C., Riethman, H., and Cremer, T., 1990. Painting of human chromosomes with probes generated from hybrid cell lines by PCR with Alu and L1 primers. *Hum Genet*, 86, 1–6.
17. Lichter, P., Cremer, T., Borden, J., Manuelidis, L., and Ward, D. C., 1988. Delineation of individual human chromosomes in metaphase and interphase cells by in situ suppression hybridization using recombinant DNA libraries. *Hum Genet*, 80, 224–234.
18. MAQC Consortium, Shi, L., Ried, L. H., Jones, W. D., Shippy, R., Warrington J. A., Baker, S. C., et al., 2006. The MicroArray Quality Control (MAQC) project shows inter- and intraplatform reproducibility of gene expression measurements. *Nat Biotechnol*, 24, 1151–1161.
19. Bartel, D. P. (2004). MicroRNAs: Genomics, biogenesis, mechanism, and function. *Cell*, 116, 281–297.
20. Olsen, P. H. and Ambros, V., 1999. The lin-4 regulatory RNA controls developmental timing in *Caenorhabditis elegans* by blocking LIN-14 protein synthesis after the initiation of translation. *Dev Biol*, 216, 671–680.
21. Ambros, V., 2003. MicroRNA pathways in flies and worms: Growth, death, fat, stress, and timing. *Cell*, 113, 673–676.
22. Bentwich, I., Avniel, A., Karov, Y., Aharonov, R., Gilad, S., Barad, O., Barzilai, A., et al., 2005. Identification of hundreds of conserved and nonconserved human microRNAs. *Nat Genet*, 37, 766–770.

23. Lim, L. P., Lau, N. C., Weinstein, E. G., Abdelhakim, A., Yekta, S., Rhoades, M. W., Burge, C. B., and Bartel, D. P., 2003. The microRNAs of *Caenorhabditis elegans. Genes Dev*, 17, 991–1008.

24. Lagos-Quintana, M., Rauhut, R., Lendeckel, W., and Tuschl, T., 2001. Identification of novel genes coding for small expressed RNAs. *Science*, 294, 853–858.

25. Nakahara, K. and Carthew, R. W., 2004. Expanding roles for miRNAs and siRNAs in cell regulation. *Curr Opin Cell Biol*, 16, 127–133.

26. Stark, A., Brennecke, J., Russell, R. B., and Cohen, S. M., 2003. Identification of drosophila microRNA targets. *PLoS Biol*, 1, E60.

27. John, B., Enright, A. J., Aravin, A., Tuschl, T., Sander, C., and Marks, D. S., 2004. Human microRNA targets. *PLoS Biol*, 2, e363.

28. Goff, L. A., Yang, M., Bowers, J., Getts, R. C., Padgett, R. W., and Hart, R. P., 2005. Rational probe optimization and enhanced detection strategy for microRNAs using microarrays. *RNA Biol*, 2(3), 93–100.

29. Mattie, M. D., Benz, C. C., Bowers, J., Sensinger, K., Wong, L., Scott, G. K., Fedele, V., Ginzinger, D., Getts, R., and Haqq, C., 2006. Optimized high-throughput microRNA expression profiling provides novel biomarker assessment of clinical prostate and breast cancer biopsies. *Mol Cancer*, 5, 24.

16 Development of an Oligonucleotide Microarray for Mutational Analysis Using Single or Multiple Sample Hybridization

Il-Jin Kim, Hio Chung Kang, and Jae-Gahb Park

CONTENTS

16.1   INTRODUCTION

In the first edition of this book, we provided detailed protocols for fabricating oligonucleotide microarrays for mutational analysis [1]. Here, in the second edition, we provide (i) improved protocols developed during the last three years; and (ii) protocols for multiple sample hybridization with competitive deoxyribonucleic acid (DNA) hybridization (CDH) along with single sample hybridization. The information provided in this chapter should allow researchers to choose between single and multiple sample hybridization with CDH.

16.1.1   Oligonucleotide Microarrays for Mutational Analysis

Microarrays for the analysis of gene expression have been mainly developed for laboratory research and commercial use. There are three major applications of gene expression analysis using oligonucleotide or complementary DNA (cDNA) microarrays: class comparison, class prediction, and class discovery [2,3]. Although classification of cancers has improved over the past 30 years, there has not been a general approach for identifying new classes of cancer (class discovery) or for assigning tumors to known classes (class prediction). Expression profiles obtained from diseased tissue using high-density microarrays have been used to predict the biological group, diagnostic category, or prognostic stage of a patient [4].

Single nucleotide polymorphism (SNP) arrays are often used for genotyping, linkage analysis, association studies, and analysis of changes in copy number [5,6]. Other types of oligonucleotide microarrays are used for detecting disease-associated mutations (e.g., in cancer, diabetes mellitus, and other hereditary diseases), but they have not been sufficiently developed compared to other types of microarrays. The development of oligonucleotide microarrays for the analysis of mutations has been slow owing to several problems. First, it is difficult to find causing genes with high penetrance in many human diseases such as cancer or hereditary syndromes because these conditions are caused by complicated or multiple genetic changes or because the mutations cause one typical phenotype [7]. Second, when the disease-causing mutations occur throughout the gene, it is difficult to detect all possible mutations. Since human cancers may have many types of mutations, including missense, nonsense, and frameshift mutations, aberrant splicing, large genomic rearrangement, and imbalances in allelic expression [8,9,10], it is difficult to design oligonucleotide microarrays that can detect all possible mutations on a single chip. For this reason, the diagnostic chips used for mutational analysis have low sensitivity. Third, the developed oligonucleotide microarrays have not been sufficiently evaluated with regard to cost effectiveness. Although the direct experimental cost of oligonucleotide microarrays is less than those of other techniques, such as direct sequencing, additional methods must be employed if oligonucleotide microarrays are unable to detect all possible mutations.

The Affymetrix GeneChip™, a representative and successful oligonucleotide microarray, uses glass wafers on which tens of thousands of short oligonucleotides are synthesized *in situ*, one nucleotide at a time, using a modification of semiconductor photolithography [11]. At least four detection arrays (*p53*, HIV, CYP450, and *BRCA1*) have been reported, and *p53* arrays are used by many cancer research groups to detect *p53* mutations [12]. The Affymetrix P450 GeneChip is used for pharmacogenetic screening. This GeneChip is an efficient and reliable tool for testing five alleles of the *CYP2D6* gene [13].

Owing to the factors outlined above, these GeneChips are used less often for mutational analysis than other GeneChips that are used for the analysis of gene expression and SNPs. Thus, many groups have attempted to produce small-scale microarrays for detecting hot-spot mutations, for example in K-*ras* [14], *RET* [15], β-catenin [16], and human papillomavirus (HPV) typing microarrays [17]. The *RET* oligonucleotide microarray can detect more than 95% of *RET* mutations in multiple endocrine neoplasia type 2A and 2B syndromes and familial medullary thyroid carcinoma [15]. The K-*ras* and β-catenin oligonucleotide microarrays were developed for detecting hot-spot mutations in various types of human cancers, especially colorectal cancers [14,16].

Although oligonucleotide microarrays must be improved for clinical use, it is certain that they will be key diagnostic tools in the near future.

16.1.2 PRINCIPLE OF CDH

In the case of GeneChips, only one sample per chip is used for the analysis of both gene expression and SNPs or mutations [5,6,11,12]. In contrast, two samples are used per microarray in the cDNA microarray format that is widely used for analysis of gene expression [18]. The cDNA microarrays contain relatively long DNA molecules immobilized on a solid surface and are mostly used for large-scale screening and expression studies. Samples are prepared by reverse transcribing ribonucleic acid (RNA) into cDNA, which is then fluorescently or radioactively labeled and used to probe a predetermined set of DNAs. Two different fluorescent dyes, usually Cy5 and Cy3, are used for cDNA microarray analysis, and a typical analysis consists of a normal tissue (control) sample labeled with a green dye and a cancer tissue (test) sample labeled with a red dye. A scanner is used to detect the fluorescence, and if both samples bind to the same target, a yellow signal is obtained [19]. Furthermore, the higher the concentration of the RNA within the original sample, the greater the hybridization signal from probe-DNA binding [20] (Figure 16.1).

For the analysis of mutations, only one sample can be analyzed per microarray. Although multiple samples can be pooled, they should be further analyzed by microarray or other methods. Thus, conventional oligonucleotide microarray analysis using one-sample hybridization does not seem to be overwhelming in the sample throughput scale to other genotyping methods, such as direct sequencing, denaturing high-performance liquid chromatography (DHPLC), and real-time polymerase chain reaction (PCR). Automatic sequencing and DHPLC are becoming more powerful and can be adapted to 96-well-based high-throughput sampling [21,22].

Development of genotyping oligonucleotide microarrays that can analyze multiple samples in a single microarray will dramatically improve the sample throughput, time required, and cost of mutation analysis. We have recently developed CDH as a method for the analysis of multiple samples in oligonucleotide microarrays [14,23]. CDH allows the analysis of several samples in a single oligonucleotide microarray without the need for further sequence confirmation or other microarray experiments. CDH was first devised to increase the sensitivity and specificity of oligonucleotide microarrays to detect mutations [14]. As the different oligonucleotide probes spotted on the slide have different abilities to hybridize their targets, some oligonucleotides give unexpectedly strong or weak signals. This can be one of the main causes of nonspecific signals and low sensitivity and specificity in oligonucleotide microarray experiments.

Nonspecific signals and low sensitivity and specificity of oligonucleotide microarrays can be resolved in several ways. First, the length and concentration of oligonucleotides can be reduced or

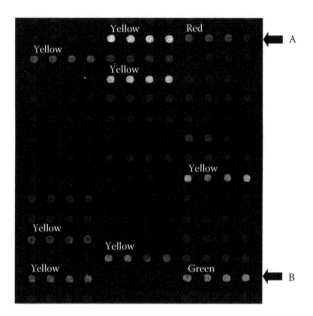

FIGURE 16.1 **(See color insert following page 138.)** Cy3- and Cy5-labeled samples were used for CDH and then visualized together as in cDNA microarray analysis. DNA from sample A with a mutation in position A was labeled with Cy5, and DNA from sample B with a mutation in position B was labeled with Cy3. When the microarray was analyzed with a ScanArray 5000 (PerkinElmer, Boston, Massachusetts, USA), the signal for mutant A appeared red, that of B appeared green, and the wild-type signals shared by the two samples appeared yellow. This cDNA-like image provides a rapid and simple method for detecting mutations.

increased, and a mismatch sequence can be inserted at specific locations to reduce very strong signals. Second, during data analysis, it is possible to use a normalization factor to adjust signals that are too low or high. These methods, however, do not provide permanent solutions or are limited to certain oligonucleotides or microarrays. For new oligonucleotide microarrays, each oligonucleotide should be carefully inspected and the normalization factor for specific codons or exons should be calculated or adjusted. Development of a generalized method for controlling very high or low signals and for reducing nonspecific signals can allow the efficient fabrication of a high-quality oligonucleotide microarray.

During our development and use of microarrays, we have discovered an important phenomenon: nonspecific signals, which are the main problem for the development of oligonucleotide microarrays, are always observed in specific regions (codons or exons). It is expected that nonspecific signals occur randomly according to the type of samples, hybridization conditions, and other experimental conditions; however, we have found that if oligonucleotide microarrays are correctly designed, fabricated, and analyzed, it is possible to observe that nonspecific results do not occur randomly in most samples.

We originally developed CDH to eliminate these nonspecific results by multiple sample hybridization. In CDH, several samples are applied together to the oligonucleotide microarrays, similar to cDNA microarray hybridization methods that require competition between two samples. If three different samples are labeled with different fluorescent dyes and hybridized together to a single oligonucleotide microarray, the DNA fragments in the three samples compete with each other to bind to the spotted oligonucleotides. All three samples are expected to bind wild-type or control spots and show the same nonspecific signals. Theoretically, signals from these spots will be reduced by one-third [23]. As all three samples rarely have the same mutations in real genetic screening, the signal from the mutation will rarely be shared and will not be reduced in a certain sample or samples. As a result, the signal for the mutations becomes threefold more prominent than the wild-type

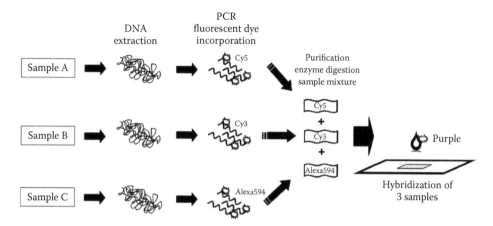

FIGURE 16.2 Principle and procedure of CDH.

or nonspecific spots. Thus, CDH is more effective at detecting mutations than conventional single sample hybridization methods.

The principle of CDH is shown in Figure 16.2. If three samples harbor the same mutation, which is very rare, the mutant spot is also shared by the three samples, and CDH is not predicted to have an increased ability to detect the mutation [23]. In this case, the result will be the same as that obtained using conventional single sample hybridization, which we define as non-CDH to distinguish it from CDH. Detailed protocols for non-CDH and CDH are described in the following sections [23].

CDH can increase not only the sample throughput but also the sensitivity and specificity of detecting mutations using oligonucleotide microarrays. Since multiple samples can be analyzed using a single oligonucleotide microarray, the cost and time needed are reduced according to the sample size. The sample size in CDH is limited by the sensitivity of the fluorescence scanner. Ten different samples may be analyzed in a single oligonucleotide micorarray if a commercial fluorescent scanner can detect 10 different fluorescent dyes. Using CDH, better results are obtained with more samples for hybridization [23]. CDH is expected to improve the sample throughput, cost effectiveness, and speed of different types of microarrays. Furthermore, because the CDH protocol is not much different from conventional hybridization, it can be easily applied to many types of oligonucleotide microarrays.

16.2 OLIGONUCLEOTIDE FABRICATION

16.2.1 OLIGONUCLEOTIDE DESIGN, SYNTHESIS, AND QUALITY CONTROL

Accurate fabrication of oligonucleotides is critical for the development of diagnostic oligonucleotide microarrays. Furthermore, oligonucleotides should be properly designed, synthesized, and stored. In general, designed oligonucleotide sequences are sent to a commercial laboratory for synthesis, modification, and purification. Oligonucleotides are usually modified to bind to slides via functional residues on their 5′- or 3′-ends. Amine modification is a common method for binding oligonucleotides to aldehyde-coated slides. Modified oligonucleotides should be purified by high-performance liquid chromatography or similarly sensitive methods. Here, we describe a protocol for using DHPLC to check the status of oligonucleotides. Although oligonucleotides are initially checked for quality, they can be degraded by repetitive freeze-thawing and become unsuitable for the fabrication of microarrays. Therefore, it is best to aliquot oligonucleotides after synthesis and to regularly check their quality by DHPLC.

16.2.1.1 Quality Control of Oligonucleotides by DHPLC (WAVE®)

WAVE is the DHPLC equipment for detecting mutations, SNPs, aberrant methylation, and other genetic changes [21,24], and is widely used for various genetic screens. DHPLC is ion-pair reversed-phase liquid chromatography using specific chemicals such as triethylammonium acetate [24], which binds to DNA and interacts with the hydrophobic column. Mutation screening or SNP analysis is performed under partially denaturing conditions. When boiled and slowly cooled, mutated oligonucleotides form hetero- and homo-duplexes. The hetero-duplexes elute from the column before homo-duplexes, allowing simple screening for mutations.

In addition, oligonucleotides can be checked or purified using WAVE. In the first edition of this book, we explained that the ultraviolet (UV) detector for DHPLC can be used to check the quality of oligonucleotides [1]. Moreover, a fluorescence detector (WAVE High Sensitivity Detection (HSD) Module) has a much higher sensitivity than the UV detector. The general protocol for using the HSD fluorescence detector is similar to that for the UV detector, but much lower amounts of oligonucleotides can be analyzed with much higher sensitivity with the HSD system. The main advantage of WAVE for checking the status of oligonucleotides is that quantity and quality can be analyzed at the same time. In the case of a correctly synthesized oligonucleotide, the chromatogram will show a single main peak, whereas for incorrectly synthesized or purified oligonucleotides, the chromatogram will show multiple peaks (Figure 16.3). The quantity of oligonucleotide can also be checked based on the peak height. Although samples are reported to contain fixed quantities of oligonucleotide, they tend to show significantly different peak heights. Since these differences in quantities can cause nonspecific signals, it is helpful to adjust the quantity according to the measured peak height. As the synthesized oligonucleotides are usually of the same length, it is easy to use the peak patterns to check both the quality and quantity of oligonucleotides. When short oligonucleotides or 6-carbon spacers are provided instead of 12-carbon spacers, early elution of peaks indicates incorrect

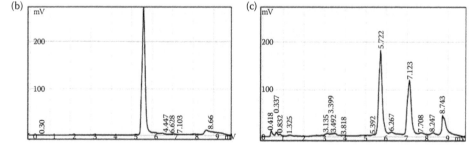

FIGURE 16.3 Analysis of oligonucleotide quality using WAVE HSD. (a) The WAVE HSD system; (b) correctly synthesized and purified oligonucleotides show a large single peak in the middle of the chromatogram; and (c) chromatograms for poor quality oligonucleotides show multiple peaks.

oligonucleotide synthesis. It is also helpful to use matrix-assisted laser desorption/ionization time-of-flight mass spectrometry (MALDI-TOF MS) to check the mass of the synthesized oligonucleotides for the purpose of obtaining detailed information on oligonucleotide quality.

16.2.2 PROTOCOL 1: OLIGONUCLEOTIDE FABRICATION

1. Oligonucleotide design and synthesis
   - Oligonucleotide length: 23 bp
   - Modification: 5′-amine
   - Spacers: 12-carbon
   - HPLC-purified oligonucleotides (Metabion, Martinsried, Germany)
2. Analysis of oligonucleotide quality check by using WAVE HSD
   - Open the Navigator™ software
   - Select "oligo purification" and "50 TO 80 BASES NO FLUORESCENT TAQ"
   - Load 4 μL (10 pmol/μL) of each oligonucleotide at 80°C
   - Set the WAVE buffer conditions as follows:

Step	Time	% A	% B	% D
Loading	0	82	18	
Start gradient	0.1	82	18	
Stop gradient	7.1	52	48	
Start clean	7.2	0	100	0
Stop clean	7.7	0	100	0
Start equilibrate	7.8	82	18	
Stop equilibrate	8.3	82	18	

   - Click the "run samples"
3. Oligonucleotide storage
   - Aliquot the oligonucleotides
   - Store oligonucleotides in a deep freezer (−70°C).

16.3 OLIGONUCLEOTIDE MICROARRAY FABRICATION (SPOTTING PROCEDURE)

16.3.1 OLIGONUCLEOTIDE SPOTTING

It is important to select the appropriate type of slide for spotting the oligonucleotide probes. As described in Section 16.2, most oligonucleotides can be modified with residues, for example amines, that can be linked to other functional residues on the slide. Amine-modified oligonucleotides can be linked to aldehyde- or epoxy-coated slides, and both amine-modified and unmodified long oligonucleotides and cDNAs can be linked to epoxy-coated slides. Both types of slides are commercially available. Aldehyde residues on the slides react with amine residues in the oligonucleotides by Schiff's base formation and create covalent attachment between the slide and the oligonucleotide [1,25]. The chemical binding between the oligonucleotides and the slides should be tight enough to withstand pretreatment (blocking), hybridization, and vigorous washing [1].

A variety of commercially available solutions can be used for spotting the oligonucleotides. Donut-shaped or distorted spots can occur due to an inadequate concentration of oligonucleotides or inappropriate spotting buffer or spotting pin selection. Spotting pins should also be selected according to the numbers of slides that will be spotted and the microarray types.

Preliminary spotting experiments must be carried out under standard experimental conditions. When many slides are spotted simultaneously, the temperature and humidity should be controlled to prevent drying of the oligonucleotides. This can be done using a cooling nest and humidifier in the microarrayer (spotter). Drying of the oligonucleotides can cause irregular or nonspecific signals.

16.3.2 PROTOCOL 2: SPOTTING PROCEDURE

1. Mix 10 μL of oligonucleotide (80 pmol/μL) with 10 μL of 2X ArrayIt™ spotting solution (TeleChem, Sunnyvale, California, USA)
2. Load a 384-well plate (ABgene, Epsom, UK) with 20–30 μL per well of oligonucleotides in spotting solution
3. Seal the plate with book tape 845 (3M, St. Paul, USA) and vortex for 10 min at room temperature
4. Centrifuge the plate at 1400 rpm using a model 5810R centrifuge (Eppendorf, Hamburg, Germany) for 2 min at room temperature
5. Slide (substrate) selection
   - CSS-100 silylated slides (aldehyde; CEL Associates, TeleChem, Sunnyvale, California, USA)
   - SuperAldehyde 2 substrates (TeleChem, Sunnyvale, California, USA)
   - SuperEpoxy 2 substrates (TeleChem, Sunnyvale, California, USA)
6. Set the humidity to 60% in a microarrayer (Cartesian MicroSys 5100; Genomic Solutions, Ann Arbor, Michigan, USA) and set the cooling nest temperature around the plate to 8°C to prevent evaporation
7. Check the status of the pins (Stealth pin; TeleChem, Sunnyvale, California, USA) by microscopy, and wash out the pins three times
8. Example of spot characteristics using stealth pins
   - Prespotting: 10 times
   - Spot spacing: 300 μm
   - Spot diameter: 130 μm
9. Store the spotted slides at 4°C until use.

16.4 SAMPLE PREPARATION (NON-CDH VERSUS CDH)

Conventional single sample hybridization (non-CDH) and multiple sample hybridization using CDH differ mainly in the sample preparation steps; there are few differences in the design, synthesis, quality control of oligonucleotides, spotting, hybridization with prepared samples, washing, or data analysis. Thus, sample preparation is the distinguishing aspect of CDH.

16.4.1 SELECTION OF FLUORESCENT DYES

16.4.1.1 Single Sample Hybridization (Non-CDH)

Only one fluorescent dye is used in conventional single sample hybridization. Which fluorescent dye depends on the labeling method, the preference of the researcher, intensity of the signals, and the lasers present in the scanner. The fluorescent dyes Cy5 and Cy3 are usually selected because they are readily available from commercial sources. Other fluorescent dyes such as Alexa594 and Texas Red can also be used, depending on the experimental conditions. The best way to select the fluorescent dye is to consider the cost effectiveness and experimental adaptability. The high cost of fluorescent dyes can be a problem for diagnostic oligonucleotide microarrays used for mutational analysis because many samples are often analyzed. Cy5 or Cy3 is highly recommended due to their adaptability to microarray experiments.

The protocol for single sample hybridization is described in this section. This protocol uses direct fluorescent dye (Cy5) labeling during PCR amplification. In this method, Cy5 is more efficiently incorporated into the sample DNA than Cy3. In the case of indirect fluorescent dye labeling, for example via biotin–streptavidin, there is no difference in the efficiency of incorporation, and Cy3 can be used for strong signals. Generally, final selection of the fluorescent dye should be made based on preliminary experiments with different types of dyes.

16.4.1.2 Multiple Sample Hybridization (CDH)

CDH analysis requires the use of multiple fluorescent dyes according to the number of samples. The numbers of samples that can be co-hybridized on one slide depends on the number and capacity of the scanner lasers. For example, if a fluorescent scanner has three different lasers, three different samples, each labeled with different fluorescent dyes, then it can be analyzed on a single microarray.

It is very important in CDH to select fluorescent dyes with different emission and excitation wavelengths so that they do not interfere with each other. Three different fluorescent dyes (Cy5, Cy3, and Alexa594) with different excitation wavelengths have been employed for CDH: Cy5 (excitation 649 nm), scanned at 632.8 nm with a red laser; Cy3 (excitation 550 nm), scanned at 543.5 nm with a green laser; and Alexa594 (excitation 590 nm), scanned at 594 nm with a yellow laser [2,23].

16.4.2 SAMPLE AMPLIFICATION AND FLUORESCENT DYE LABELING

16.4.2.1 Single Sample Hybridization (Non-CDH)

This part is a process for PCR amplification and direct labeling with a fluorescent dye (Cy5). There are few differences with conventional PCR, although a custom-made fluorescent deoxyribonucleotide triphosphate (dNTP) mixture is used in place of the standard dNTP mixture. The ratio between fluorescent and unlabeled dNTPs must be adjusted so that sufficient fluorescently labeled amplicons are generated. Since Cy5 dye is usually provided as Cy5-deoxycytidine triphosphate (dCTP), the ratio between Cy5-dCTP and dCTP should be considered when preparing the dNTP mixture. Excessive dCTP produces a dense PCR band that lacks sufficient fluorescent dye incorporation. In contrast, excessive Cy5-dCTP can result in a weak PCR band that does not produce a sufficient hybridization signal. Thus, several ratios of fluorescent and unlabeled dNTPs should be tested. In the protocol presented here, equal amounts of Cy5-dCTP and dCTP are used, which results in the generation of fluorescently labeled PCR amplicons suitable for hybridization. To obtain clean hybridization results, the amplicons should be purified to remove the unincorporated Cy5-dCTP. The purification will be described in a later section.

16.4.2.2 Multiple Sample Hybridization (CDH)

The general PCR procedure is the same as that for single sample hybridization (non-CDH) except that the PCRs are performed using fluorescent dNTP mixtures. If Cy5, Cy3, and Alexa594 are selected for three-sample CDH, three independent PCR steps are required with each dNTP mixture. Individual fluorescent dyes have different molecular sizes and efficiencies in direct labeling, resulting in different PCR band densities. Since Cy5 and Cy3 are linked to dCTP, the ratio between Cy5(3)-dCTP and dCTP in the PCR is critical. Likewise, Alexa594 is linked to dUTP, so that the ratio between Alexa594-dUTP and dTTP is important for the success of amplification and labeling. After checking the PCR bands by 1–3% agarose gel electrophoresis, the PCR amplicons can be purified.

16.4.3 SAMPLE PURIFICATION

16.4.3.1 Single Sample Hybridization (Non-CDH)

To obtain clean results for both non-CDH and CDH, unused or remaining fluorescent dye, PCR buffers, dNTP mixtures, and other reagents should be removed. This purification should be performed in the dark to avoid degradation of the fluorescent dye. Commercial column purification kits (described in the non-CDH method) are more effective than ethanol precipitation.

FIGURE 16.4 The 96-well plate purification system. The PCR Cleanup kit is used to remove excess primers and salts from a PCR mixture and provides sufficient recovery of the DNA fragments. The process is performed using a vacuum manifold. This protocol includes a single filtration step, followed by resuspension and recovery of the sample.

16.4.3.2 Multiple Sample Hybridization (CDH)

As CDH uses multiple sample amplicons with different fluorescent dyes, adequate purification is essential. It is possible to purify individual samples separately or as a pooled sample in one column or well. Although these two methods both work well, it is simpler and more cost- and time-effective to pool several samples before purification. After verifying the presence of the PCR bands in each sample, each PCR amplicon is collected in a tube for purification. Cy5 dye shows blue color (red 632.8 nm excitation laser) and Cy3 shows red color (green 543.5 nm). Alexa954 shows pink color (yellow 594 nm excitation laser). A purple solution is obtained when all three dyes are combined for purification.

A 96-well vacuum manifold system can be used for the purification of large numbers of samples. Although the single-column method described for non-CDH also works well for CDH, the plate purification may be more suitable for CDH using a large number of samples. The mixed purple solution is simply applied to one well of a plate and then vacuum is applied (Figure 16.4). The unused fluorescent dyes and buffers pass through the well and are discarded. Fluorescently labeled PCR amplicons remain in the well and can be eluted with distilled water. The eluted samples are then ready for the next step (enzymatic digestion).

16.4.4 SAMPLE FRAGMENTATION (DNASE I-BASED METHOD)

For both CDH and non-CDH, purified amplicons are fragmented by DNase I enzyme to allow them to more easily bind the spotted oligonucleotides and therefore raise the hybridization efficiency. For CDH, the time of digestion or amount of digestion must be adjusted if more than three samples are mixed. In the case of three samples, an additional 5 min is required for the sufficient fragmentation of the samples.

16.4.5 PROTOCOL 3: SAMPLE PREPARATION

16.4.5.1 Non-CDH

1. Prepare dNTP/fluorescent dye mixture
   - 25 μL of deoxyadenosine triphosphate (dATP), deoxyguanosine triphosphate (dGTP), and deoxythymidine triphosphate (dTTP) at 5 mM each
   - 25 μL of dCTP and Cy5-dCTP (Amersham Pharmacia Biotech Ltd., Piscataway, USA) at 1 mM each
2. Perform PCRs for sample amplification and Cy5-labeling in a volume of 35 μL using a thermocycler (GeneAmp PCR System 9700; Applied Biosystems Inc., Foster City, California, USA)
   - 2 μL containing 100 ng of genomic DNA
   - 10 pmol of each primer

- 1.6 µL of premixed dNTP mixture
- 0.5 unit of Taq polymerase (Qiagen Inc., Valencia, California, USA)
- PCR conditions are as follows: initial denaturation at 94°C for 5 min, followed by 35 cycles of amplification (94°C for 30 s, 50–65°C for 30 s, and 72°C for 1 min), and a final denaturation at 72°C for 7 min

3. Check the quality of the PCR products by gel electrophoresis
4. Purify 35 µL of Cy5-labeled PCR product using a commercial purification kit (Qiagen Inc.) and elute it with 45 µL of distilled water. To cut the PCR product into random fragments, add 0.25 unit of DNase I (Takara, Shiga, Japan) and 4.5 µL of 10X DNase I buffer and incubate at 25°C for 10 min
5. Heat-inactivate at 80°C for 10 min and centrifuge
6. Dry the pellet in a vacuum concentrator (Eppendorf)
7. Before proceeding to the hybridization step, store the dried pellet at −20°C.

16.4.5.2   CDH

1. Prepare a dNTP mixture for each fluorescent dye
   - 25 µL of dATP, dGTP, and dTTP at 5 mM each
   - 25 µL of dCTP and Cy5 or Cy3-dCTP (Amersham Pharmacia Biotech Ltd.) at 1 mM each
   - 25 µL of dCTP, dGTP, and dATP at 5 mM each
   - 25 µL of dTTP and Alexa594-dUTP (Molecular Probes, Eugene, Oregon, USA) at 1 mM each
2. Perform PCRs for each sample with each fluorescent dye (Cy5, Cy3, and Alexa594) in a volume of 35 µL using a thermocycler (GeneAmp PCR System 9700; Applied Biosystems Inc.)
   - 2 µL containing 100 ng of genomic DNA
   - 10 pmol of each primer
   - 1.6 µL of premixed dNTP mixture
   - 0.5 unit of Taq polymerase (Qiagen Inc.)
   - PCR conditions are as follows: initial denaturation at 94°C for 5 min, followed by 35 cycles of amplification (94°C for 30 s, 50–65°C for 30 s, and 72°C for 1 min), and a final denaturation at 72°C for 7 min
3. Check the quality of the PCR products by gel electrophoresis
4. Mix the three samples (solution will be purple)
5. Apply the mixture to one well of a 96-well plate (Montage PCR96 Cleanup Kit; Millipore) on a vacuum manifold (MultiScreen® Resist vacuum manifold; Millipore)
6. Apply vacuum for 45–60 min. Ensure that no solution remains in the wells
7. Apply 45 µL of distilled water to the well and seal the plate with book tape 845 (3M)
8. Carefully shake the plate on a vortex mixer for 10 min
9. Remove 45 µL of the eluted sample and place it in a 1.5 mL microcentrifuge tube
10. To cut the PCR product into random fragments, add 0.25 unit of DNase I (Takara) and 4.5 µL of 10X DNase I buffer and incubate at 25°C for 15 min
11. Heat-inactivate at 80°C for 10 min and centrifuge
12. Dry the pellet completely in a vacuum concentrator
13. Before proceeding to the hybridization step, store the dried pellet at −20°C.

16.5   HYBRIDIZATION AND WASHING

16.5.1   Slide Pretreatment

Hybridization is divided into three steps: slide pretreatment, hybridization, and washing of the hybridized slide. Slide pretreatment is important for fabrication of diagnostic oligonucleotide microarrays. As described above, the amine residues on oligonucleotides and aldehyde residues on the slide form a Schiff base, leading to the formation of a covalent bond. The remaining aldehyde residues that fail to bind oligonucleotides should be inactivated with sodium borohydride in ethanol as part of a slide pretreatment prewashing step. Before the slide is dipped into the sodium borohydride solution, the spotted slide is heated at 90°C to denature the oligonucleotides. Following this pretreatment step, the spotted slide is ready for hybridization with prepared samples.

16.5.2   HYBRIDIZATION

The hybridization steps for non-CDH and CDH are the same because the samples are prepared as a dry pellet in a single tube, after digestion with DNase I. Different oligonucleotides have different melting or hybridization temperatures due to different adenine and thymine (AT) or guanine and cytosine (GC) contents. Therefore, the effect of the hybridization on the melting temperature should be considered. The different melting temperatures of oligonucleotides can be a substantial problem for hybridization and data analysis and should therefore be controlled during oligonucleotide design or selection of the hybridization solution. It is difficult to control this problem during oligonucleotide design because the target sequence (mutation sequence) should be located in the middle of a 23-mer oligonucleotide. Thus, it is more effective to select an appropriate hybridization solution to equalize the melting temperatures of oligonucleotides with different GC contents. Commercial hybridization solutions, for example UniHyb® solution (TeleChem), are available for controlling the melting temperature [1].

The hybridization time for diagnostic oligonucleotide microarrays is much shorter than for other oligonucleotide microarrays or cDNA microarrays for the analysis of gene expression. Approximately 3 h of hybridization is sufficient to obtain data on mutations. Longer hybridization times can increase nonspecific signals more than the signals for mutant or wild-type spots. The hybridization time should be selected based on the preliminary experiments and can be reduced to <2 or 3 h according to the characteristics of the gene.

Pretreated slides and cover slides should be washed and dried at least 1 h before hybridization and stored in a dust-free container. Incomplete drying of the cover slide or pretreated slide can interfere with the hybridization.

16.5.3   WASHING OF HYBRIDIZED SLIDES

Postwashing with detergent-containing solutions is carried out to remove fluorescent debris or unhybridized sample DNA. In addition to specific hybridization, a wash step to eliminate nonspecific binding is essential for obtaining accurate mutation data. Too high a temperature or too low a salt concentration in the washing solution may cause the detachment of specifically hybridized DNA from oligonucleotides.

16.5.4   HYBRIDIZATION AND WASHING PROCEDURE

16.5.4.1   Protocol 4.1: Pretreatment of Slides

1. Wash the spotted slides twice in 500 mL of 0.5% sodium dodecyl sulfate (SDS) with vigorous stirring for 2 min
2. Wash the slides in distilled water with vigorous stirring for 2 min
3. Rinse the slides in 90°C distilled water for 2 min without stirring
4. Leave the slides at room temperature for 1 min
5. Prepare the sodium borohydride solution by dissolving 1.3 g of $NaBH_4$ in 375 mL of phosphate buffered saline (PBS) and add 125 mL of ethanol
6. Wash the slides in the sodium borohydride solution with mild stirring for 5 min
7. Wash the spotted slides twice in 500 mL of 0.5% SDS with vigorous stirring for 1 min
8. Wash the slides in distilled water with vigorous stirring for 1 min

9. Centrifuge the slide in a 50 mL centrifuge tube at 1400 rpm for 2 min
10. Slightly open the cap of a 50 mL tube to allow evaporation of the water at the bottom of the tube.

16.5.4.2 Protocol 4.2: Hybridization

1. Add 4 μL of distilled water and 1 μL of prewarmed (60°C) 5X hybridization solution (ArrayIt; TeleChem) to the prepared target samples.
2. Vortex the samples vigorously.
3. Heat the samples in a water bath at 95°C for 2 min and place them on ice.
4. Wash the cover slides with detergent.
5. Load 3.5 μL of the prepared sample on the prewashed slide.
6. Place a cleaned cover slide over the sample, avoiding bubbles.
7. Place the slide into a 50 mL centrifuge tube using a Kim wipe wetted with 350 μL of distilled water.
8. Wrap the tube with foil and place it in the hybridization chamber at the designed temperature for 3 h.

16.5.4.3 Protocol 4.3: Washing

1. Remove the cover slide by dipping up and down in hybridization solution (5 mL 10% SDS, 10 mL 20X (solution consisting of trisodium citrate and sodium chloride) SSC, and 535 mL distilled water).
2. Wash the slide in hybridization solution (5 mL 10% SDS, 10 mL 20X SSC, and 535 mL distilled water) with stirring in the dark for 15 min.
3. Wash the slide in hybridization solution (2.5 mL 10% SDS, 5.5 mL 20X SSC, and 542 mL distilled water) with stirring in the dark for 15 min.
4. Wash the slide in hybridization solution (3 mL 20X SSC and 547 mL distilled water) with stirring in the dark for 5 min.
5. Wash the slide in distilled water for 1 min.
6. Centrifuge the slide in the 50 mL tube at 1400 rpm for 2 min.

16.6 SCANNING AND DATA ANALYSIS

Scanning and data analysis are the same for non-CDH and CDH except that scans at multiple wavelengths (according to the number of samples) are used for CDH.

16.6.1 SCANNING

16.6.1.1 Single Sample Hybridization (Non-CDH)

Since a single sample labeled with a single fluorescent is hybridized to a single slide, only one scan with the appropriate laser is necessary.

16.6.1.2 Multiple Sample Hybridization

For CDH, different samples are mixed and hybridized together. These samples can be easily distinguished by scanning with different lasers. When Cy5, Cy3, or Alexa594 are used for CDH, red, green, or yellow laser sources, respectively, are required. As these three fluorescent dyes have different excitation wavelengths, there is no risk of confusion in the data analysis. The images obtained by scanning with the different lasers are analyzed in the next step.

16.6.2 DETERMINATION OF MUTATIONS (M/W RATIO)

As discussed in Section 16.1.1, diagnostic oligonucleotide microarrays for detecting mutations are not as well developed as microarrays for assessing gene expression or for SNP genotyping. One major

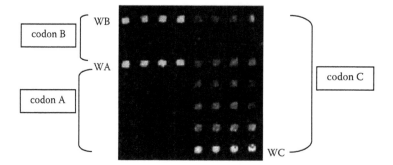

FIGURE 16.5 Spots in a specific area (codon) generally show higher signals than those in other areas (codons). WA, wild-type of codon A; WB, wild-type of codon B; and WC, wild-type of codon C.

reason for this is the lack of appropriate criteria for determining mutations or software for custom-made small-scale oligonucleotide microarrays. If different criteria for determining mutations, genes, hybridization times, and fabrication times are used by different researchers, it will not be possible to generalize oligonucleotide microarrays for the detection of mutations. To resolve this problem, we introduced the mutation/wild-type (M/W) ratio to determine mutations using oligonucleotide microarrays. The M/W ratio for each spot is calculated using the "signal" or "signal-to-noise" values from the Quant Array software. The concept of the M/W ratio is based on codon-by-codon or exon-by-exon analysis. In our previous study, all signal values were calculated in a statistical manner, and any signal of more than the 99% confidence interval was regarded as a mutation. We set the cutoff level for significant signals according to the following formula: cutoff level = mean background + 2.58 (standard deviation of the background) [15]. This method, however, sometimes caused some nonspecific signals to appear specific or caused some true mutations to be missed. Oligonucleotides for specific codons or exons show very high hybridization efficiencies and strong signals. However, some oligonucleotides that hybridize poorly to other codons show very low hybridization efficiencies and relatively lower signals. Thus, all oligonucleotides should not be compared without considering their locations.

In Figure 16.5, spots in codon C area show very strong signals compared to those in codon areas A and B. Thus, the possibility of mutations should be checked codon by codon or specific area by area. Spots with an M/W ratio above the cutoff level (0.5) indicate a true mutation in a codon or exon. The more samples hybridized in CDH, the higher the M/W ratio [23]. For example, we previously reported that the M/W ratio for three-sample CDH (0.97) is threefold higher than that of non-CDH (0.32) [23]. Since the M/W ratio is used for data analysis in its own codon, it can be easily used for other types of oligonucleotide microarrays.

16.6.3 PROTOCOL 5: SCANNING AND DATA ANALYSIS

16.6.3.1 M/W Ratio Determination

1. Prerun the scanner (ScanArray5000; PerkinElmer) and laser device before the slides are scanned.
2. Scan the microarray with the appropriate photomultiplier tube (PMT) values and laser power.
3. Correct the scanned images and quantify the signals using the commercial analysis software (QuantArray, Packard Instrument Co., Meriden, Connecticut, USA).
4. Calculate "signal" or "signal-to-noise" values for every spot using QuantArray.
5. Calculate M/W ratios for each codon.
6. Set an appropriate cutoff level (such as 0.5) after considering statistical significance.
7. Regard any spot over the cutoff level as a mutation.

ACKNOWLEDGMENTS

This work was supported by a research grant 2006 from the National Cancer Center, South Korea and the BK21 project for Medicine, Dentistry, and Pharmacy.

REFERENCES

1. Kim IJ, Kang HC, Park JH, Shin Y, Ku JL, Yoo BC, and Park JG. Development and application of an oligonucleotide microarray for mutational analysis. In: G. Hardiman (ed.), *Microarrays Methods and Applications*, Vol. I, pp. 249–271, 2003. Eagleville, Pennsylvania: DNA Press.
2. Kim IJ, Kang HC, and Park JG. Microarray applications in cancer research. *Cancer Res Treat* 2004; 36: 207–213.
3. Golub TR, Slonim DK, Tamayo P, Huard C, Gaasenbeek M, Mesirov JP, Coller H, et al. Molecular classification of cancer: class discovery and class prediction by gene expression monitoring. *Science* 1999; 286: 531–537.
4. Simon R. Diagnostic and prognostic prediction using gene expression profiles in high-dimensional microarray data. *Br J Cancer* 2003; 89: 1599–1604.
5. Kennedy GC, Matsuzaki H, Dong S, Liu WM, Huang J, Liu G, Su X, et al. Large-scale genotyping of complex DNA. *Nat Biotechnol* 2003; 21: 1233–1237.
6. Sawcer SJ, Maranian M, Singlehurst S, Yeo T, Compston A, Daly MJ, De Jager PL, et al. Enhancing linkage analysis of complex disorders: An evaluation of high-density genotyping. *Hum Mol Genet* 2004; 13: 1943–1949.
7. Vogelstein B, Fearon ER, Hamilton SR, Kern SE, Preisinger AC, Leppert M, Nakamura Y, White R, Smits AM, and Bos JL. Genetic alterations during colorectal-tumor development. *N Engl J Med* 1988; 319: 525–532.
8. Papadopoulos N, Leach FS, Kinzler KW, and Vogelstein B. Monoallelic mutation analysis (MAMA) for identifying germline mutations. *Nat Genet* 1995; 11: 99–102.
9. Renkonen ET, Nieminen P, Abdel-Rahman WM, Moisio AL, Jarvela I, Arte S, Jarvinen HJ, and Peltomaki P. Adenomatous polyposis families that screen APC mutation-negative by conventional methods are genetically heterogeneous. *J Clin Oncol* 2005; 23: 5651–5659.
10. den Dunnen JT and Antonarakis SE. Mutation nomenclature extensions and suggestions to describe complex mutations: a discussion. *Hum Mutat* 2000; 15: 7–12.
11. Macgregor PF and Squire JA. Application of microarrays to the analysis of gene expression in cancer. *Clin Chem* 2002; 48: 1170–1177.
12. Warrington JA, Dee S, and Trulson M. Large-scale genomic analysis using Affymetrix GeneChip® probe arrays. In: M. Schena (ed.), *Microarray Biochip Technology*, pp. 201–220, 2000. Natick, Massachusetts: Eaton Publishing.
13. Chou WH, Yan FX, Robbins-Weilert DK, Ryder TB, Liu WW, Perbost C, Fairchild M, de Leon J, Koch WH, and Wedlund PJ. Comparison of two CYP2D6 genotyping methods and assessment of genotype–phenotype relationships. *Clin Chem* 2003; 49: 542–551.
14. Park JH, Kim IJ, Kang HC, Shin Y, Park HW, Jang SG, Ku JL, Lim SB, Jeong SY, and Park JG. Oligonucleotide microarray-based mutation detection of the K-ras gene in colorectal cancers with use of competitive DNA hybridization. *Clin Chem* 2004; 50: 1688–1691.
15. Kim IJ, Kang HC, Park JH, Ku JL, Lee JS, Kwon HJ, Yoon KA, et al. RET oligonucleotide microarray for the detection of RET mutations in multiple endocrine neoplasia type 2 syndromes. *Clin Cancer Res* 2002; 8: 457–463.
16. Kim IJ, Kang HC, Park JH, Shin Y, Ku JL, Lim SB, Park SY, Jung SY, Kim HK, and Park JG. Development and applications of a beta-catenin oligonucleotide microarray: β-catenin mutations are dominantly found in the proximal colon cancers with microsatellite instability. *Clin Cancer Res* 2003; 9: 2920–2925.
17. An HJ, Cho NH, Lee SY, Kim IH, Lee C, Kim SJ, Mun MS, Kim SH, and Jeong JK. Correlation of cervical carcinoma and precancerous lesions with human papillomavirus (HPV) genotypes detected with the HPV DNA chip microarray method. *Cancer* 2003; 97: 1672–1680.
18. DeRisi J, Penland L, Brown PO, Bittner ML, Meltzer PS, Ray M, Chen Y, Su YA, and Trent JM. Use of a cDNA microarray to analyse gene expression patterns in human cancer. *Nat Genet* 1996; 14: 457–460.
19. Kurian KM, Watson CJ, and Wyllie AH. DNA chip technology. *J Pathol* 1999; 187: 267–271.

20. Glanzer JG, Haydon PG, and Eberwine JH. Expression profile analysis of neurodegenerative disease: Advances in specificity and resolution. *Neurochem Res* 2004; 29: 1161–1168.

21. Kim IJ, Shin Y, Kang HC, Park JH, Ku JL, Park HW, Park HR, Lim SB, Jeong SY, Kim WH, and Park JG. Robust microsatellite instability (MSI) analysis by denaturing high-performance liquid chromatography (DHPLC). *J Hum Genet* 2003; 48: 525–530.

22. Kang HC, Kim IJ, Park JH, Kwon HJ, Won YJ, Heo SC, Lee SY, et al. Germline mutations of BRCA1 and BRCA2 in Korean breast and/or ovarian cancer families. *Hum Mutat* 2002; 20: 235.

23. Kim IJ, Kang HC, Jang SG, Ahn SA, Yoon HJ, and Park JG. Development and applications of a BRAF oligonucleotide microarray. *J Mol Diagn* 2007; 9: 55–63.

24. Xiao W and Oefner PJ. Denaturing high-performance liquid chromatography: A review. *Hum Mutat* 2001; 17: 439–474.

25. Hermanson GT. Introduction of primary amine groups. In: *Bioconjugate Techniques*, pp. 100–112, 1996. San Diego, California: Academic Press.

17 Development of an Integrated Molecular Diagnostic Test to Identify Respiratory Viruses

Frédéric Raymond, Whei-Kuo Wu, and Jacques Corbeil

CONTENTS

17.1 INTRODUCTION

This chapter describes the development of an integrated diagnostic assay to identify respiratory viruses. From the analysis of the problem to the design of the assay and its validation, this chapter aims to guide the reader through all the design steps of a molecular diagnostic test involving primer design, polymerase chain reaction (PCR), primer extension, and microarray hybridization. The system chosen for the integration of this assay is the AutoGenomics Infiniti System.

17.1.1 BURDEN OF RESPIRATORY VIRUS DISEASE

From the avian flu to severe acute respiratory syndrome (SARS), respiratory viruses are currently
the cause of great concern worldwide. Fear of an influenza pandemic is present through all strata of
the population and the media devotes a substantial amount of airtime to every new case of a suspicious respiratory disease. However, these highly talked about viruses are only the tip of the iceberg
when speaking in terms of viral respiratory infections. They are, first and foremost, the main causes
of the common cold. Non-influenza-related viral respiratory tract infections (VRTIs) had an estimated cost of $39.5 billion for the year 2000 in just the United States, a number that includes $17
billion of direct costs and $22.5 billion of indirect costs [1]. According to the same study, there are
approximately 500 million noninfluenza-related VRTI episodes per year in the United States alone.
Some respiratory virus illnesses result in death, primarily among young infants and the elderly.

17.1.2 ADVANTAGES OF TESTING

It is often difficult to distinguish between the different possible causes of respiratory infections.
Many viruses have similar symptoms and their precise diagnosis requires microbiological laboratory testing. Tests allowing the detection and identification of the most important viruses within one
assay would have a positive impact on patient management [2]. Such assays could also test for
uncommon respiratory viruses such as SARS or avian influenza [3]. Dual respiratory virus infections involving human respiratory syncytial virus (HRSV) have been shown to reduce interferon
(INF) γ response, which is associated with a more severe illness [4].

Rapid testing of VRTI has been shown to have many clinical and financial benefits. Woo and
collaborators observed that rapid testing for VRTI could lead to a significant reduction in the length
of hospital stays in the case of influenza or parainfluenza virus infections [5]. They also observed a
decrease in antibiotic use when influenza virus, parainfluenza virus, and adenovirus were tested.
Barenfanger and colleagues estimated that an average $5716 per patient could be saved by appropri-
ately treating the patients for VRTI after a rapid diagnosis of respiratory viruses, mostly because of
a decrease in the length of mean hospital stays [6]. Those studies suggest that rapid, in-clinic testing
in case of VRTI is advisable, providing additional financial advantages.

Unnecessary use of antibiotics can be linked to parents often expecting clinicians to prescribe
their children antibiotics. In a study carried out between 1996 and 2000, it was found that 45% of
consultations for VRTI lead to the prescription of antibiotics. With the rise of antibiotic-resistant
bacteria, it is of utmost importance to limit the misuse of antibiotics [7,8]. The rapid molecular
diagnostics of VRTI would give clinicians a new tool to decrease the prescription of antibiotics and
increase the appropriate use of the anti-virals available for many virus species [9].

17.1.3 CURRENT TESTS AVAILABLE

The gold standard for respiratory virus detection is cell culture. However, this technique requires
between 5 and 10 days before the results are available to the clinician [10]. In the last 15 years, many
other techniques have been devised to detect respiratory viruses. Those techniques include antigen
detection, PCR, and microarray, among others. Antigen detection allows for a rapid detection of any
respiratory virus. However, this technique has a limited sensitivity and is not appropriate for multiple
virus detection. Over 10 antigen detection tests are available commercially for influenza A detection
[10]. Multiplex PCR tests for the detection of respiratory viruses have been described in the literature
[11–14], but with the limitation of requiring gel electrophoresis for amplicon visualization. Real-time
PCR is limited in its multiplexing capabilities. Moreover, in use are the more technologically complex DNA hybridization-based techniques. Such techniques include reverse-line blot [15], semiconductor-based hybridization [16], microfluidic flow-thru microarray [17], and tiling microarray [18].
All these techniques have the disadvantage of either requiring highly skilled personnel, of taking

TABLE 17.1
Viruses to be Detected Using the Respiratory Virus Assay, Including Results from Published Prevalence Studies

Families	Virus Species	Types	$n = 200$ [12]	Children <4 Years Old $n = 536$ [20]	$n = 446$ [21]
			Prevalence Studies (%)		
Adenoviridae	Adenovirus	A, B, C, D, E, and F	12.9	16.6	2.3
Coronaviridae	Coronavirus	NL63, 229E, HKU1, OC43, and SARS	NA	NA	3.4
Orthomyxoviridae	Influenza	A and B	22.8	14.6	8.8
Paramyxoviridae	Parainfluenza	1, 2, 3, and 4	6.3	28.3	3.2
Paramyxoviridae	HRSV	A and B	37.5	29.3	43.6
Paramyxoviridae	HMPV	A and B	NA	NA	4.4
Picornaviridae	Rhinovirus	A and B	NA	59.6	31.8
Picornaviridae	Enterovirus	A, B, C, and D	10.6	NA	2.1

many hours of laboratory work, or of being based on a recent technology that has not yet been commercialized. There is an unmet need to automate a well-controlled detection assay for respiratory viruses that is easy to use and provides results in a timely manner.

17.1.4 RESPIRATORY VIRUSES

Common cold symptoms can be caused by many viruses. Most of them are ribonucleic acid (RNA) viruses, except for the adenoviruses, which are deoxyribonucleic acid (DNA) viruses. For inclusion in our respiratory virus assay, we selected viruses that produce common cold-like symptoms. These viruses include enteroviruses, influenzaviruses, coronaviruses, rhinoviruses, parainfluenzaviruses, metapneumoviruses, and respiratory syncytial virus. Most of these virus lead to different peaks in the level of infection among children throughout the year, but some are endemic [19]. However, the peak time of each virus may depend on geographical location. Adults and, particularly, chronic obstructive pulmonary disease patients are also at risk of contracting respiratory viruses. Influenza virus infection is one of the most important threats related to respiratory viruses. Pandemic influenza caused million of deaths in the past hundred years and epidemic influenza causes the death of many young children and elderly people each year. The viruses to be detected in our assay are listed in Table 17.1, along with results from published prevalence studies [20].

17.2   PRINCIPLE OF THE ASSAY

17.2.1 SUMMARY

The Infiniti respiratory virus assay has eight main steps, five of which are performed on the AutoGenomics Infiniti System. Only three steps need to be done by laboratory technicians: sample extraction, reverse transcription, and multiplex PCR. Figure 17.1 outlines the steps involved in testing samples using the respiratory virus assay.

17.2.2 COMPLETE DESCRIPTION

A nasopharyngeal aspirate is obtained from the patient and the viral RNA is extracted using the QIAamp Viral RNA Mini Kit (cat# 52906) from Qiagen (Missisauga, Ontario). A reverse

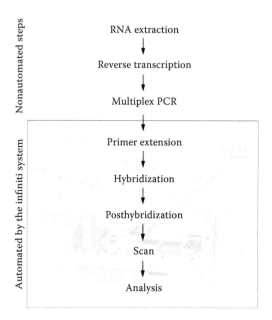

FIGURE 17.1   Steps involved in the respiratory virus assay and integration using the Infiniti platform.

transcription step (using Superscript II protocol from Invitrogen, Canada) is conducted upon the extracted RNA and a small volume of the product is added to the multiplex PCR reagents in a 24-well PCR plate. The multiplex PCR for the respiratory virus assay contains 37 different oligonucleotides, some of which are degenerated. The 40-cycle PCR is done at an annealing temperature of 55°C in a conventional thermocycler. The PCR plate is then transferred into the Infiniti System, where all the remaining steps of the assay are performed.

The primer extension reagents are added to each well and the thermal cycling for primer extension is done by the Infiniti System. During the design stage, a unique oligonucleotide tag sequence was added to each detection primer. The detection primers are composed of two parts: a unique detection tag sequence and a target-specific primer sequence. During primer extension, the target-specific sequence hybridizes to the amplicons and generates a linear amplification. This primer amplification step has three purposes:

1. Incorporation of fluorescently labeled nucleotides
2. Specific amplification of target sequences
3. Increase in sensitivity.

After primer extension, hybridization solution is added to each well and the mix is added to the chip for hybridization. The tags in 5′ of the detection primers will hybridize to the specific complementary probe spotted on the biochip's surface. Chips will then be washed, dried, and scanned using a confocal scanner integrated in the Infiniti apparatus. The results are automatically normalized and a diagnosis is proposed to the user.

17.3   ASSAY DESIGN

17.3.1   Requirements of the Assay

The quality of a molecular diagnostic assay can be estimated using many criteria [21]. During the development of the respiratory virus assay, we developed the test according to the following criteria.

We wanted the respiratory virus diagnostic assay to follow many guidelines that assess the quality of a molecular diagnostic test. To be useful in clinics, a diagnostic test must be easy to implement and to conduct in a standard clinical setting. The Infiniti System allows for simplified in-clinic procedures, by automating all steps following PCR. Some steps, such as RNA extraction and reverse transcription, are not included in the current test, as they must be performed by laboratory personnel. However, we wanted to keep the test to the fewest steps possible, so we optimized the assay as a one-tube multiplex PCR, in order to reduce the need for manipulation and pipetting. Additionally, having one PCR tube reduced the danger for cross-contamination, tube mislabeling, or other undesirable handling errors. To achieve this goal, the multiplex PCR contains as few PCR primers as possible for each virus to be detected.

Three important qualities of molecular diagnostic testing—sensitivity, specificity, and ubiquity—were also strictly followed and evaluated. Sensitivity is the capability of a test to detect as few viruses as possible in order to eliminate false negatives caused by low viral load. Specificity is the ability to distinguish the different viruses among each other and without false positives. Ubiquity is the capacity of a test to detect all possible strains of a targeted species without generating a false negative. Qualities such as specificity and ubiquity can be evaluated theoretically during the analysis of known virus sequences. However, they can only truly be assessed by testing many samples and by conducting statistical analyses. Similar approaches need to be implemented to test other qualities such as sensitivity, robustness, and reliability.

17.3.2   BIOINFORMATIC TOOLS

Many bioinformatic tools were used in the design of the respiratory virus assay. Sequence alignment was done using CLUSTAL W [22] and visualized using MEGA version 3.1 [23]. Phylogenetic and molecular evolutionary analyses were also carried out using MEGA version 3.1. FastPCR was used to estimate some primer properties and to manipulate sequences [24]. The RNAsoft tool Pairfold was used for primer–dimer analysis [25]. Sequence comparison was done using blast and blast2seq [26,27]. Some sequence analysis tools were created in-house using the Perl language to perform simple repetitive bioinformatic tasks.

17.3.3   GENERAL SEQUENCE ANALYSIS

In the first steps of sequence analysis, we downloaded the reference sequences for each virus' genome. To determine if it was possible to use common PCR primers for groups of viruses, we aligned the reference genomes of all viruses and grouped the viruses with the most similarities using CLUSTAL W. Using this procedure, we were able to determine that rhinoviruses and enteroviruses could have common PCR primers. Metapneumovirus and respiratory syncytial virus could also be analyzed together, but further analysis suggested that we should use different primers for both species. These analyses also allowed us to group viruses within species to achieve the lowest possible quantity of PCR primers required for amplification. Thus, two groups of virus types were determined for adenoviruses, parainfluenzaviruses, and coronaviruses. Following this analysis, alignment of reference sequences was carried out, with the goal of finding a common region, suitable for primer design for each virus.

17.3.4   SEQUENCES USED FOR DESIGN

Initially, we made first attempts at PCR primer design based solely on virus genomes found on RefSeq [28]. However, when we compared the reference sequences to other sequences available on the National Center for Biotechnology Information (NCBI) database, we noticed that some reference sequences had many polymorphisms when compared to related sequences found in the NCBI database. For those viruses, there was more homogeneity between nonreference sequences than between reference sequences and nonreference sequences. Thus, we concluded that better sequence analysis would be done using reference sequences along with all the other sequences available in the database for a given virus.

For viruses with high sequence variability, such as type A Influenza, for which more than 1500 sequences for each virus segment are available on the influenza (FLU) database [29], a random sample of 100 sequences was aligned using CLUSTAL W [22]. This alignment was analyzed to identify regions that were suitable for primer design. We aimed at regions that would minimize the number of primers necessary, while allowing us to detect as many influenza variants as possible. The same process was used with all viruses and virus types targeted by this assay. For many viruses, the number of available sequences was not sufficiently high to allow a randomized selection of sequences used in the design, so all available sequences were used instead. With better epidemiological studies, we may be able to determine the prevalence of certain types that will assist in improving the coverage in a clinical setting.

17.3.5 SPECIFICITY

The nature of the respiratory virus assay warranted the design of low specificity PCR primers and high specificity detection primers in order to maximize detection. As discussed earlier, we wished to amplify the viruses by using as few PCR primers as possible. Towards this end, we needed to increase the number of sequences amplified by the PCR primers. This leads to less specific detection primers. However, what is lost in specificity during multiplex PCR is not problematic if the detection primers are properly designed with appropriate stringency. To achieve an adequate level of specificity, the sequences targeted by the detection primers should be as different as possible between the virus types amplified by the same PCR primers. However, the most important parameter to insure specificity is the 3′ nucleotide of the detection primer. Since the DNA polymerase requires that the 3′ nucleotide of the detection primer be hybridized to the target DNA in order to begin DNA synthesis, a mismatch at this base would not allow primer extension. Thus, a careful choice of a detection primer's 3′ nucleotides allows the design of very specific detection primers. All primers were blasted on the NCBI database to insure their specificity.

17.3.6 UBIQUITY

As discussed in Section 17.3.4, for many of the viruses included in the respiratory virus assay, all related sequences found on the NCBI database were used for both PCR and detection primer design. Thus, their theoretical ubiquity was insured at the primer design step. For primers for which more than 100 sequences were available, we selected a random 100 samples to align and to use in primer design. After primer design, the sequences of the primers were blasted on all the sequences available for their target virus to insure a near-perfect theoretical ubiquity. Still, we must keep in mind that the theoretical ubiquity of primers may not reflect the ubiquity observed with real specimens. We plan to investigate, through DNA sequencing, any sample found to be positive for a virus using another method. We would then use these results to refine our design process, in order to increase the ubiquity of the PCR primers and that of the detection primers.

17.3.7 PRIMER DESIGN

For each virus, PCR primers and detection primers were designed to carefully optimize its thermodynamic properties while keeping the primers specific and ubiquitous. One degenerate nucleotide was allowed for each primer. We observed that two or more degenerate bases decreased PCR efficiency. Target amplicon size was ~200 nucleotides and the melting temperature for both types of primers was between 58°C and 60°C. Secondary structures of ΔG (a measure of free energy; a high negative value of ΔG suggests the increased possibility of secondary structure, an undesirable attribute for any primer) of more negative than −9 kcal/mol were rejected, especially if the 3′ end of the

TABLE 17.2
Summary of the Primer Sets Included in the Respiratory Virus Assay

Primer Sets		Amplicon Length (nucleotides)	PCR Primers (Nb primers)	Detection Primers (Nb primers)
Virus	**Type**			
Adenovirus	A, B, and C	174	4	4
Adenovirus	D, E, and F	174	3	3
Coronavirus	SARS	168	2	1
Coronavirus	HKU1 and OC43	123	2	2
Coronavirus	NL63 and 229E	149	2	2
Enterovirus	A, B, C, and D	197	3	4 + 1[a]
Rhinovirus	A and B			2 + 1[a]
Influenza	A	201	3	2
Influenza	B	201	3	1
HMPV	A and B	148	3	2
HRSV	A and B	245	3	2
HPIV	1	272	2	1
HPIV	3	243	2	1
HPIV	2, 4A, and 4B	275	5	3
Total			37	31

[a] Detection primer targeting both rhinovirus and enterovirus.

primers were involved. Primer–dimers were checked using tools described in Section 17.3.2. Primers were systematically rejected if their 3′ end hybridized to any other primer with a ΔG more negative than −9 kcal/mol. To insure proper labeling of the primer extension products, detection primers were oriented so that many labeled nucleotides could be incorporated during primer extension.

17.3.8   CONTENT OF THE ASSAY

The respiratory virus assay's multiplex PCR is composed of 37 primers. The primer extension mix contains 32 detection primers, each with a different 5′ tag. A summary of the number of primers used to detect each virus type is shown in Table 17.2.

17.4   VALIDATION OF THE ASSAY

17.4.1   ABOUT VALIDATION

Careful design of PCR primers and detection primers is of utmost importance, but the quality of a test can only be verified by performing the test on actual samples. The validation stage comprises three steps:

1. Simplex PCR validation
2. Multiplex PCR validation
3. Primer extension validation.

During the validation stage of the assay, some PCR primers or detection primers may be replaced by new designs. It is important to carefully track all primer sets. A primer set is a group of primers that are used together to amplify and detect one or more virus types. A primer set may contain more than two primers.

17.4.2   Simplex PCR Validation

Simplex PCR validation consists of testing all PCR primer sets separately to insure that all viruses can be detected using their respective PCR primers. Since the real test for the PCR primers is their behavior in multiplex PCR, it is not necessary to optimize the PCR conditions and primer concentrations at this step. However, if some primer sets yield no or low amplification, it may be useful to redesign the troublesome primers. Sometimes, a small modification of primer sequence can lead to great improvement. In the case of the respiratory virus assay, all viruses yielded a significant band of appropriate length in gel electrophoresis.

17.4.3   Multiplex PCR Validation

Multiplex PCR validation requires the creation of a multiplex primer mix that includes all PCR primer sets. This multiplex mix is then used to amplify specimens from each virus type, in order to insure the proper amplification properties of each primer set. Results of this step for the respiratory virus assay are shown in Figure 17.2. At this step, it may be important to qualify each primer set according to the amplification yield. Primer sets that give good amplification can be kept as such, but primer sets that give no or low amplification require attention. Optimization of two parameters is useful at this step. First, the annealing temperature of the thermocycling program can be either increased or decreased, or a gradient PCR can be added to the 6–12 first PCR cycles, starting from a higher temperature and decreasing in increment at each step. Second, the concentration of primer sets yielding lower results can be increased by a factor of two or higher, if necessary. If none of those parameters allow a primer

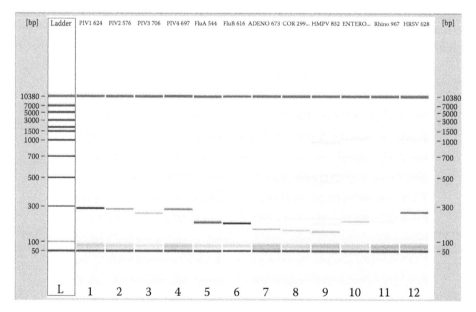

FIGURE 17.2 Multiplex PCR validation of the respiratory virus assay on Bioanalyser. Twelve different viruses were amplified using the respiratory virus multiplex PCR. The results were run on the Agilent Bioanalyser to quantify the yield of amplification and to verify the length of PCR products for each virus. In this experiment, only rhinovirus gave low amplification. This problem was later solved by increasing the concentration of rhinovirus PCR primers in the multiplex PCR.

set to yield significant amplification, it may be necessary to modify the primer set, or to design a new primer set. However, before concluding that a primer set does not work, it may be advisable to try the multiplex on a different specimen, to insure that the low or absent amplification is actually due to the primer set and not due to the low quality of a specimen, low DNA concentration, or a variant DNA sequence at the targeted region. Thus, the multiplex PCR validation may also include a test for ubiquity. This test is conducted by amplifying many strains of all virus types to insure that all are detected. This step will also be needed for primer extension testing, but performing this experiment at the stage of multiplex PCR evaluation gives insight on the assay's ubiquity.

17.4.4 PRIMER EXTENSION VALIDATION

Primer extension validation requires that one or more specimens of each virus type be amplified by multiplex PCR and then submitted to primer extension. Primer extension is done using a mix of all detection primers at the same concentration. However, before initiating primer extension, it may be useful to perform an exonuclease and alkaline phosphatase step on the amplicons in order to destroy the remaining free nucleotides and to inactivate the remaining PCR primers. After the primer extension step, the products are hybridized to the Infiniti chips, which are processed using the same protocol as the automated assay. The chips are then read by the Infiniti System and the results are analyzed. Figure 17.3 shows results obtained for four viruses using the Infiniti respiratory virus assay.

For this first step of primer extension validation, we need to identify four types of signals associated with each virus:

1. Significant true-positive signals
2. Nonsignificant true-positive signals
3. False-negative signals
4. False-positive signals.

The first case, where the amplified virus gives a significant positive signal on its targeting probe, is the expected result, and conditions relating to such results should not be modified at this stage.

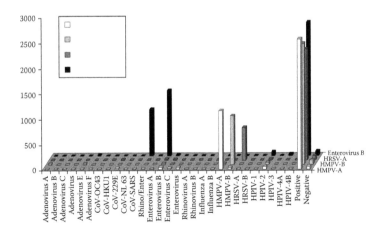

FIGURE 17.3 Results from virus strains analyzed using the respiratory virus assay. The viruses HMPV type A, HMPV type B, HRSV type A, and enterovirus type B were amplified using the respiratory virus multiplex PCR. The products were then subjected to alkaline phosphatase and to 3′ exonuclease. Primer extension was conducted upon the treated amplicons and the products were hybridized to the microarray. The results were read on the Infiniti platform. The x-axis refers to the different microarray probes. The y-axis refers to the fluorescence intensity of each spot. The z-axis refers to the target virus.

The second case, where the amplified virus gives a nonsignificant signal on its targeting probe, suggests that this probe set requires optimization. A first step in the optimization of primer sets giving this type of result is to increase the concentration of primers targeting this virus type in the mix of detection primers. It is also important to compare the signal obtained on the chip to the PCR amplification yield of this virus type. Optimization of the PCR step may be needed for such a primer set.

The third case, where the amplified virus gives no signal on its target probe, is problematic. In this case, the first step is to try the same troubleshooting technique as described above for a non-significant true-positive signal. However, it may be important to re-analyze the sequence of the detection primers to insure that their sequences are correct. Sequencing of the amplicon may be warranted and redesign of the primers may be necessary.

The fourth case, where a false-positive signal is observed on a probe for which the target virus type was not added before PCR, is a problem that will most certainly require a redesign of the detection primer. The most frequent cause of false positives is primer–dimers between detection primers, especially dimers that involve the 3′ nucleotides of the detection primer. Thus, the first step in troubleshooting this type of result is to verify the presence of primer–dimers between the false-positive detection primer and the other detection primers. If primer–dimers involving the 3′ extremity of the detection primer are found, the primers will need to be redesigned. The false-positive detection primer can also be compared to the different amplicons to insure that no false-positive result is caused by cross-hybridization.

17.5 RESPIRATORY VIRUS ASSAY

The respiratory virus assay was designed and validated following the guidelines described in Sections 17.3 and 17.4, respectively. Upon further validation and optimization of the respiratory virus assay, it will be used as an epidemiological tool in a prospective study on a cohort of over 200 children between 1 and 3 years of age who are seen at our children's hospital in Quebec City, Canada. During this study, we will compare the results obtained with the Infiniti respiratory virus assay with results obtained with standard laboratory methods and with a real-time PCR assay that we are currently developing. This study will allow us to estimate the usefulness of this assay in a clinical setting. Furthermore, we aim to use this assay to conduct a similar study over 3 consecutive years, on children who are seen at our hospital and at a pediatric clinic in Quebec City. The results will help us to understand the epidemiology of the viruses tested and to gain insight in their treatment and control. We also hope to correlate the viruses infecting the children with the severity of the symptoms. As respiratory viruses become more and more of a worldwide concern, such a tool will have an increased usefulness for diagnostic and treatment.

ACKNOWLEDGMENTS

The authors wish to thank Dr. Guy Boivin for the initial impetus to start the project and the Canadian Research Chair Program (JC). FR acknowledges the support of the Canadian Institute of Health Research.

REFERENCES

1. Fendrick AM, Monto AS, Nightengale B, and Sarnes M. The economic burden of non-influenza-related viral respiratory tract infection in the United States. *Arch Intern Med* 2003, 163: 487–494.
2. Wallace LA, Collins TC, Douglas JD, McIntyre S, Millar J, and Carman WF. Virological surveillance of influenza-like illness in the community using PCR and serology. *J Clin Virol* 2004, 31: 40–45.
3. Speers DJ. Clinical applications of molecular biology for infectious diseases. *Clin Biochem Rev* 2006, 27(1): 39–51.
4. Aberle JH, Aberle SW, Pracher E, Hutter HP, Kundi M, and Popow-Kraupp T. Single versus dual respiratory virus infections in hospitalized infants: Impact on clinical course of disease and interferon-gamma response. *Pediatr Infect Dis J* 2005, 24: 605–610.

5. Woo PC, Chiu SS, Seto WH, and Peiris M. Cost-effectiveness of rapid diagnosis of viral respiratory tract infections in pediatric patients. *J Clin Microbiol* 1997, 35: 1579–1581.
6. Barenfanger J, Drake C, Leon N, Mueller T, and Troutt T. Clinical and financial benefits of rapid detection of respiratory viruses: An outcomes study. *J Clin Microbiol* 2000, 38: 2824–2828.
7. Nyquist A, Gonzales R, Steiner JF, and Sande MA. Antibiotic prescribing for children with colds, upper respiratory tract infections, and bronchitis. *JAMA* 1998, 279: 875–877.
8. Mainous AG, 3rd, Hueston WJ, and Clark JR. Antibiotics and upper respiratory infection: Do some folks think there is a cure for the common cold. *J Fam Pract* 1996, 42: 357–361.
9. Abed Y and Boivin G. Treatment of respiratory virus infections. *Antiviral Res* 2006, 70: 1–16.
10. Vega R. Rapid viral testing in the evaluation of the febrile infant and child. *Curr Opin Pediatr* 2005, 17: 363–367.
11. Osiowy C. Direct detection of respiratory syncytial virus, parainfluenza virus, and adenovirus in clinical respiratory specimens by a multiplex reverse transcription-PCR assay. *J Clin Microbiol* 1998, 36: 3149–3154.
12. Grondahl B, Puppe W, Hoppe A, Kuhne I, Weigl JA, and Schmitt HJ. Rapid identification of nine microorganisms causing acute respiratory tract infections by single-tube multiplex reverse transcription-PCR: Feasibility study. *J Clin Microbiol* 1999, 37: 1–7.
13. Gruteke P, Glas AS, Dierdorp M, Vreede WB, Pilon JW, and Bruisten SM. Practical implementation of a multiplex PCR for acute respiratory tract infections in children. *J Clin Microbiol* 2004, 42: 5596–5603.
14. Bellau-Pujol S, Vabret A, Legrand L, Dina J, Gouarin S, Petitjean-Lecherbonnier J, Pozzetto B, Ginevra C, and Freymuth F. Development of three multiplex RT-PCR assays for the detection of 12 respiratory RNA viruses. *J Virol Methods* 2005, 126: 53–63.
15. Coiras MT, Lopez-Huertas MR, Lopez-Campos G, Aguilar JC, and Perez-Brena P. Oligonucleotide array for simultaneous detection of respiratory viruses using a reverse-line blot hybridization assay. *J Med Virol* 2005, 76: 256–264.
16. Lodes MJ, Suciu D, Elliott M, Stover AG, Ross M, Caraballo M, Dix K, et al. Use of semiconductor-based oligonucleotide microarrays for influenza A virus subtype identification and sequencing. *J Clin Microbiol* 2006, 44: 1209–1218.
17. Kessler N, Ferraris O, Palmer K, Marsh W, and Steel A. Use of the DNA flow-thru chip, a three-dimensional biochip, for typing and subtyping of influenza viruses. *J Clin Microbiol* 2004, 42: 2173–2185.
18. Wang D, Coscoy L, Zylberberg M, Avila PC, Boushey HA, Ganem D, and DeRisi JL. Microarray-based detection and genotyping of viral pathogens. *Proc Natl Acad Sci USA* 2002, 99: 15687–15692.
19. Makela MJ, Puhakka T, Ruuskanen O, Leinonen M, Saikku P, Kimpimaki M, Blomqvist S, Hyypia T, and Arstila P. Viruses and bacteria in the etiology of the common cold. *J Clin Microbiol* 1998, 36: 539–542.
20. Monto AS. Epidemiology of viral respiratory infections. *Am J Med* 2002, 112(Suppl. 6A): 4S–12S.
21. Freymuth F, Vabret A, Cuvillon-Nimal D, Simon S, Dina J, Legrand L, Gouarin, S, Petitjean J, Eckart P, and Brouard J. Comparison of multiplex PCR assays and conventional techniques for the diagnostic of respiratory virus infections in children admitted to hospital with an acute respiratory illness. *J Med Virol* 2006, 78: 1498–1504.
22. Thompson JD, Higgins DG, and Gibson TJ. CLUSTAL W: Improving the sensitivity of progressive multiple sequence alignment through sequence weighting, positions-specific gap penalties and weight matrix choice. *Nucleic Acids Res* 1994, 22: 4673–4680.
23. Kumar S, Tamura K, and Nei M. MEGA3: Integrated software for molecular evolutionary genetics analysis and sequence alignment. *Brief Bioinform* 2004, 5: 150–163.
24. Ruslan Kalendar R. 2006. FastPCR, PCR primer design, DNA and protein tools, repeats and own database searches program. Available at www.biocenter.helsinki.fi/bi/programs/fastpcr.htm.
25. Andronescu M, Aguirre-Hernandez R, Condon A, and Hoos HH. RNAsoft: A suite of RNA secondary structure prediction and design software tools. *Nucleic Acids Res* 2003, 31: 3416–3422.
26. Altschul SF, Gish W, Miller W, Myers EW, and Lipman DJ. Basic local alignment search tool. *J Mol Biol* 1999, 215: 403–410.
27. Tatusova TA and Madden TL. Blast 2 sequences—a new tool for comparing protein and nucleotide sequences. *FEMS Microbiol Lett* 1999, 174: 247–250.
28. Pruitt KD, Tatusova T, and Maglott DR. NCBI Reference Sequence (RefSeq): a curated non-redundant sequence database of genomes, transcripts and proteins. *Nucleic Acids Res* 2005, 33(Database issue): D501–D504.
29. National Center for Biotechnology Information. 2004. Influenza virus resource. Available at http://www.ncbi.nlm.nih.gov/genomes/FLU/FLU.html.

18 Microarrays in Neuroscience

*Massimo Ubaldi, Wolfgang Sommer, Laura Soverchia,
Anbarasu Lourdusamy, Barbara Ruggeri, Roberto
Ciccocioppo, and Gary Hardiman*

CONTENTS

18.1 INTRODUCTION

In the past few years, microarray technology has matured into a commonly used laboratory tool in biomedical research, including neuroscience. A variety of topics in neuroscience have now been investigated using microarray-based approaches. Although the initial microarray publications appeared in the early 1990s, the first neuroscience microarray study dates to just 1999. To date, several deoxyribonucleic acid (DNA) array studies have been published with a constant and significant increase throughout the years (Figure 18.1). Microarray-based studies in neuroscience encompass neurological, and psychiatric disorders, drug addiction (see later), memory and learning [1,2], and circadian rhythms [3,4]. Arrays have been used to address the topic of human brain evolution comparing the gene expression patterns of *Homo sapiens*, to those of the close relative *Pan troglodytes* and other primates (for a review see Ref. [5]).

Recent years have brought increased quality in microarray production, streamlined protocols for sample processing, improvement in signal reproducibility, and improved data analysis [6]. Despite these advances in microarray technology, the specific methodological issues encountered in the study of the nervous tissue and the consequent production of high-quality data with general reproducibility and true biological relevance remain a significant challenge for most researchers [7].

Neurobiology poses several challenges to the application of microarray technology. The compartmentalization of the brain and the multitude of highly specialized cells result in highly

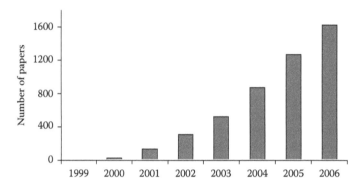

FIGURE 18.1   Increase in neuroscience microarray publications since 1999.

heterogeneous samples. Consequently expression data from brain samples represent averages of a variety of different transcriptional profiles. Furthermore, the majority of genes are expressed at a low copy number, on the order of 10 or less copies per cell. In addition, many genes show a highly restricted pattern of expression, sometimes confined to only a few specialized cell types. The physiology of the neurons requires strong adaptive mechanisms for the generation of action potentials, thus many genes are regulated within tight boundaries. As a result, any possible effect of a treatment or physiological reaction is dampened by dilution in highly heterogeneous samples. Therefore small changes in expression are expected and may be of biological importance. Differences in transcript levels between groups are usually <30% in behavioral or neuropharmacological microarray experiments, and fold changes exceeding a doubling in transcript levels are rarely seen. This poses considerable demands on assay sensitivity and reproducibility.

18.2   MICROARRAY METHODOLOGICAL APPROACHES IN NEUROSCIENCE

18.2.1   Designing the Microarray Experiment

There are several important issues one needs to consider when moving from microarray experimental design to the generation of a list of differentially expressed genes. Careful experimental design will systematically identify possible sources of variability and, by minimizing their impact, reduce the noise level in the experiment. We can divide variability according to three typical sources: biological, experimental, and technical.

Biological variability is caused by the spectrum of responses of the individual organism. A seemingly obvious implication would be that, if available, inbred strains should be used for microarray experiments, which have reduced individual variability in gene expression. This is likely true in most cases. However, different lines of experimental animals differ in their "behavioral reaction norm." This term describes the amount of change in a quantitative phenotypic trait displayed in response to a certain amount of environmental challenge or pressure [8]. An inbred line or a line generated through the pressure of selective breeding is "locked into" a particular phenotype and it may have a very narrow reaction norm. This will make it less suitable for studies addressing changes in behavioral phenotype and gene expression patterns in relation to perturbations (e.g., pharmacological or environmental challenge).

Experimental variability is related to both the phase of the actual animal experiment as well as the preparation of samples before hybridization to the microarray. It may appear trivial to point out that good pharmacological or behavioral practice and design are at the core of any successful animal experiment including microarray studies. In our experience this all too often fails due to lack of communication and interaction between animal experimenters and molecular biologists. Most specialists are only aware of a few of the many details that comprise an experiment, and systematic variation in any of them may compromise the entire experiment. In our experience it is difficult to know beforehand what factors may influence gene expression. We recommend that animal experimenters and molecular biologists record the conditions of the experiments as exhaustively as possible. To facilitate access to this important information, which may affect data analysis of the microarrays, it is good practice to record all observations in simple text or Excel format. We were able, in one large microarray study, to trace the major source of variance to the specific day in which the RNA was extracted. Since the experimental groups were poorly balanced for this procedural step, the bias if undetected, would have skewed the entire data analysis (Figure 18.2).

Technical details should be made available together with the publication of the entire microarray experiment. This will allow other laboratories to revisit the data in search of sources of systematic error if discrepant results arise in the literature. Among the most common confounding factors in performing an animal experiment are exposure to perturbations (e.g., novel compounds) and time of the experiment. One has to keep in mind that stress and circadian rhythm are powerful regulators of gene expression. It is well documented that immediate early genes respond within minutes to animal handling [9]. Therefore a single experienced animal keeper should work with animals during the entire experiment and the euthanasia. Control groups are often treated differently, for example by much less exposure to the experimental environment or different time points of sample collection. If an experiment needs to be spread out over several occasions for sample collection, this effect needs to be balanced across all groups. Both animal and microarray researchers must develop awareness of such potential pitfalls and discuss their potential impact on the experiment.

We use the term "technical variability" to group all sources of variability which may come about during the generation of the expression profiles. Several steps of the Affymetrix protocol are fairly standardized; hence technical variability is usually low relative to biological and experimentally induced variability. However, small differences in technique, which would hardly impact most optimized bench-top procedures, provide significant differences when it comes to the measurement

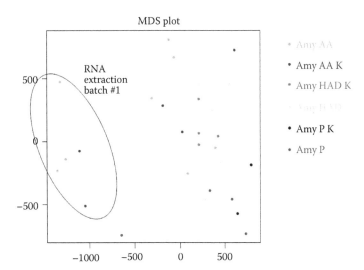

FIGURE 18.2 (See color insert following page 138.) The multidimensional scaling plot (MDS) displays 30 samples from 6 different experimental groups (color coded). Distances within the plot represent approximately the sum of absolute differences over all genes between any two samples. Overlaying the plot with the experimental information reveals that RNA extraction is a major source of variance.

of many genes on a microarray; although it is impossible to optimize the measure of thousands of genes simultaneously. In addition, such small systematic differences are magnified by statistical procedures, such as clustering, that are commonly used in microarray studies. Therefore, as discussed under experimental variability, procedures should be kept consistent throughout an experiment and records should be kept for most variable factors, including researcher, day of the experiment, batch numbers of kits, solutions and reagents, and so on.

18.2.2 TISSUE PRESERVATION AND POSTMORTEM BRAIN

RNA quality has a direct influence on messenger RNA (mRNA) levels and for microarray studies, high RNA integrity is an essential prerequisite. It is relatively easy to obtain high-quality mRNA from frozen tissue. However, frozen tissue is not always available and in some circumstances tissue samples are available from different storage conditions [10,11]. One important source of samples, often provided with related clinical histories and prognosis, is represented by fixed and paraffin-embedded specimens. They are often used for immunohistochemical preparations and are widely available in many brain banks and pathology departments [12,13]. Extraction of genomic DNA from formalin-fixed paraffin-embedded (FFPE) tissue is well documented and is now a routine diagnostic process [14]. However extraction of RNA from archived tissue is not trivial; RNA extraction from fixed paraffin-embedded tissue strongly depends on the type of fixative and the experimental protocol used for fixation [15,16]. A number of different fixatives are routinely used for histology and tissue preservation. Simple organic coagulants such as ethanol, methanol, or acetone have precipitative effects. In contrast, aldehyde-based fixatives (such as formalin or paraformaldehyde) function as chemical cross-linking agents between proteins or between protein and nucleic acid, involving hydroxymethylene bridges [17].

Formalin is the most commonly used fixative in routine pathological practice, including brain tissue. Gene expression analysis in FFPE tissues has been limited to techniques such as polymerase chain reaction (PCR) and in situ hybridization analyses that evaluate only a few genes simultaneously [18,19]. There have been a few reports of microarray analysis of RNA extracted from FFPE tissue [20,21], but no direct comparison between FFPE and unfixed snap-frozen tissue samples. In the neuroscience field, Karsten et al. performed complementary DNA (cDNA) microarray experiments using brain RNA samples from frozen tissue and tissue fixed in ethanol or formalin. The results revealed that RNA from frozen tissue yielded very reproducible expression data, whereas RNA from fixed tissue generated results characterized by higher variability and lower correlation coefficients. Formalin-fixed tissue was more severely affected than ethanol fixed tissue in all conditions [22].

To improve the yield of RNA from FFPE tissues a number of procedures have been proposed [23–26]. The most commonly used method is proteinase K digestion of the (FFPE) tissues [27,28]. This enzyme weakens the strong chemical bonds formed during fixation, facilitating subsequent RNA extraction. RNA extraction is highly problematic from samples preserved in stronger cross-linking fixatives such as gluteraldehyde, modified formalins containing mercuric chloride, and Bouin's fixative [11].

During the last decade concerns have arisen about RNA stability in postmortem brain samples. Often, brain tissue (especially human) cannot be sampled, frozen, or fixed before a certain postmortem interval (PMI); the longer the PMI, the higher the risk of RNA degradation [29,30]. In this regard the literature is rather confusing and recent studies report that many factors contribute to overall RNA integrity. A successful RNA extraction from postmortem tissues is correlated to the nature of tissue [20,31]. Focusing the attention on brain, a further difference was observed between animal and human postmortem brain tissues. In animal studies, where agonal factors are controlled, the stronger predictor of RNA degradation is PMI. A study conducted on postmortem mouse brain tissue demonstrated that increasing PMI (6–48 h time points) was associated with RNA degradation as revealed by the analysis of 28S/18S ribosomal RNA ratios [32]. A significant correlation between PMI and mRNA degradation could not be found for human postmortem brain tissue [33]. A number of reports indicate that human brain tissue pH rather than PMI is a better predictor of RNA degradation [34–36]. Intact mRNA is associated with a brain pH in the range between 6.1 and 7.0; tissues with low pH (<6) contain fragmented mRNAs. Death subsequent to prolonged agony causing hypoxia may result in a drop in tissue pH and a subsequent reduction of mRNA integrity [37,38].

18.2.3 SAMPLE COLLECTION

The brain possesses many structures with different evolutionary origins and functions, each containing a large variety of cell types (i.e., neurons, astrocytes, oligodendrocytes, and microglia) with specific functions and characterized by different gene expression profiles.

Microarray-based studies have helped to assess the distinctive transcriptional profiles of many brain areas including the cortex, hippocampus, amygdala, entorhinal cortex, midbrain, and cerebellum [39,40]. An example of such brain region-specific expression patterns is given in Figure 18.3. With the arrival of genome-wide brain expression maps (e.g., www.brain-map.org), region-specific gene expression is easily verified and can be used to assess the quality of a microarray experiment. Moreover the hippocampus differences in gene expression were analyzed among sub-regions (CA1, CA3, and dentate gyrus (DG)). The study revealed sets of unique genes that were specifically expressed in sub-regions, supporting on a molecular basis, the previously defined sub-divisions of this brain area as assessed by anatomical studies [41].

The studies reported above illustrate the extreme precision and accuracy required in the tissue collection to avoid undesired sampling from surrounding areas. Different techniques can be used for brain tissue dissection. A common procedure when large brain areas need to be sampled consists of the use of Brain Matrix Blocks; chambers designed for hands-free slicing of discrete regions. The Brain Matrix allows slicing of either coronal or sagittal sections through the brain at intervals as small as 1 mm. When small brain areas need to be sampled, an alternative to Brain Matrix Blocks is the cryostat. In this case, brains previously stored at −80°C can be sliced at variable thickness (as thin as 5 μm) and at different angles (coronal, sagittal, or horizontal), and material can be punched out from selected brain areas, collected, and used for subsequent RNA extraction. A high level of accuracy can be reached with laser-capture microdissection (LCM), a technology originally developed at the National Institutes of Health in 1996 [42]. LCM permits the selection and capture of cells, cell aggregates, and discrete morphological structures deriving from thin tissue sections. Cells are visualized through a thermoplastic film, which is attached to the bottom of an optically clear microfuge tube cap. A laser pulse, directed onto the target cells through the film, melts the film and allows it to flow onto the targeted area where upon cooling it binds with the underlying cell(s). The film including the adhered cells or clusters is then lifted and the captured cells may be used for gene expression profiling. LCM has been a very useful tool in the study of differential gene expression of small and large neurons in the rat dorsal root ganglia (DRG). Large (diameter of >40 μm) and small (diameter <25 μm) neurons were cleanly and individually captured by LCM from Nissl-stained sections of DRG. This system was chosen because small and large neurons are adjacent to each other and numerous differentially expressed genes (small versus large cells) have been reported [43]. LCM was used to demonstrate the feasibility of gene expression mapping studies in seven rat brain nuclei, locus coeruleus (LC), dorsal raphe nucleus (DR), parvocellular division (PA), and magnocellular division (MG) of the hypothalamic paraventricular nucleus (PVN) and CA1, CA3, and DG divisions of the hippocampal formation [44]. CA1 and CA3 pyramidal neurons were also isolated by LCM in fresh-frozen normal postmortem human brains [45].

The effective use of LCM to harvest specific cellular phenotypes requires visual guidance by the distinctive anatomy, morphology, or specific labeling of the target cells [46]. In the study conducted by Yao et al. [47], quick staining with cresyl violet acetate, a technique routinely used to identify the dissection area, was substituted by a fluorescent neuronal tracer to specifically label rat midbrain

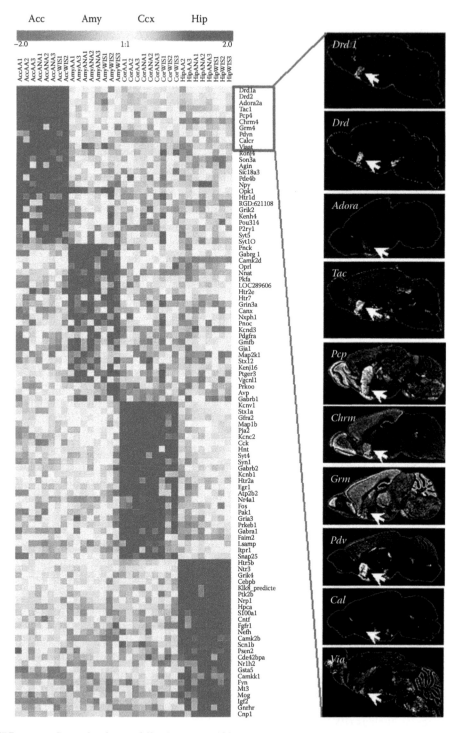

FIGURE 18.3 (See color insert following page 138.) Clustering of 33 microarray experiments from rat forebrain. Left panel: Four brain regions (Acc: nucleus accumbens, Amy: amygdala, Ccx: cingulate cortex, Hip: hippocampus) from three rat lines (AA, ANA, Wistar) were analyzed for common region-specific gene expression patterns. Only the top 25 genes per cluster are shown. Data are reanalyzed from a previously published experiment [125]. Right panel: Verification of accumbens-specific gene expression using the Allen Brain Atlas (http://mouse.brain-map.org; [135]). For 8 of the 10 selected genes, the atlas shows increased *in situ* hybridization signals in the ventral striatum including the nucleus accumbens. Two genes (red labels) could not be verified.

dopamine neurons and nondopamine cortical neurons. The labeled cells were used to visually guide harvesting of the cells by LCM.

A novel approach to analyze cell-specific expression profiles is to use transgenic mice that express a fluorogenic marker in a distinct subset of neurons. Using this technique Sugino et al. analyzed the expression profile in 12 populations of neurons, including several types of GABAergic inter-neurons and pyramidal neurons from cingulate cortex, as well as neurons derived from hippocampus, amygdala, and thalamus. They found highly heterogeneous gene expression patterns between different classes of neurons even within a single brain region [48,49].

18.2.4 SAMPLE POOLING

A strong constraint for microarray studies in neurobiology is the limited amount of tissue. This can be dealt with in two principal ways, pooling or RNA amplification. This should be decided early during the experimental design because inconsistently processed samples will not be comparable; this could happen if for example the researcher tries to "rescue" a subset of samples with additional amplification steps.

There is disagreement about the use of pooling to overcome limited sample size among practitioners and also among statisticians, but in general the field has moved away from this design. Some statisticians advocate pooling to reduce variance, even when it is not necessary to deal with small samples. In theory, if the variation of a gene among different individuals is "reasonable" (has a minimum number of outliers), then pooling-independent samples result in reduction of variance. We could further reduce the variation by making technical replicates of the pooled RNA. Since technical variation is usually less than individual variation, this strategy would in theory give us more accurate estimates of the group means for each gene. In practice, the expression levels of many genes among individuals show frequent outliers. Some individual samples have levels of stress–response proteins and immunoglobulins 5–10-fold higher than levels typical of the group. This can be due to many factors unrelated to the experimental treatment: for example, individual animals or subjects may be infected, or some tissue samples may be anoxic for long periods before preservation, which allows cells to respond to stress. It is easier to detect, and compensate for such anomalies, if individual samples are hybridized. A comparison of samples with pooled and unpooled designs found that the majority of genes that were identified as differently expressed between two groups in the pooled study turned out to be extreme in only one individual in the unpooled study. Furthermore, examining individual samples separately allows one to estimate variation between individuals, which is sometimes important and often interesting. Therefore, we generally recommend against pooling from only a small number of individuals ($n < 20$).

18.2.5 RNA AMPLIFICATION

Standard protocols for microarray hybridization require RNA amounts corresponding to approximately 10^7 cells or several milligrams of tissue. RNA extracted from cells harvested by needle biopsy, cell sorting, or LCM necessitates an amplification step. Two main approaches, signal amplification and global mRNA amplification, have been developed to overcome the hurdle of working with small tissue samples. Signal amplification methodologies, such as tyramide signal amplification (TSA) [22] and dendrimer technology [50], increase the fluorescence signal emitted per mRNA molecule. Global mRNA amplification increases the number of transcripts [51]. To obtain adequate RNA from small tissue samples, global amplification techniques based either on exponential PCR [52] or isothermal linear RNA polymerase amplification [53] are employed. PCR-based amplification approaches can affect initial quantitative relationship between transcripts of a population due to exponential amplification, whereas linear RNA amplification has been used successfully to generate enough input RNA for robust hybridization [54,55]. Several studies carried out with limited starting material employed the classical T7 RNA polymerase amplification method, also termed amplified antisense RNA (aRNA) amplification [53]. This method utilizes a synthetic oligo (dT) primer linked to a phage T7 RNA polymerase promoter to prime the synthesis of first strand cDNA by reverse transcription of the polyA⁺ RNA component of total RNA. Second strand cDNA synthesis begins with RNase H degrading the polyA⁺ RNA strand, followed by second strand synthesis using *Escherichia coli* DNA polymerase I. Amplified aRNA is synthesized via *in vitro* transcription of the double-stranded cDNA (ds cDNA) template using T7 RNA polymerase.

Modification of the initial aRNA procedure and development of strategies combining PCR with aRNA and isothermal RNA amplification have been reported in several neuroscience studies [56–60]. These methods have been reviewed elsewhere [7,61].

18.2.6 MICROARRAY FOLLOW-UP STUDIES

Microarray-based studies offer the possibility to perform a system-wide transcriptome analysis of the brain. However, this technique gives a high level of false positives requiring follow-up studies to validate the results. Most microarray experiments in neuroscience applications are viewed as a screen to identify potential candidate genes that are then confirmed with alternative complementary methodologies.

The process of validation requires the assessment of the expression of specific genes using independent RNA samples. In practice, especially in the case of human postmortem studies in which the samples are limited, microarray findings are often validated with RNA from the same samples used for the microarray assay.

Many techniques can be used for confirming the expression pattern as assessed by gene expression profiling studies. These include *in situ* hybridization, quantitative RT-PCR (Q-PCR) and *in vivo* experiments.

18.2.6.1 *In Situ* Hybridization

In situ hybridization examines the cellular localization of expression changes allowing confirmation that the changes observed in the microarray study are not due to differences in the tissue dissection process. *In situ* histochemistry has been used to confirm the presence of relevant transcripts in a specific region of the central nervous system (CNS) or more precisely, in nuclei or layers of that region. It can be used as a follow-up to a microarray experiment to examine if the observed gene expression changes are due to specific steps of neuro-differentiation [62], the consequence of a treatment [63], or the genes associated with schizophrenia (SZ) [64].

18.2.6.2 Quantitative RT-PCR

Q-PCR is routinely used to assess the reliability of microarray studies. In almost every microarray paper published today, some of the most significant genes found differentially expressed are checked by Q-PCR for confirmation. Q-PCR is more sensitive than microarrays in detecting differentially expressed low-abundance transcripts. The finding that apolipoprotein L1 (*ApoL1*) is up-regulated in schizophrenic subjects was confirmed by Q-PCR. In addition, the up-regulation of other family members, in particular *ApoL2*, was observed. While with Q-PCR these results were reproducibly determined, the microarray analysis did not detect differential expression of *ApoL2* because it was below the detection threshold [65]. This example exemplifies how an integrated use of complementary techniques represents a powerful approach to neurobiology studies.

Often Q-PCR, *in situ* hybridization, and microarray are used together. For example, these three approaches were used to study the pathophysiology of Alzheimer's disease (AD) in order to more precisely identify the genes differentially expressed between the frontal cortices of Alzheimer-type dementia (ATD) subjects compared to controls. They showed that the autotaxin mRNA levels were considerably higher in ATD than in controls suggesting that this gene may be involved in the pathology of ATD [66].

18.2.6.3 *In vivo* Experiments

In some cases the microarray results can be validated by *in vivo* experiments. The analysis of gene expression changes that underlie memory formation has led to the identification of many neurobiologically relevant transcripts including glutamate receptors, ion channel, and trafficking protein. One gene, identified in the course of this analysis, fibroblast growth factor 18 (FGF18), was injected into the hippocampus demonstrating an improvement in rat spatial memory when animals were tested in a water maze [1]. The memory formation process was also studied with the complementary use of microarrays and mutagenesis in *Drosophila*. Convergent findings from both approaches identified candidate memory genes and the involvement of a specific pathway in memory. Behavioral tests carried out after the disruption of proteins, crucial in this pathway, provided evidence of a deficit in long-term memory [2].

18.3 DATA ANALYSIS

A well-defined scientific hypothesis followed by a clear experimental design will make the subsequent analysis and interpretation as simple and as powerful as possible. Microarray experiments are constructed using one of several design paradigms including direct, reference, loop, and saturated [67,68]. Each of these designs specifies the number of samples needed and the manner in which the samples should be compared in order to obtain a desired level of accuracy and reproducibility. The number of samples should include both biological and technical replicates, which are necessary to provide a reliable estimate of gene expression levels.

The experience gained from the microarray studies carried out so far highlights the importance of the quality assessment of chips used in the data analysis. Low-quality chips are identified and removed from subsequent analysis. Excluding low-quality chips from the analysis should increase the number of differentially expressed genes detected, while simultaneously reducing false positives.

One important issue in microarray data analysis is the normalization of the data across all the arrays used in the experiment [69,70]. The need for this arises from the observation that measurements from different hybridizations may occur on different scales and in order to properly compare datasets, normalization is required. Normalization can be performed based on a set of predefined genes ("housekeeping genes" or "control genes") that are believed to have the same expression across all arrays used in the experimental series or based on overall microarray signal intensity, assuming that the overall microarray signal intensity is constant across all arrays. Furthermore, normalization across arrays may be performed using linear or nonlinear methods. In linear normalizations, all microarray signals are adjusted to a common mean by a single, array-specific scaling factor. In contrast, nonlinear normalizations recognize that a single, array-specific scaling factor may not be ideal for all gene targets on a single array. For spotted cDNA microarrays, a two-step normalization strategy can be employed to correct for systematic variation of the data: nonlinear normalization by LOWESS local regression for within-slide bias correction followed by the linear normalization for between-slide bias correction. The LOWESS (locally weighted scatter plot smoothing) normalization is an iterative method that works on the assumption that the average of up-regulated expression profiles is approximately the same as the down-regulated expression profiles at each intensity level [70]. The quantile normalization method is commonly used for Affymetrix arrays, where it first ranks data on each array and then substitutes data of the same rank across all arrays by using the mean of the data [71].

Once normalized the data are statistically analyzed to detect genes that show reproducible differences in mRNA abundance between sample classes. A huge variety of methods have been used for this task, and when sufficient replicates are available all of them work well. For a simple two-group comparison, parametric (e.g., *t*-test) or nonparametric (e.g., Mann–Whitney) methods can be applied to test the hypothesis of each gene. In addition, many methods based on modified *t*-test such as SAM [72], Limma [73], and BayesT [74] can also effectively be used for differential gene expression analysis. When the normality assumption for hypothesis testing is violated in microarray experiments, one can use permutation or bootstrapping to calculate the *p*-values. Since the hypothesis testing to identify differentially expressed genes involves thousands of genes, the calculated *p*-values should be corrected for multiple hypothesis testing. The family-wise error rate (FWER) corrects the *p*-values by controlling the false positives in the differentially expressed genes. The FWER is defined as the probability of at least one false positive in the list of differentially expressed genes [75]. In general, FWER is conservative, as it restricts the number of differentially expressed genes and misses a significant number of truly differentially expressed genes. An alternative approach to deal with the multiple hypothesis testing problem is to use the false discovery rate (FDR) correction method proposed by Benjamini and Hochberg [76]. The FDR is the expected proportion of false positives among total rejected hypotheses. The concept of FDR has been widely accepted in large-scale microarray analysis [77–79].

18.4   MICROARRAY-BASED STUDIES OF NEUROLOGICAL AND PSYCHIATRIC DISORDERS

Many datasets are accumulating from microarray studies focusing on neurological and psychiatric disorders. Among the psychiatric disorders most of the studies have dealt with SZ; however, studies have also been performed on major depression [80–82], bipolar disorder [83–85], and autism [86,87]. DNA microarray transcriptome profiling of SZ points to altered glial functions in the pathophysiology of this disease [85,88–90]. In addition, other important classes of transcripts were found altered in the SZ brain when compared to control. Significant up-regulation of genes encoding high-density lipoproteins were identified in SZ brains [65], and several studies reported the involvement of genes related to neuronal development and plasticity [88] and presynaptic functions [91,92]. DNA microarray-based studies have also identified specific metabolic pathways that are altered in SZ and that can contribute to the disease process [93]. Genome-wide expression studies have identified genes and neurobiological pathways not previously associated with SZ. Moreover the application of this technology will contribute a broader picture of the neurological systems implicated in SZ. For example, microarray analysis coupled with traditional genome-wide linkage analysis has identified candidate genes and regulatory sites for SZ [94].

Among the neurological disorders AD, Parkinson's disease (PD), and multiple sclerosis (MS) have been extensively studied using microarray expression profiling. AD is the most common form of neurodegenerative disease in humans. In the last few years, many microarray studies have tried to elucidate the molecular mechanism associated with this neuropathological disease. The signs of degeneration typical of AD include β amyloid-containing plaques, intracellular neurofibrillary tangles (NFTs), and degeneration of cholinergic neurons. These markers are widespread in different brain regions. Consequently, microarray studies have examined a large number of brain regions including hippocampus, amygdala, cingulate cortex, and parietal cortex. Comparison of hippocampal neurons in AD subjects and controls confirmed the expression of genes encoding for proteins previously associated with the disease as well as identifying genes not previously implicated in AD. Furthermore, in the AD brain a general down-regulation of transcription factor signaling is observed. Additionally, transcripts of proteins involved in ion homeostasis, neurotrophic support, and synaptic plasticity are down-regulated. In contrast, genes involved in apoptosis and neuroinflammation were up-regulated [95].

Transcriptional profiling on a genome-wide scale has also been useful in monitoring the gene expression changes that take place during the different stages of AD development. The functional categorization of the genes differentially expressed between AD and control cases revealed that specific biological processes are involved in the different phases of AD dementia [96–99]. Of particular interest is the finding that genes related to synaptic vesicle function are dysregulated. The decline of cognitive functions is one of the most significant symptoms of AD and it has been associated with loss of synapses. Frontal cortex samples from AD cases revealed a significant down-regulation of genes involved in synaptic vesicle trafficking, whereas other synaptic genes not involved in vesicle trafficking were unchanged, suggesting a possible role of the synaptic vesicle trafficking pathway in the synapse malfunction in AD [100].

PD is a neurodegenerative disorder that leads to the impairment of motor skills. Many factors contribute to the development of the disease including genetic susceptibility and the effect of endocrine and environmental neurotoxins. Studies carried out with the purpose of correlating genetic variations with PD have found many candidate genes including *α-synuclein* [101] and *parkin* [102]. However none of these genes seem to play a crucial role in the development of the disease. Recent studies suggest an involvement of multifactorial processes and a number of distinct pathways in the PD pathogenesis [101,103]. The environmental exposure to toxins such as 1-methyl 4-phenyl 1,2,3,6-tetrahydro-pyridine (MPTP) and 6-hydroxydopamine (6-OHDA) induce the dopaminergic neurodegeneration causing most of the symptoms observed in PD, thus the selective use of these toxins has allowed the reproduction of PD in animal models. Microarray analysis has also been applied to assess gene expression changes in cultured cell models of dopaminergic (DA) neuron death [104–106]. Transcriptional profiling of human postmortem samples identified signatures unique to PD and elucidated the molecular pathways that play a role in disease pathology [107,108]. Moreover, microarray-based studies on human brain samples have played an important role in assessing the homology existing between animal models and human PD. For example, the significant down-regulation in synaptic gene expression observed in substantia nigra (SN) and the striatum of individuals affected by PD have also been observed in MPTP-treated mice [109].

MS is an autoimmune disease of the central nervous system which affects 1 million people worldwide. The hallmark of this disease is a series of plaques representing demyelinating lesions in the brain. It is not yet clear whether the degeneration of axons and myelin observed in MS depends on the inflammatory process, or if the immune response is secondary to the neurodegenerative process. Transcriptome studies of MS brain taken both early and late in the disease progression can help to elucidate the pathogenesis of MS. DNA microarray was applied to compare the gene expression changes in MS specific lesions of subjects affected by MS and in controls. Gene expression changes in specific lesions of subjects affected by the disease were compared with control individuals. The study was extended to the alterations occurring in the mouse brain following induction of autoimmune encephalomyelitis (EAE), an animal model of MS. The most important finding was the up-regulation of certain genes involved in the inflammatory activity that is a crucial component in MS [110]. The comparison of acute and chronic lesions observed in MS brains revealed various differentially transcribed genes that represent promising targets for therapy [111].

Genome-wide expression approaches were also employed to analyze gene expression changes induced by drugs of abuse including cocaine [112–114], morphine [115], and amphetamine [116]. Moreover, expression profiling with DNA microarrays has identified alcohol-responsive genes in human and animal brain [117–124] and characterized the gene expression patterns of rats selectively bred for differences in ethanol preference [125–130].

The results show that many biological systems and pathways are involved in ethanol preference and in maintaining ethanol dependence. However the data collectively gathered from all these experiments have failed thus far to identify a specific set of candidate genes directly responsible for alcoholism. In animal studies this can be in part due to the different phenotypes used. Combining genome-wide expression profiling and quantitative trait locus (QTL) analysis, Saba et al. [131] have elegantly demonstrated that the genetic networks underlying two commonly assessed alcoholism-related phenotypes, that is, ethanol preference and tolerance, are very different from each other. Furthermore, selective breeding for a particular phenotype, for example ethanol preference, also randomly uncovers alleles that are not directly associated with the phenotype under selection. Thus, not all the gene expression changes observed can be directly associated with a preference for ethanol consumption. In this context new approaches are of interest, such as meta-analysis to combine expression data of various animal models to extract candidate genes directly involved with alcoholism [132]. It is clear however that that the genome/transcriptome studies have identified pathways that may play significant roles in the risk of alcohol abuse and dependence.

18.5 CONCLUSIONS

In the near future a rapid development of genomics technologies and data analysis methods is expected. Improvements that will facilitate more sensitive and standardized analysis as well as more powerful tools for statistical analysis and data mining are anticipated. This will permit better integration of datasets from different sources.

The combination of knowledge derived from genetic, genomic, and proteomic studies will expand our understanding of brain disorders, improving early disease diagnosis, and allowing more effective and tailored treatments. For example, treatment for AD currently commences when the disease is in a moderate or mild state as assessed by cognitive observations. Preliminary studies suggest that biomarkers associated with the early stages of AD could allow timely diagnosis and therapeutic intervention. Routine screening however, remains to be implemented [133,134].

Microarray studies have produced long lists of genes associated with different neurobiological conditions and sources. The results have stimulated many multidisciplinary studies such as behavioral, anatomic, and pharmacogenomics research, which will ultimately increase our global understanding of the brain's biology.

REFERENCES

1. Cavallaro S, D'Agata V, Manickam P, Dufour F, and Alkon DL. 2002. Memory-specific temporal profiles of gene expression in the hippocampus. *Proc Natl Acad Sci USA* 99(25): 16279–16284.
2. Dubnau J, Chiang AS, Grady L, Barditch J, Gossweiler S, McNeil J, Smith P, Buldoc F, Scott R, Certa U, Broger C, and Tully T. 2003. The staufen/pumilio pathway is involved in *Drosophila* long-term memory. *Curr Biol* 13(4): 286–296.
3. Ueda HR, Matsumoto A, Kawamura M, Iino M, Tanimura T, and Hashimoto S. 2002. Genome-wide transcriptional orchestration of circadian rhythms in *Drosophila*. *J Biol Chem* 277(16): 14048–14052.
4. Duffield GE, Best JD, Meurers BH, Bittner A, Loros JJ, and Dunlap JC. 2002. Circadian programs of transcriptional activation, signaling, and protein turnover revealed by microarray analysis of mammalian cells. *Curr Biol* 12(7): 551–557.
5. Preuss TM, Cáceres M, Oldham MC, and Geschwind DH. 2004. Human brain evolution: insights from microarrays. *Nat Rev Genet* 5: 850–860.
6. Hardiman G. 2004. Microarray platforms—comparisons and contrasts. *Pharmacogenomics* 5(5): 487–502.
7. Soverchia L, Ubaldi M, Ciccocioppo R, and Hardiman G. 2005. Microarrays and the interrogation of brain gene expression. *Addict Biol* 10: 5–13.
8. Levine L, Grossfield J, and Rockwell RF. 1979. Functional relationships between genotypes and environments in behavior: effects of different kinds of early social experience on interstrain fighting in male mice. *J Hered* 70: 317–320.
9. Asanuma M and Ogawa N. 1994. Pitfalls in assessment of c-fos mRNA expression in the brain: effects of animal handling. *Rev Neurosci* 5: 171–178.
10. Coombs NJ, Gough AC, and Primrose JN. 1999. Optimisation of DNA and RNA extraction from archival formalin-fixed tissue. *Nucleic Acids Res* 27(16): e12.
11. Van Deerlin VM, Gill LH, and Nelson PT. 2002. Optimizing gene expression analysis in archival brain tissue. *Neurochem Res* 27: 993–1003.
12. Duyckaerts C, Sazdovitch V, Seilhean D, Delaere P, and Hauw JJ. 1993. A brain bank in a neuropathology laboratory (with some emphasis on diagnostic criteria). *J Neural Transm* 39: 107–118.
13. Hulette CM. 2003. Brain banking in the United States. *J Neuropathol Exp Neurol* 62: 715–722.
14. Mies C. 1992. Molecular pathology of paraffin-embedded tissue: current clinical applications. *Diagn Mol Pathol* 1: 206–211.
15. Greer CE, Lund JK, and Manos MM. 1991. PCR amplification from paraffin-embedded tissues: recommendations on fixatives for long-term storage and prospective studies. *PCR Methods Appl* 1: 46–50.
16. Foss RD, Guha-Thakurta N, Conran RM, and Gutman P. 1994. Effects of fixative and fixation time on the extraction and polymerase chain reaction amplification of RNA from paraffin-embedded tissue. Comparison of two housekeeping gene mRNA controls. *Diagn Mol Pathol* 3: 148–155.
17. Werner M, Chott A, Fabiano A, and Battifora H. 2000. Effect of formalin tissue fixation and processing on immunohistochemistry. *Am J Surg Pathol* 24(7): 1016–1019.

18. Qian X, Bauer RA, Xu HS, and Lloyd RV. 2001. In situ hybridization detection of calcitonin mRNA in routinely fixed, paraffin-embedded tissue sections: a comparison of different types of probes combined with tyramide signal amplification. *Appl Immunohistochem Mol Morphol* 9(1): 61–69.

19. Macabeo-Ong M, Ginzinger DG, Dekker N, McMillan A, Regezi JA, Wong DT, and Jordan RC. 2002. Effect of duration of fixation on quantitative reverse transcription polymerase chain reaction analyses. *Mod Pathol* 15(9): 979–987.

20. Lee CH, Bang SH, Lee SK, Song KY, and Lee IC. 2005. Gene expression profiling reveals sequential changes in gastric tubular adenoma and carcinoma in situ. *World J Gastroenterol* 11(13): 1937–1945.

21. Onken MD, Worley LA, Ehlers JP, and Harbour JW. 2004. Gene expression profiling in uveal melanoma reveals two molecular classes and predicts metastatic death. *Cancer Res* 64(20): 7205–7209.

22. Karsten SL, Van Deerlin VM, Sabatti C, Gill LH, and Geschwind DH. 2002. An evaluation of tyramide signal amplification and archived fixed and frozen tissue in microarray gene expression analysis. *Nucleic Acids Res* 30: E4.

23. Koopmans M, Monroe SS, Coffield LM, and Zaki SR. 1993. Optimization of extraction and PCR amplification of RNA extracts from paraffin-embedded tissue in different fixatives. *J Virol Methods* 43: 189–204.

24. Krafft AE, Duncan BW, Bijwaard KE, Taubenberger JK, and Lichy JH. 1997. Optimization of the isolation and amplification of RNA from formalin-fixed, paraffin-embedded tissue: The Armed Forces Institute of Pathology Experience and Literature Review. *Mol Diagn* 2: 217–230.

25. Godfrey TE, Kim SH, Chavira M, Ruff DW, Warren RS, Gray JW, and Jensen RH. 2000. Quantitative mRNA expression analysis from formalin-fixed, paraffin-embedded tissues using 5′ nuclease quantitative reverse transcription-polymerase chain reaction. *J Mol Diagn* 2: 84–91.

26. Specht K, Richter T, Muller U, Walch A, Werner M, and Hofler H. 2001. Quantitative gene expression analysis in microdissected archival formalin-fixed and paraffin-embedded tumor tissue. *Am J Pathol* 158: 419–429.

27. Masuda N, Ohnishi T, Kawamoto S, Monden M, and Okubo K. 1999. Analysis of chemical modification of RNA from formalin-fixed samples and optimization of molecular biology applications for such samples. *Nucleic Acids Res* 27: 4436–4443.

28. Lewis F, Maughan NJ, Smith V, Hillan K, and Quirke P. 2001. Unlocking the archive—gene expression in paraffin-embedded tissue. *J Pathol* 195: 66–71.

29. Shyu AB, Greenberg ME, and Belasco JG. 1989. The c-fos transcript is targeted for rapid decay by two distinct mRNA degradation pathways. *Genes Dev* 3: 60–72.

30. Schramm M, Falkai P, Tepest R, Schneider-Axmann T, Przkora R, Waha A, Pietsch T, Bonte W, and Bayer TA. 1999. Stability of RNA transcripts in post-mortem psychiatric brains. *J Neural Transm* 106: 329–335.

31. Heinrich M, Matt K, Lutz-Bonengel S, and Schmidt U. 2007. Successful RNA extraction from various human postmortem tissues. *Int J Legal Med* 121(2): 136–142.

32. Catts VS, Catts SV, Fernandez HR, Taylor JM, Coulson EJ, and Lutze-Mann LH. 2005. A microarray study of post-mortem mRNA degradation in mouse brain tissue. *Brain Res Mol Brain Res* 138(2): 164–177.

33. Preece P and Cairns NJ. 2003. Quantifying mRNA in postmortem human brain: influence of gender, age at death, postmortem interval, brain pH, agonal state and inter-lobe mRNA variance. *Brain Res Mol Brain Res* 118(1–2): 60–71.

34. Kingsbury AE, Foster OJ, Nisbet AP, Cairns N, Bray L, Eve DJ, Lees AJ, and Marsden CD. 1995. Tissue pH as an indicator of mRNA preservation in human post-mortem brain. *Brain Res Mol Brain Res* 28: 311–318.

35. Hynd MR, Lewohl JM, Scott HL, and Dodd PR. 2003. Biochemical and molecular studies using human autopsy brain tissue. *J Neurochem* 85: 543–562.

36. Li JZ, Vawter MP, Walsh DM, Tomita H, Evans SJ, Choudary PV, Lopez JF, et al. 2004. Systematic changes in gene expression in postmortem human brains associated with tissue pH and terminal medical conditions. *Hum Mol Genet* 13(6): 609–616.

37. Harrison PJ, Procter AW, Barton AJ, Lowe SL, Najlerahim A, Bertolucci PH, Bowen DM, and Pearson RC. 1991. Terminal coma affects messenger RNA detection in post mortem human temporal cortex. *Brain Res Mol Brain Res* 9: 161–164.

38. Harrison PJ, Heath PR, Eastwood SL, Burnet PW, McDonald B, and Pearson RC. 1995. The relative importance of premortem acidosis and postmortem interval for human brain gene expression studies: selective mRNA vulnerability and comparison with their encoded proteins. *Neurosci Lett* 200: 151–154.

39. Sandberg R, Yasuda R, Pankratz DG, Carter TA, Del Rio JA, Wodicka L, Mayford M, Lockhart DJ, and Barlow C. 2000. Regional and strain-specific gene expression mapping in the adult mouse brain. *Proc Natl Acad Sci USA* 97: 11038–11043.

40. Zirlinger M, Kreiman G, and Anderson DJ. 2001. Amygdala-enriched genes identified by microarray technology are restricted to specific amygdaloid subnuclei. *Proc Natl Acad Sci USA* 98: 5270–5275.

41. Zhao X, Lein ES, He A, Smith SC, Aston C, and Gage FH. 2001. Transcriptional profiling reveals strict boundaries between hippocampal subregions. *J Comp Neurol* 441(3): 187–196.

42. Emmert-Buck MR, Bonner RF, Smith PD, Chuaqui RF, Zhuang Z, Goldstein SR, Weiss RA, and Liotta LA. 1996. Laser capture microdissection. *Science* 274: 998–1001.

43. Luo L, Salunga RC, Guo H, Bittner A, Joy KC, Galindo JE, Xiao H, et al. 1999. Gene expression profiles of laser-captured adjacent neuronal subtypes. *Nat Med* 5: 117–122.

44. Bonaventure P, Guo H, Tian B, Liu X, Bittner A, Roland B, Salunga R, et al. 2002. Nuclei and subnuclei gene expression profiling in mammalian brain. *Brain Res* 943: 38–47.

45. Torres-Munoz JE, Van Waveren C, Keegan MG, Bookman RJ, and Petito CK. 2004. Gene expression profiles in microdissected neurons from human hippocampal subregions. *Brain Res Mol Brain Res* 127(1–2): 105–114.

46. Ginsberg SD and Che S. 2004. Combined histochemical staining, RNA amplification, regional, and single cell cDNA analysis within the hippocampus. *Lab Invest* 84(8): 952–962.

47. Yao PJ, Zhu M, Pyun EI, Brooks AI, Therianos S, Meyers VE, and Coleman PD. 2003. Defects in expression of genes related to synaptic vesicle trafficking in frontal cortex of Alzheimer's disease. *Neurobiol Dis* 12(2): 97–109.

48. Sugino K, Hempel CM, Miller MN, Hattox AM, Shapiro P, Wu C, Huang ZJ, and Nelson SB. 2006. Molecular taxonomy of major neuronal classes in the adult mouse forebrain. *Nat Neurosci* 9: 99–107.

49. Nelson SB, Hempel C, and Sugino K. 2006. Probing the transcriptome of neuronal cell types. *Curr Opin Neurobiol* 16: 571–576.

50. Stears RL, Getts RC, and Gullans SR. 2000. A novel, sensitive detection system for high-density microarrays using dendrimer technology. *Physiol Genomics* 3: 93–99.

51. Nygaard V, Loland A, Holden M, Langaas M, Rue H, Liu F, Myklebost O, Fodstad O, Hovig E, and Smith-Sorensen B. 2003. Effects of mRNA amplification on gene expression ratios in cDNA experiments estimated by analysis of variance. *BMC Genomics* 4: 11.

52. Lukyanov K, Diatchenko L, Chenchik A, Nanisetti A, Siebert P, Usman N, Matz M, and Lukyanov S. 1997. Construction of cDNA libraries from small amounts of total RNA using the suppression PCR effect. *Biochem Biophys Res Commun* 230: 285–288.

53. Van Gelder RN, von Zastrow ME, Yool A, Dement WC, Barchas JD, and Eberwine JH. 1990. Amplified RNA synthesized from limited quantities of heterogeneous cDNA. *Proc Natl Acad Sci USA* 87: 1663–1667.

54. McClain KL, Cai YH, Hicks J, Peterson LE, Yan XT, Che S, and Ginsberg SD. 2005. Expression profiling using human tissues in combination with RNA amplification and microarray analysis: assessment of Langerhans cell histiocytosis. *Amino Acids* 28(3): 279–290.

55. Ginsberg SD. 2005. RNA amplification strategies for small sample populations. *Methods* 3: 229–237.

56. Xiang CC, Chen M, Ma L, Phan QN, Inman JM, Kozhich OA, and Brownstein MJ. 2003. A new strategy to amplify degraded RNA from small tissue samples for microarray studies. *Nucleic Acids Res* 31(9): e53.

57. Wang E, Miller LD, Ohnmacht GA, Liu ET, and Marincola FM. 2000. High-fidelity mRNA amplification for gene profiling. *Nat Biotechnol* 18: 457–459.

58. Luzzi V, Holtschlag V, and Watson MA. 2001. Expression profiling of ductal carcinoma in situ by laser capture microdissection and high-density oligonucleotide arrays. *Am J Pathol* 158(6): 2005–2010.

59. Ginsberg SD and Che S. 2002. RNA amplification in brain tissues. *Neurochem Res* 10: 981–992.

60. Lefebvre d'Hellencourt C and Harry GJ. 2005. Molecular profiles of mRNA levels in laser capture microdissected murine hippocampal regions differentially responsive to TMT-induced cell death. *J Neurochem* 93(1): 206–220.

61. Ruggeri B, Soverchia L, Ubaldi M, Ciccocioppo R, and Hardiman G. 2007. Amplification strategies and DNA biochips. In: *Biochips as Pathways to Discovery*, A. Carmen and G. Hardiman (eds), pp. 253–260. New York: Taylor & Francis.

62. Yu S, Zhang JZ, and Xu Q. 2006. Genes associated with neuronal differentiation of precursors from human brain. *Neuroscience* 141(2): 817–825.

63. Conti B, Maier R, Barr AM, Morale MC, Lu X, Sanna PP, Bilbe G, Hoyer D, and Bartfai T. 2006. Region-specific transcriptional changes following the three antidepressant treatments electro convulsive therapy, sleep deprivation and fluoxetine. *Mol Psychiatry* 12(2): 167–189.

64. Vawter MP, Shannon Weickert C, Ferran E, Matsumoto M, Overman K, Hyde TM, Weinberger DR, Bunney WE, and Kleinman JE. 2004. Gene expression of metabolic enzymes and a protease inhibitor in the prefrontal cortex are decreased in schizophrenia. *Neurochem Res* 29(6): 1245–1255.

65. Mimmack ML, Ryan M, Baba H, Navarro-Ruiz J, Iritani S, Faull RL, McKenna PJ, et al. 2002. Gene expression analysis in schizophrenia: reproducible up-regulation of several members of the apolipoprotein L family located in a high-susceptibility locus for schizophrenia on chromosome 22. *Proc Natl Acad Sci USA* 99(7): 4680–4685.

66. Umemura K, Yamashita N, Yu X, Arima K, Asada T, Makifuchi T, Murayama S, et al. 2006. Autotaxin expression is enhanced in frontal cortex of Alzheimer-type dementia patients. *Neurosci Lett* 400(1–2): 97–100.

67. Yang YH and Speed T. 2002. Design issues for cDNA microarray experiments. *Nat Rev Genet* 3(8): 579–588.

68. Churchill GA. 2002. Fundamentals of experimental design for cDNA microarrays. *Nat Genet* 32(Suppl.): 490–495.

69. Quackenbush J. 2002. Microarray data normalization and transformation. *Nat Genet* 32(Suppl.): 496–501.

70. Yang YH, Dudoit S, Luu P, Lin DM, Peng V, Ngai J, and Speed TP. 2002. Normalization for cDNA microarray data: a robust composite method addressing single and multiple slide systematic variation. *Nucleic Acids Res* 30(4): e15.

71. Bolstad BM, Irizarry RA, Astrand M, and Speed TP. 2003. A comparison of normalization methods for high density oligonucleotide array data based on variance and bias. *Bioinformatics* 19(2): 185–193.

72. Tusher V, Tibshirani R, and Chu C. 2001. Significance analysis of microarrays applied to ionizing radiation response. *Proc Natl Acad Sci USA* 98: 5116–5121.

73. Smyth GK. 2004. Linear models and empirical Bayes methods for assessing differential expression in microarray experiments. *Stat Appl Genet Mol Biol* 3(1): article 3.

74. Baldi P and Long AD. 2001. A Bayesian framework for the analysis of microarray expression data: regularized *t*-test and statistical inferences of gene changes. *Bioinformatics* 17(6): 509–519.

75. Westfall PH and Young SS. 1993. *Resampling-Based Multiple Testing: Examples and Methods for P-Value Adjustment*. New York: Wiley.

76. Benjamini Y and Hochberg Y. 1995. Controlling the false discovery rate: a practical and powerful approach to multiple testing. *J Roy Statist Soc Ser B* 57: 289–300.

77. Reimers M. 2005. Statistical analysis of microarray data. *Addict Biol* 10(1): 23–35.

78. Reimers M, Heilig M, and Sommer WH. 2005. Gene discovery in neuropharmacological and behavioral studies using Affymetrix microarray data. *Methods* 37(3): 219–228.

79. Reimers M and Weinstein JN. 2005. Quality assessment of microarrays: visualization of spatial artifacts and quantitation of regional biases. *BMC Bioinformatics* 6: 166.

80. Aston C, Jiang L, and Sokolov BP. 2005. Transcriptional profiling reveals evidence for signaling and oligodendroglial abnormalities in the temporal cortex from patients with major depressive disorder. *Mol Psychiatry* 10(3): 309–322.

81. Evans SJ, Choudary PV, Neal CR, Li JZ, Vawter MP, Tomita H, Lopez JF, et al. 2004. Dysregulation of the fibroblast growth factor system in major depression. *Proc Natl Acad Sci USA* 101(43): 15506–15511.

82. Sibille E, Arango V, Galfalvy HC, Pavlidis P, Erraji-Benchekroun L, Ellis SP, and John Mann J. 2004. Gene expression profiling of depression and suicide in human prefrontal cortex. *Neuropsychopharmacology* 29(2): 351–361.

83. Bezchlibnyk YB, Wang JF, McQueen GM, and Young LT. 2001. Gene expression differences in bipolar disorder revealed by cDNA array analysis of post-mortem frontal cortex. *J Neurochem.* 79(4): 826–834.

84. Konradi C, Eaton M, MacDonald ML, Walsh J, Benes FM, and Heckers S. 2004. Molecular evidence for mitochondrial dysfunction in bipolar disorder. *Arch Gen Psychiatry* 61(3): 300–308.

85. Tkachev D, Mimmack ML, Ryan MM, Wayland M, Freeman T, Jones PB, Starkey M, Webster MJ, Yolken RH, and Bahn S. 2003. Oligodendrocyte dysfunction in schizophrenia and bipolar disorder. *Lancet* 362(9386): 798–805.

86. Purcell AE, Jeon OH, Zimmerman AW, Blue ME, and Pevsner J. 2001. Postmortem brain abnormalities of the glutamate neurotransmitter system in autism. *Neurology* 57(9): 1618–1628.

87. Samaco RC, Nagarajan RP, Braunschweig D, and LaSalle JM. 2004. Multiple pathways regulate MeCP2 expression in normal brain development and exhibit defects in autism-spectrum disorders. *Hum Mol Genet* 13(6): 629–639.

88. Hakak Y, Walker JR, Li C, Wong WH, Davis KL, Buxbaum JD, Haroutunian V, and Fienberg AA. 2001. Genome-wide expression analysis reveals dysregulation of myelination-related genes in chronic schizophrenia. *Proc Natl Acad Sci USA* 98(8): 4746–4751.

89. Aston C, Jiang L, and Sokolov BP. 2004. Microarray analysis of postmortem temporal cortex from patients with schizophrenia. *J Neurosci Res* 77(6): 858–866.

90. Pongrac J, Middleton FA, Lewis DA, Levitt P, and Mirnics K. 2002. Gene expression profiling with DNA microarrays: advancing our understanding of psychiatric disorders. *Neurochem Res* 27(10): 1049–1063.

91. Mirnics K, Middleton FA, Marquez A, Lewis DA, and Levitt P. 2000. Molecular characterization of schizophrenia viewed by microarray analysis of gene expression in prefrontal cortex. *Neuron* 28(1): 53–67.

92. Vawter MP, Crook JM, Hyde TM, Kleinman JE, Weinberger DR, Becker KG, and Freed WJ. 2002. Microarray analysis of gene expression in the prefrontal cortex in schizophrenia: a preliminary study. *Schizophr Res* 58(1): 11–20.

93. Middleton FA, Mirnics K, Pierri JN, Lewis DA, and Levitt P. 2002. Gene expression profiling reveals alterations of specific metabolic pathways in schizophrenia. *J Neurosci* 22(7): 2718–2729.

94. Vawter MP, Atz ME, Rollins BL, Cooper-Casey KM, Shao L, and Byerley WF. 2006. Genome scans and gene expression microarrays converge to identify gene regulatory loci relevant in schizophrenia. *Hum Genet* 119(5): 558–570.

95. Colangelo V, Schurr J, Ball MJ, Pelaez RP, Bazan NG, and Lukiw WJ. 2002. Gene expression profiling of 12633 genes in Alzheimer hippocampal CA1: transcription and neurotrophic factor down-regulation and up-regulation of apoptotic and pro-inflammatory signaling. *J Neurosci Res* 70(3): 462–473.

96. Pasinetti GM and Ho L. 2001. From cDNA microarrays to high-throughput proteomics. Implications in the search for preventive initiatives to slow the clinical progression of Alzheimer's disease dementia. *Restor Neurol Neurosci* 18(2–3): 137–142.

97. Blalock EM, Geddes JW, Chen KC, Porter NM, Markesbery WR, and Landfield PW. 2004. Incipient Alzheimer's disease: microarray correlation analyses reveal major transcriptional and tumor suppressor responses. *Proc Natl Acad Sci USA* 101(7): 2173–2178.

98. Loring JF, Wen X, Lee JM, Seilhamer J, and Somogyi R. 2001. A gene expression profile of Alzheimer's disease. *DNA Cell Biol* 20(11): 683–695.

99. Emilsson L, Saetre P, and Jazin E. 2006. Alzheimer's disease: mRNA expression profiles of multiple patients show alterations of genes involved with calcium signaling. *Neurobiol Dis* 21(3): 618–625.

100. Yao F, Yu F, Gong L, Taube D, Rao DD, and MacKenzie RG. 2005. Microarray analysis of fluoro-gold labeled rat dopamine neurons harvested by laser capture microdissection. *J Neurosci Methods* 143(2): 95–106.

101. Spillantini MG, Schmidt ML, Lee VM, Trojanowski JQ, Jakes R, and Goedert M. 1997. α-Synuclein in Lewy bodies. *Nature* 388(6645): 839–840.

102. Shimura H, Hattori N, Kubo SI, Mizuno Y, Asakawa S, Minoshima S, Shimizu N, Iwai K, Chiba T, Tanaka K, and Suzuki T. 2000. Familial Parkinson disease gene product, parkin, is a ubiquitin-protein ligase. *Nat Genet* 25(3): 302–305.

103. Mandel S, Grünblatt E, and Youdim M. 2000. cDNA microarray to study gene expression of dopaminergic neurodegeneration and neuroprotection in MPTP and 6-hydroxydopamine models: implications for idiopathic Parkinson's disease. *J Neural Transm* 60(Suppl.): 117–124.

104. Holtz WA and O'Malley KL. 2003. Parkinsonian mimetics induce aspects of unfolded protein response in death of dopaminergic neurons. *J Biol Chem* 278(21): 19367–19377.

105. Yoo MS, Chun HS, Son JJ, DeGiorgio LA, Kim DJ, Peng C, and Son JH. 2003. Oxidative stress regulated genes in nigral dopaminergic neuronal cells: correlation with the known pathology in Parkinson's disease. *Brain Res Mol Brain Res* 110(1): 76–84.

106. Chun HS, Gibson GE, DeGiorgio LA, Zhang H, Kidd VJ, and Son JH. 2001. Dopaminergic cell death induced by MPP(+), oxidant and specific neurotoxicants shares the common molecular mechanism. *J Neurochem* 76(4): 1010–1021.

107. Hauser MA, Li YJ, Xu H, Noureddine MA, Shao YS, Gullans SR, Scherzer CR, et al. 2005. Expression profiling of substantia nigra in Parkinson disease, progressive supranuclear palsy, and frontotemporal dementia with parkinsonism. *Arch Neurol* 62(6): 917–921.

108. Grunblatt E, Mandel S, Jacob-Hirsch J, Zeligson S, Amariglo N, Rechavi G, Li J, et al. 2004. Gene expression profiling of parkinsonian substantia nigra pars compacta; alterations in ubiquitin-proteasome, heat shock protein, iron and oxidative stress regulated proteins, cell adhesion/cellular matrix and vesicle trafficking genes. *J Neural Transm* 111(12): 1543–1573.

109. Miller RM, Kiser GL, Kaysser-Kranich TM, Lockner RJ, Palaniappan C, and Federoff HJ. 2006. Robust dysregulation of gene expression in substantia nigra and striatum in Parkinson's disease. *Neurobiol Dis* 21(2): 305–313.
110. Whitney LW, Ludwin SK, McFarland HF, and Biddison WE. 2001. Microarray analysis of gene expression in multiple sclerosis and EAE identifies 5-lipoxygenase as a component of inflammatory lesions. *J Neuroimmunol* 121(1–2): 40–48.
111. Lock C, Hermans G, Pedotti R, Brendolan A, Schadt E, Garren H, Langer-Gould A, et al. 2002. Gene-microarray analysis of multiple sclerosis lesions yields new targets validated in autoimmune encephalomyelitis. *Nat Med* 8(5): 500–508.
112. Albertson DN, Pruetz B, Schmidt CJ, Kuhn DM, Kapatos G, and Bannon MJ. 2004. Gene expression profile of the nucleus accumbens of human cocaine abusers: evidence for dysregulation of myelin. *J Neurochem* 88(5): 1211–1219.
113. Yuferov V, Kroslak T, Laforge KS, Zhou Y, Ho A, and Kreek MJ. 2003. Differential gene expression in the rat caudate putamen after "binge" cocaine administration: advantage of triplicate microarray analysis. *Synapse* 48(4): 157–169.
114. Economidou D, Mattioli L, Ubaldi M, Lourdusamy A, Soverchia L, Hardiman G, Campolongo P, Cuomo V, and Ciccocioppo R. 2007. Role of cannabinoidergic mechanisms in ethanol self-administration and ethanol-seeking in rat adult offspring following perinatal exposure to Δ9-tetrahydrocannabinol. *Toxicol Appl Pharmacol* 223(1): 73–85.
115. Ammon S, Mayer P, Riechert U, Tischmeyer H, and Hollt V. 2003. Microarray analysis of genes expressed in the frontal cortex of rats chronically treated with morphine and after naloxone precipitated withdrawal. *Brain Res Mol Brain Res* 112(1–2): 113–125.
116. Gonzalez-Nicolini V and McGinty JF. 2002. Gene expression profile from the striatum of amphetamine-treated rats: a cDNA array and in situ hybridization histochemical study. *Brain Res Gene Expr Patterns* 1(3–4): 193–198.
117. Rimondini R, Arlinde C, Sommer W, and Heilig M. 2002. Long-lasting increase in voluntary ethanol consumption and transcriptional regulation in the rat brain after intermittent exposure to alcohol. *FASEB J* 16: 27–35.
118. Saito M, Smiley J, Toth R, and Vadasz C. 2002. Microarray analysis of gene expression in rat hippocampus after chronic ethanol treatment. *Neurochem Res* 27(10): 1221–1229.
119. Daniels GM and Buck KJ. 2002. Expression profiling identifies strain-specific changes associated with ethanol withdrawal in mice. *Genes Brain Behav* 1(1): 35–45.
120. Sommer W, Arlinde C, and Heilig M. 2005. The search for candidate genes of alcoholism: evidence from expression profiling studies. *Addict Biol* 10(1): 71–79.
121. Bjork K, Saarikoski ST, Arlinde C, Kovanen L, Osei-Hyiaman D, Ubaldi M, Reimers M, Hyytia P, Heilig M, and Sommer WH. 2006. Glutathione-S-transferase expression in the brain: possible role in ethanol preference and longevity. *FASEB J* 20(11): 1826–1835.
122. Mayfield RD, Lewohl JM, Dodd PR, Herlihy A, Liu J, and Harris RA. 2002. Patterns of gene expression are altered in the frontal and motor cortices of human alcoholics. *J Neurochem* 81(4): 802–813.
123. Covarrubias MY, Khan RL, Vadigepalli R, Hoek JB, and Schwaber JS. 2005. Chronic alcohol exposure alters transcription broadly in a key integrative brain nucleus for homeostasis: the nucleus tractus solitarius. *Physiol Genomics* 24(1): 45–58.
124. Hoffman P and Tabakoff B. 2005. Gene expression in animals with different acute responses to ethanol. *Addict Biol* 10(1): 63–69.
125. Arlinde C, Sommer W, Bjork K, Reimers M, Hyytia P, Kiianmaa K, and Heilig M. 2004. A cluster of differentially expressed signal transduction genes identified by microarray analysis in a rat genetic model of alcoholism. *Pharmacogenomics J* 4: 208–218.
126. Worst TJ, Tan JC, Robertson DJ, Freeman WM, Hyytia P, Kiianmaa K, and Vrana KE. 2005. Transcriptome analysis of frontal cortex in alcohol-preferring and nonpreferring rats. *J Neurosci Res* 80: 529–538.
127. Edenberg HJ, Strother WN, McClintick JN, Tian H, Stephens M, Jerome RE, Lumeng L, Li TK, and McBride WJ. 2005. Gene expression in the hippocampus of inbred alcohol-preferring and -nonpreferring rats. *Genes Brain Behav* 4: 20–30.
128. Rodd ZA, Bertsch BA, Strother WN, Le-Niculescu H, Balaraman Y, Hayden E, Jerome RE, et al. 2006. Candidate genes, pathways and mechanisms for alcoholism: an expanded convergent functional genomics approach. *Pharmacogenomics J* 7(4): 222–256.
129. Sommer W, Hyytia P, and Kiianmaa K. 2006. The alcohol-preferring AA and alcohol-avoiding ANA rats: neurobiology of the regulation of alcohol drinking. *Addict Biol* 11(3–4): 289–309.

130. Ciccocioppo R, Economidou D, Cippitelli A, Cucculelli M, Ubaldi M, Soverchia L, Lourdusamy A, and Massi M. 2006. Genetically selected Marchigian Sardinian alcohol-preferring (msP) rats: an animal model to study the neurobiology of alcoholism. *Addict Biol* 11(3–4): 339–355.
131. Saba L, Bhave SV, Grahame N, Bice P, Lapadat R, Belknap J, Hoffman PL, and Tabakoff B. 2006. Candidate genes and their regulatory elements: alcohol preference and tolerance. *Mamm Genome* 17: 669–688.
132. Mulligan MK, Ponomarev I, Hitzemann RJ, Belknap JK, Tabakoff B, Harris RA, Crabbe JC, et al. 2006. Toward understanding the genetics of alcohol drinking through transcriptome meta-analysis. *Proc Natl Acad Sci USA* 103(16): 6368–6373.
133. Palotas A, Puskas LG, Kitajka K, Palotas M, Molnar J, Pakaski M, Janka Z, Penke B, and Kalman J. 2004. Altered response to mirtazapine on gene expression profile of lymphocytes from Alzheimer's patients. *Eur J Pharmacol* 497(3): 247–254.
134. Davidsson P and Sjogren M. 2005. The use of proteomics in biomarker discovery in neurodegenerative diseases. *Dis Markers* 21(2): 81–92.
135. Lein ES, Hawrylycz MJ, Ao N, Ayres M, Bensinger A, Bernard A, Boe AF, et al. 2007. Genome-wide atlas of gene expression in the adult mouse brain. *Nature* 445(7124): 168–176.

19 Optimization of Protein Array Fabrication for Establishing High-Throughput Ultra-Sensitive Microarray Assays for Cancer Research

*Gary Hardiman, C. Ramana Bhasker,
and Richard J.D. Rouse*

CONTENTS

19.1 INTRODUCTION

The complexity of cancer and its basic underlying principles are slowly beginning to unfold. Neoplasia is an intricate process that involves a variety of defects that impair normal cell proliferation and homeostasis. Hanahan et al. [1] reported that of the more than 100 distinct types of cancer, there are six essential changes that can occur in cells that will determine malignancy: (a) self-sufficiency in growth signals; (b) insensitivity to anti-growth signals; (c) evasion of apoptosis; (d) potential of limitless cellular replication; (e) angiogenesis; and (f) metastasis. The proteomic network related to these changes is becoming clearer and therapeutic development is being targeted at distinct proteins that play key roles in these pathways. Therefore, the need for high-throughput multiplexed proteomic techniques with the capability of accurately measuring how these pathways are altered and how treatments can modify these networks effectively is warranted [2,3]. Molecular profiling technologies are rapidly advancing and individualized patient treatment is becoming a reality.

The microarray format is well suited for measuring biomarkers since it is a sensitive multiplex assay that requires small amounts of material. Antibody-based arrays, referred to as forward phase assays, however, require extensive optimization, because these molecules have variable affinities and specificities [4]. Typically, antibody specificity is determined by the Immuno (Western) blot method where a single band at the matched molecular weight is an indication of specificity. Antibody affinity is a more complicated parameter since it determines the linear range of the assay. In multiplex assays using antibodies with variable affinities, linearity would be difficult to determine for every feature unless the readout can be carried out *in solution* over the course of multiple washing steps.

Reverse phase arrays involve spotting cell lysates or antigen targets on the surface of slides and probing with specific antibodies. This array format is designed for high-throughput protein quantification since the lysates are dispensed in a dilution series ensuring that part of the array will be in the linear antibody-binding range. Since it is a sensitive, robust, and quantitative high-throughput assay, reverse phase arrays are well adapted for paneling cellular responses for preclinical therapeutic development.

The reverse phase array technology has been validated for both monitoring the expression of disease-related proteins and investigating the effects of pharmaceutical agents. It was first implemented for profiling 60 human cancer cell lines (NCI-60) in a search for screening potential compounds for anti-cancer activity [5]. That study accurately measured 52 proteins across the NCI-60 cell lines. Paweletz et al. [6] used reverse phase arrays to analyze pro-survival checkpoint proteins at the transition stage from normal prostate epithelium to prostate intraepithelial neoplasia. This assay format has also been set up to map signaling pathways for prostate cancers, gliomablastomas, and ovarian cancers [7–9].

In addition to engineering specific and consistent antigen–antibody-binding characteristics, translating this onto a microarray format can be challenging. There are many ways by which this application can be improved, some of which should focus on the characterization of antibodies. Optimizing protein microarray assay techniques are evolving also due to increasingly sophisticated instrumentation. Piezoelectric drop-on-demand microarrayers are capable of dispensing a diverse volume across broad viscosity ranges. This when combined with highly sensitive imaging systems that are amendable for high-throughput processing can provide a well-suited pipeline for developing robust proteomic assays. We describe elements of a high capacity and sensitive protein microarray assay platform that consists of the GeSiM Nano-Plotter piezoelectric spotter (produces spots down to 50 pl) and the Zeptosens planar waveguide (PWG) imager (detects 600 molecules in ~300 pl samples).

19.2 TECHNICAL CHALLENGES FOR OPTIMIZING ANTIBODY ASSAYS FOR ONCOLOGY DIAGNOSTIC APPLICATIONS

Multiplexed diagnostic bioassays have received some profitable success in recent years. Genomic Health has established a United States Food and Drug Administration (FDA)-validated laboratory test (Oncotype DX™) that predicts the likelihood of a recurrence in women after being diagnosed with early-stage invasive breast cancer. This genetic test measures the expression of 21 breast cancer-associated genes. A similar assay, MammaPrint® (developed by Agendia, Amsterdam), is used to determine the 10-year survival of patients. This FDA-approved test utilizes its relatively high multiplexing (70 genes) to produce very accurate data (having a significance level of 96.7% [10]). These tests are based on gene expression profiling using ribonucleic acid (RNA). It should be noted that messenger RNA (mRNA) abundance in a cell often correlates poorly with the amount of protein synthesized. Expression analysis using deoxyribonucleic acid (DNA) microarrays measures only the transcriptome and important regulation occurs during translation and enzymatic activities.

The progress of implementing proteomics technologies into clinical diagnostic assays is slow. Most of the current clinical immuno-applications are still based on the immunohistochemistry (IHC) principle, with limited capacity for multiplexing. For example, the HeceptTest™ (developed by Dako, Carpinteria, California) is an IHC assay used to identify patients' *ErbB2* expression so that they may benefit from antibody-mediated treatment against *ErbB2*.

Developing a multiplexed protein assay is more challenging compared to its genomic counterpart [11,12]. Protein samples are more complex, comprising diverse compositions. Antibody–antigen interactions can therefore be quite promiscuous when molecules with similar epitopes can occur even though the genetic sequence is entirely different. In addition, each antibody–antigen interaction can be influenced by (assay) buffers based on the charge and hydrophobicity owing to the diverse nature of protein structure. Antibody-binding characteristics are based on their intrinsic monovalent binding affinity (true affinity) and their avidity (functional affinity) [13]. It is important to consider aspects related to true affinity and avidity. True affinity refers to the interaction between one arm of a multivalent antibody (IgG is a bivalent antibody) and an epitope. Avidity involves the interaction of the antibody as a whole molecule to the antigen. In addition avidity can be affected by a variety of factors that include true affinity, antibody valence, antigen density/concentration, and steric factors.

Creating arrays with well-defined specificity and binding affinity is one of the technical challenges of implementing this type of assay since both primary and secondary antibodies should be assessed for these properties. A number of factors can affect molecular binding characteristics, including the interaction of the binding motifs to the target, the number of motifs (valence), array feature density, and steric factors. There is thus a need to select and compile monoclonal antibodies that both are specific to the target protein under study, and have matched binding affinities in microarray assays. To effectively screen these proteins, it is important to implement a system capable of evaluating a number of assay conditions and one that is sufficiently flexible to be adapted for manufacturing [14].

19.3 ENSURING ACCURATE SPOTTING IN THE ARRAY FABRICATION PROCESS

As opposed to genomic-based microarrays where the most widely used platform (Affymetrix, Santa Clara, California) is produced by an *in situ* synthesis technique, protein arrays are typically manufactured using either a contact or noncontact printing process that transfers samples from microwell plates to the microarray surface. To assure high-quality printing, it is advantageous to establish systems that check for fabrication errors that can be used to repair "miss-spotting" problems. The GeSiM (GeSiM GmbH, Grosserkmannsdorf, Germany) Nano-Plotter piezoelectric non-contact dispensing platform has features that can be implemented to provide quality control before, during, and after array fabrication.

Before spotting, it is important to evaluate whether the pipettes can successfully dispense samples. Issues related to inadequate sample aspiration and/or sample composition can cause tips to "misfire." The Nano-Plotter utilizes an infrared stroboscope camera that visualizes droplets firing from the piezoelectric nozzle. If stable dispensing is achieved, a still image of droplets can be seen while the nozzle is dispensing in real-time, shown in Figure 19.1. Preset parameters can be programmed to evaluate the droplet number, size, and deviation from vertical to mark that the nozzle is "functional" to spot on chips or "nonfunctional," where a new sample is taken up or the process is aborted completely. A correction for missing spots can be taken into account later. The stroboscope droplet measurement can be conducted repeatedly to verify stable dispensing over time. Furthermore, both raw images and computed statistical data can be collected for every aspiration event. Data recording is also possible in real-time, and the system can be fitted with a head camera positioned to visualize the tips while they are spotting. This is shown in Figure 19.2, where the Nano-Plotter is spotting on Whatman (Whatman plc, Maidstone, Kent, UK) FAST nitrocellulose-coated slides. When the droplet hits the nitrocellulose, the color change is clearly visible. These images can be saved for each sample dispensed for every array and every slide and be used to verify whether misfiring occurred during the spotting run.

After spotting, the Nano-Plotter can be set up to evaluate microarray quality using a video microscope for optical detection. Integration of the video microscope onto the print head requires

(a)   (b)

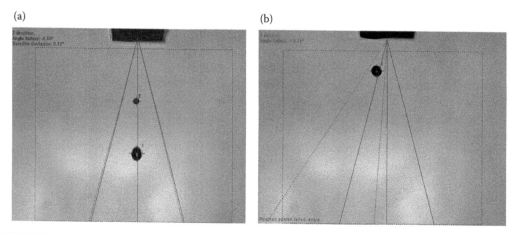

FIGURE 19.1 Demonstration of the stroboscope pre-spot check utility. (a) A droplet that passed the inspection criteria; and (b) a droplet that that failed the inspection criteria.

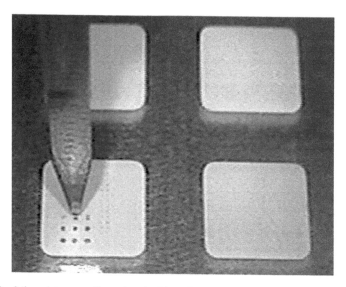

FIGURE 19.2 Real-time data recording using the Nano-Plotter head camera.

the implementation of image processing software. To enable sufficient contrast for automatic recognition features such as circular spots, the software utilizes Gaussian smoothing, edge detection, and cutoff filters. These procedures work well with surfaces having optical properties that allow for sufficient contrast and its efficacy will vary depending on the composition of the spotting buffer. Demonstrated in Figure 19.3 shows an image of 0.25mg/mL antibody in phosphate buffered saline (PBS) spotted on Gentel PATH ultra-thin nitrocellulose slides (Gentel Biosciences, Madison, Wisconsin, USA). Implementing the video microscope to check the array quality makes it possible to correct for problems with spotting without having to take the slides off the Nano-Plotter. This is convenient since loading slides off and on the spotting deck introduces the need for realignment.

"Miss-spotting" can occur in arbitrary locations. For example, one replicate spot could be missing while other adjacent spots are present. In other cases, the "spots" could be missing from a selected number of arrays out of the total spotting run. Therefore when repairing arrays, it is

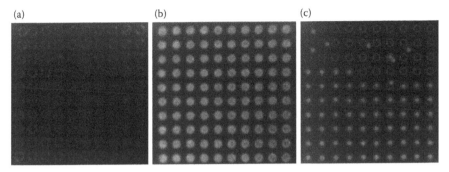

FIGURE 19.3 Detection of array spots with a video microscope. (a) Low contrast image of IgG (2 mg/mL) in a low salt buffer (20 mM sodium phosphate, 30 mM NaCl); (b) high contrast image of IgG (0.25 mg/mL) spotted in PBS (phosphate buffered saline); and (c) automated spot-finding to identify "miss-spots."

important to work with a flexible software. The Nano-Plotter system implements transfer lists that are files that map source plates to specific array locations. The transfer list syntax can be simple as

P1 S1 1:10,4:7

where P1 is the plate number 1, S1 the sample position number, 1:10 the rows 1–10, and 4:7 the columns 4–7. From one aspiration, it is possible to spot in array positions that are not adjacent. For example,

P1 S1 1:10,4:7 14:20,29

This enables the possibility to spot in two locations within one array that are not next to one another. The design of the transfer list is also set up to handle complicated syntax:

H-1 P1 S15 TG1 TA[tR1/tC1, tR2/tC2] BL7 TI1 1:5,1:5 = 10 1:5,6

where H-1 means plate holder position 1 and TG1 the target group one. Target groups consist of one more spotting targets. So it is possible to spot on targets having different dimensions (i.e., such as slides and microwell plates) or different array patterns (one group having 0.5 mm center-to-center spacing and another group having 0.75 mm center-to-center spacing) in one spotting run. TA[tR1/tC1, tR2/tC2] makes it possible to spot on a selected number of targets within a target group, per aspiration event. BL7 enables the operator to select a specific array block. TI1 makes it possible to select a specific tip to dispense while turning the other tips off. The 1:5,1:5 = 10 means the tip will spot from rows 1 to 5 and columns 1 to 5 with 10 drops per spot.

19.4 SPOTTING BUFFER OPTIMIZATION

Although in principle, noncontact spotters can function with diverse spotting buffers, it is important to determine the appropriate buffer for a given application. Forward-based antibody array spotting with high salt buffers containing no surface-acting agents are more prone towards "miss-spotting" events. Conducting experiments to determine the most appropriate buffer formulation requires an assessment of array quality over multiple sample aspirations. Close attention is required to ensure how well a particular spotting buffer works in the binding assays. Furthermore, it is recommended to conduct carry over tests to assure that there is no cross-contamination between samples.

Figures 19.4 and 19.5 demonstrate how spotting buffer compositions can produce very different droplet characteristics. Low salt buffers produce relatively transparent spots. Although they are

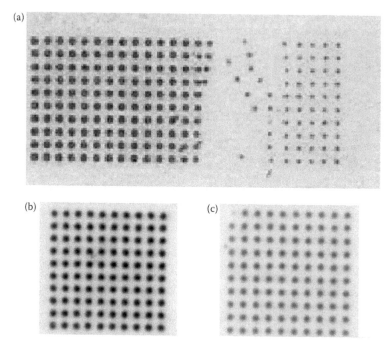

FIGURE 19.4 Demonstration of high salt effects on spotting on water-sensitive paper. (a) IgG (1 mg/mL) in PBS. This highlights how "miss-spotting" can occur, (b) is an array of 1 × PBS and (c) is a 10 × PBS array. The higher the salt content, the faster the sample dries. Spotting buffers with high salt content can dry in the tip during the spotting run. Therefore close attention to both the buffer composition and spotting environment is important for quality production runs.

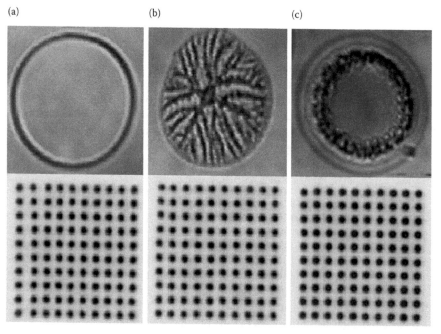

FIGURE 19.5 Array patterns and droplet appearance after spotting. (a) 2 mg/mL mouse IgG in 20 mM sodium phosphate and 30 mM NaCl; (b) 1 mg/mL mouse IgG in PBS (137 mM NaCl, 2.7 mM KCl, and 11.9 mM phosphate buffer); and (c) 1 mg/mL bovine serum albumin 2% Pluronic F68 in PBS.

amendable for robust spotting, proteins can be less stable. PBS is a relatively high salt buffer and the salt crystals are clearly apparent in the droplet. Owing to the high salt content, the samples can dry in the tip during spotting and so it is important to pay close attention to "environmental" factors. To reduce drying, it is possible to add either sucrose or glycerol into the buffer [15,16].

The Nano-Plotter has both humidity and temperature control. Temperature is controlled using a chilled thermal bath. Both the source plate and spotting target holders can be chilled to a temperature that raises the humidity sufficiently to reduce excess evaporation of the sample. For example, reducing the temperature of the target holder down to 18°C can raise the relative humidity to 55%. This allows for very robust array fabrication using even high salt buffers.

Spotting cell lysates typically contain a surfactant (such as Triton X) that acts as a surface-acting agent. Nondenaturing detergents such as Pluronic, Tween, and NP-40 (popular for producing cell lysates) are molecules possessing both a hydrophobic carbon tail and a hydrophilic region. As a result, proteins become encapsulated in micelles in these buffers, although it is possible to "wash out" the detergent after spotting.

19.5   DISPENSE PARAMETER OPTIMIZATION

The voltage and pulse width are adjustable and this will affect the dispense volume per droplet. The voltage setting controls the magnitude of the piezopulse, whereas the pulse width controls the length of the pulse. Increasing one or both settings will in turn increase the droplet volume. Figure 19.6 demonstrates the effect of voltage on droplet volume. This shows the need to pay close attention towards fine-tuning of the dispense parameters for each tip tested.

The Nano-Plotter software allows for individually setting the dispense parameters for each tip. Determining the best settings can be challenging, however. Ideally, calibration studies should be done in assays that are comparable to the tips being used. There are spotting programs available that allow for programming dispensing settings for every spot. This makes it possible to aspirate a sample and spot at various voltage and pulse width settings in order to identify tip settings that have matched dispensed volumes. Such a calibration is encouraged for the collection of tips used in the microarray fabrication run. Given that the Nano-Plotter can operate up to 16 tips simultaneously this is a convenient feature that is amenable to automation.

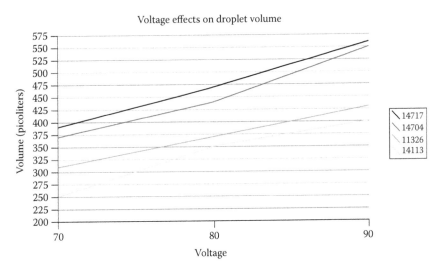

FIGURE 19.6   Increasing the voltage will increase the droplet volume. This shows the need to individually set the parameters for dispensing with each tip in order to improve spotting accuracy.

19.6 APPLICATION OF SURFACE PLASMON RESONANCE TO MEASURE ANTIBODY-BINDING KINETICS

As microarrays are multiplex assays, one of the challenges is to characterize the binding kinetics of the ligands. A number of factors can affect molecular binding characteristics, which can include the interaction of the binding motifs to the target, the number of motifs (valence), array feature density, and steric factors. This can involve extensive testing and development. Implementing imaging systems that enable label-free detection in real-time will significantly improve the efficiency of this type of assay optimization and validation.

Surface plasmon resonance (SPR) is a technique that is capable of dynamically measuring molecular interactions occurring on surfaces. SPR works by targeting a path of photons against a surface that has been coated with a material capable of "resonating" upon contact. The coating is typically a conductive metal layer (like gold). Using either a prism or a grating, resonance can be generated by coupling the photons to electrons in the metal.

It is possible to accurately measure changes in resonance in real-time. GWC Technologies SPRimager™ (GWC Technologies, Inc., Madison, Wisconsin) utilizes a charge-coupled device (CCD) camera that measures subtle changes of reflectivity upon exposure to a fixed angle of incidence. An increased reflectivity change occurs when more material binds to the array surface. A percent reflectivity change, $\Delta\%R$, is calculated and used as the parameter to quantify the amount of array feature binding.

This example demonstrates how this SPR method works with high-density microarrays. Using the Nano-Plotter, three different proteins (Protein G, beta-2-microglobulin, and streptavidin) were

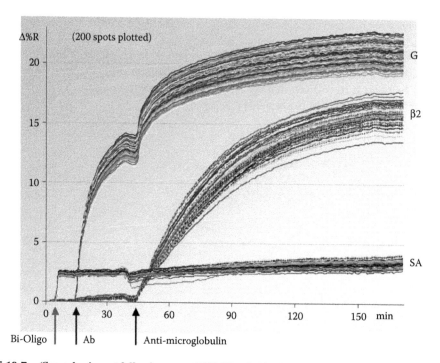

FIGURE 19.7   (See color insert following page 138.) Kinetic binding assay as a function of percent change in reflectivity in a high-density SPR microarray assay. The change in reflectivity is calculated based on a comparison to time zero. In a flow cell attached to the microarray, a biotinylated oligonucleotide was independently added until it reached a binding equilibrium as indicated by the flat line. Afterwards IgG was introduced, until reflectivity was stable, then a microglobulin-specific IgG was added which caused an increase of reflectivity change for both Protein G and beta-2-microglobulin.

spotted 200 times each across an 8 by 8 mm^2 area at 400 pl per spot. These three proteins were used to demonstrate both specificity and binding affinity. Protein G binds to the IgG antibody with high affinity. Beta-2-microglobulin is an antigen commonly expressed by mononucleated eukaryotic cells. Streptavidin exhibits highly specific binding affinity for the protein biotin. The platform is prepared by assembling the spotted array into a 1-cm flow cell that allows for known quantities of analytes to be administered into the array. Figure 19.7 plots the sequence of analyte administration and binding as a function of time. The first sample that was applied was a biotinylated oligonucleotide that quickly reached binding equilibrium. The second analyte was an IgG which binds to Protein G. The third analyte was an IgG specific to beta-2-microglobulin; this molecule will bind with both the Protein G and beta-2-microglobulin spots as indicated in Figure 19.7.

This simple experiment demonstrates how both specificity and binding affinity can be analyzed using a microarray-based SPR assay. The Protein G and beta-2-microglobulin assessment is an example which compares a specific antibody-binding interaction (Protein G to IgG) to cross-reactivity (anti-beta-2-microglobulin to Protein G). Binding affinity can be measured by determining both the level and rate of reflectivity change upon administration of a specific concentration of analyte. A less-pronounced reflectivity change was observed with the biotinylated oligonucleotide compared with what was observed using the IgG antibodies. Furthermore, Figure 19.7 demonstrates how Protein G had a greater affinity to IgG than the beta-2-microglobulin interaction since the rate of reflectivity change was higher.

19.7   IMPLEMENTING OPTIMIZED ARRAY FABRICATION FOR ULTRASENSITIVE PROTEIN ARRAY ASSAYS

The Zeptosens ZeptoREADER imaging platform is a high-throughput PWG detection system. It has lasers compatible for both Cy3 and Cy5 excitation and a high-resolution CCD camera for array imaging. The reader can load upto 60 ZeptoCHIP slides. It enables streamlined image quantification thereby allowing for automated calculation of up to 95,040 data points.

PWG sensors provide a surface-confined evanescent field excitation which allows for a limit of detection down to 600 labeled molecules in one ~200 µm microarray spot. This is about 75 times more sensitive than a confocal scanner. Furthermore, PWGs enable solution phase microarray detection so that the end user can do multiple reads over the course of higher stringency wash steps. Another advantage of solution phase detection is that the fluorophore dyes are not as prone to bleaching.

To demonstrate the assay performance with this setup, a simple study was performed to compare the effects of doxorubicin (DOX) exposure and ultraviolet (UV) irradiation on two colon carcinoma cell lines. The p53 tumor suppressor is a transcription factor functioning in a pathway that becomes activated upon genomic DNA damage. This can have an impact on tumor cell formation and viability since p53 prevents cells from undergoing neoplasia through apoptosis [16]. DNA damage can be caused by a variety of factors, which may include hypoxia, oncogenes, and telomere abnormalities.

Since the p53 pathway is inactivated in most human neoplasms, established tumor cell lines are being used to study drugs that can overexpress phosphorylated p53 to apoptosis-causing levels. In controlled cell culture conditions, UV treatment very efficiently activates this pathway and it is used as a benchmark for evaluating potential therapeutics. DOX is a DNA intercalating agent that is frequently used as a chemotherapeutic agent for cancer treatment. DOX is an anthracycline that is known to block topoisomerase II, the protein involved with unwinding DNA for DNA replication and/or RNA synthesis [17]. For this example, the effect of DOX was compared with that of UV treatment on two colon carcinoma cell lines, HT-29 and HCT-15.

The abundance of the following four DNA damage response proteins was measured:

- P-ATM/ATR substrate—phosphorylates Chk2 upon response to DNA damage
- P-Chk2—phosphorylated cell cycle regulator (in response to DNA damage) regulates p53 through phosphorylation

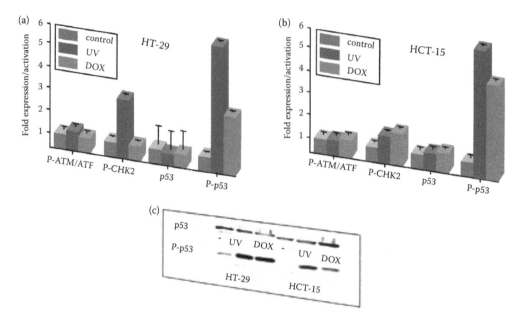

FIGURE 19.8 **(See color insert following page 138.)** Biomarker detection using reverse phase arrays. Protein levels measured across three conditions (control—blue, UV—red, DOX—orange). (a) Expression profile in HT-29 cells; (b) expression profile in HCT-15 cells; and (c) Western blot data used to validate the specificity of p53 and P-p53 antibodies.

- p53—tumor-suppressor protein 53
- P-p53—phosphorylated p53 (serine residues) that activates apoptosis.

The reverse phase assay demonstrates how DOX treatment compares with UV irradiation on the expression levels of these response proteins in HT-29 and HCT-15 cells. The levels of P-p53 significantly increased in treated cells compared to untreated cells. For HT-29, P-p53 abundance was three- and fourfold higher in HT-29 and HCT-15 cells, respectively, than in untreated cells. These data demonstrate a cell-dependent effect of DOX since it did not induce the overexpression of P-Chk2 in HT-29 cells, but did so in the HCT-15 line.

This reverse phase array assay was verified by Western blot as indicated in Figure 19.8c. The blot clearly showed stronger bands for P-p53 after UV and DOX treatment. The advantage of the reverse phase array assay is that it is a more quantitative assay (detecting 10–20% differences in abundance), and is capable of measuring biomarkers in samples sets that are as low as 0.25 fg per array spot.

REFERENCES

1. Hanahan D and Weinberg RA. The hallmarks of cancer. *Cell* 2000; 100(1): 57–70.
2. Cadd VA, Hogg PJ, Harris AL, and Feller SM. Molecular profiling of signalling proteins for effects induced by the anti-cancer compound GSAO with 400 antibodies. *BMC Cancer* 2006; 6: 155.
3. Kirmiz C, Li B, An HJ, Clowers BH, Chew HK, Lam KS, Ferrige A, Alecio R, Borowsky AD, Sulaimon S, Lebrilla CB, and Miyamoto S. A serum glycomics approach to breast cancer biomarkers. *Mol Cell Proteomics* 2007; 6(1): 43–55.
4. Sanchez-Carbayo M. Antibody arrays: technical considerations and clinical applications in cancer. *Clin Chem* 2006; 52(9): 1651–1659 (Epub 2006 Jun 29).
5. Nishizuka S, Charboneau L, Young L, Major S, Reinhold WC, Waltham M, Kouros-Mehr H, Bussey KJ, Lee JK, Espina V, Munson PJ, Petricoin E, 3rd, Liotta LA, and Weinstein JN. Proteomic profiling of the NCI-60 cancer cell lines using new high-density reverse-phase lysate microarrays. *Proc Natl Acad Sci USA* 2003; 100(24): 14229–14234.

6. Grubb RL, Calvert VS, Wulkuhle JD, Paweletz CP, Linehan WM, Phillips JL, Chuaqui R, Valasco A, Gillespie J, Emmert-Buck M, Liotta LA, and Petricoin EF. Signal pathway profiling of prostate cancer using reverse phase protein arrays. *Proteomics* 2003; 3(11): 2142–2146.

7. Jiang R, Mircean C, Shmulevich I, Cogdell D, Jia Y, Tabus I, Aldape K, Sawaya R, Bruner JM, Fuller GN, and Zhang W. Pathway alterations during glioma progression revealed by reverse phase protein lysate arrays. *Proteomics* 2006; 6(10): 2964–2971.

8. Wulfkuhle JD, Aquino JA, Calvert VS, Fishman DA, Coukos G, Liotta LA, and Petricoin EF, 3rd. Signal pathway profiling of ovarian cancer from human tissue specimens using reverse-phase protein microarrays. *Proteomics* 2003; 3(11): 20852090.

9. Dmitriev DA, Massino YS, and Segal OL. Kinetic analysis of interactions between bispecific monoclonal antibodies and immobilized antigens using a resonant mirror biosensor. *J Immunol Methods* 2003; 280(1–2): 183–202.

10. Fan C, Oh DS, Wessels L, Weigelt B, Nuyten SAD, Nobel AB, van 't Veer LJ, and Perou CM. Concordance among gene expression-based predictors for breast cancer. *N Engl J Med* 2006; 355: 560–569.

11. Talapatra A, Rouse R, and Hardiman G. Protein arrays and biochips. In: *Microarray Methods and Applications* (G. Hardiman, ed.), 2003, Chap. 8, pp. 141–154. Eagleville, PA: DNA Press Inc.

12. Talapatra A, Mahant V, Kureshy F, Vairavan R, and Hardiman G. (2004). Proteomic analysis using the Infiniti AnalyzerTM, a fully automated multiplexing microarray platform. In: *Protein Microarrays* (M. Schena, ed), Chap. 18, pp. 353–364. Sudbury, MA: Jones and Bartlett Publishers.

13. Pawlak M, Schick E, Bopp MA, Schneider MJ, Oroszlan P, and Ehrat M. Zeptosens' protein microarrays: a novel high performance microarray platform for low abundance protein analysis. *Proteomics* 2002; 2(4): 383–393.

14. Talapatra A, Rouse R, and Hardiman G. Protein microarrays: challenges and promises. *Pharmacogenomics* 2002; 3: 507–516.

15. Delehanty JB and Ligler FS. A microarray immunoassay for simultaneous detection of proteins and bacteria. *Anal Chem* 2002; 74(21): 5681–5687.

16. Haab BB, Dunham MJ, and Brown PO. Protein microarrays for highly parallel detection and quantitation of specific proteins and antibodies in complex solutions. *Genome Biol* 2001; 2(2): research0004–research 4.13.

17. Le Gac G, Esteve PO, Ferec C, and Pradhan S. DNA damage-induced down-regulation of human Cdc25C and Cdc2 is mediated by cooperation between p53 and maintenance DNA (cytosine-5) methyltransferase 1. *J Biol Chem* 2006; 281(34): 24161–24170 (Epub 2006 Jun 28).

Index

Milton Keynes UK
Ingram Content Group UK Ltd.
UKHW050453071024
449327UK00015B/363